Advances in Crop Breeding

Advances in Crop Breeding

Edited by Ayden Spears

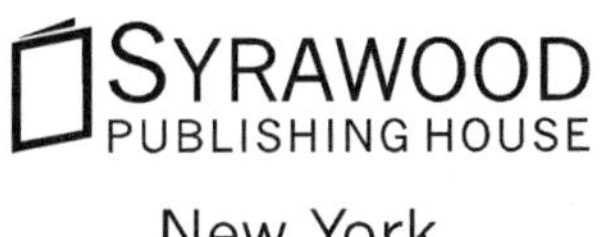

New York

Published by Syrawood Publishing House,
750 Third Avenue, 9th Floor,
New York, NY 10017, USA
www.syrawoodpublishinghouse.com

Advances in Crop Breeding
Edited by Ayden Spears

International Standard Book Number: 978-1-68286-694-8 (Hardback)

Cataloging-in-Publication Data

Advances in crop breeding / edited by Ayden Spears.
p. cm.
Includes bibliographical references and index.
ISBN 978-1-68286-694-8
1. Plant breeding. 2. Crops. 3. Breeding. I. Spears, Ayden.
SB123 .A38 2019
631.53--dc23

TABLE OF CONTENTS

PREFACE

The world is advancing at a fast pace like never before. Therefore, the need is to keep up with the latest developments. This book was an idea that came to fruition when the specialists in the area realized the need to coordinate together and document essential themes in the subject. That's when I was requested to be the editor. Editing this book has been an honour as it brings together diverse authors researching on different streams of the field. The book collates essential materials contributed by veterans in the area which can be utilized by students and researchers alike.

The science of crop breeding delves into the study of genetics of plants to develop desirable characteristics. Modern crop breeding builds on the principles of molecular biology and genetics to replace existing and time-intensive classical approaches of selective breeding. This is achieved by developing innovative practices of marker-assisted selection, double haploids and genetic modification. There has been a lot of progress in this field and its applications are being widely used and studied. Some of the diverse topics covered in this book address the multiple branches that fall under this category. The various studies that are constantly contributing towards advancing technologies and evolution of this field are examined in detail. The book includes contributions of experts and scientists, which will provide innovative insights into this field.

Each chapter is a sole-standing publication that reflects each author's interpretation. Thus, the book displays a multi-facetted picture of our current understanding of application, resources and aspects of the field. I would like to thank the contributors of this book and my family for their endless support.

Editor

RNAi-mediated resistance to SMV and BYMV in transgenic tobacco

Lo Thi Mai Thu[1], Vi Thi Xuan Thuy[1], Le Hoang Duc[2], Le Van Son[2], Chu Hoang Ha[2] and Chu Hoang Mau[3*]

***Corresponding author:**
E-mail: chuhoangmau@tnu.edu.vn

[1] Taybac University, Viet Nam
[2] Institute of Biotechnology, VAST, Viet Nam
[3] Thainguyen University of Education, Viet Nam

Abstract: *Soybean mosaic virus (SMV) and bean yellow mosaic virus (BYMV) are two typical types of viruses that cause mosaic in soybean plants. Multiple viral infections at the same site can lead to 66% to 80% yield reduction. We have aimed to improve SMV and BYMV resistance in Vietnamese soybeans using gene transfer techniques under the mechanism of RNAi. In this study, we present newly generated transgenic tobacco plants carrying RNAi [CPi (SMV-BYMV)] resistance to the two types of viruses; 73.08% of transgenic tobacco lines proved to be fully resistant to SMV and BYMV. In addition, the number of virus copies in transgenic tobacco plants was reduced on average by more than 51% compared to the control plants (wild type). This promising result shows the potential of transerring the CPi (SMV-BYMV) structure in soybean to increase resistance of soybean to SMV and BYMV and advance the aims of antiviral soybean breeding in Vietnam.*

Key words: *Antiviral soybean breeding, gene transfer, Glycine max, qRT-PCR, mosaic virus.*

INTRODUCTION

Soybean mosaic virus (SMV) and Bean yellow mosaic virus (BYMV) belong to the Potyvirus genus, which causes mosaic in legume crops. SMV and BYMV can infect the same plant simultaneously and can be very difficult to distinguish. Multiple infections from different types of viruses can reduce soybean yields from 66% to 80% (Hartman et al. 1999). Currently, the main methods for reducing losses from these two viruses are still traditional preventive measures, such as selecting clean lines for seed banks, field sanitation, crop rotation, cleanup of diseased crops, pest and disease control, and so on. These methods are very time-consuming, inefficient, and unsustainable. An effective measure to prevent SMV and BYMV is to use virus-resistant cultivars. However, natural strains resistant to SMV and BYMV are very limited. Therefore, a new approach is to generate transgenic plants resistant to viruses. One promising approach is utilizing an RNA interference (RNAi) mechanism. RNAi is considered an effective technique to improve plant resistance against viruses. RNAi is a process of post-transcriptional gene silencing (PTGS) triggered by double-stranded RNA (dsRNA) (Baulcombe 2004). Numerous groups have successfully applied RNAi strategies in generating transgenic plants resistant to viruses (Asad et al. 2003, Kasai et al. 2012). Transgenic plants have slowed the accumulation of disease symptoms or reduced them. In 2007, using the RNAi approach, Bonfim et al. obtained a transgenic soybean line with high resistance to Bean golden mosaic virus (up to 93%). Another study (Furutani et al. 2007) also tried to improve SMV resistance in soybean using genetic engineering techniques.

We designed an RNAi-based construct to silence genes encoding a coat protein (CP) of SMV and BYMV, thus limiting viral infection. From conserved regions of CP coding sequences of SMV and BYMV, we cloned a 573 bp cDNA fragment, called CPi (SMV-BYMV), which is composed of a 294 bp fragment from the SMV CP gene and a 279 bp fragment from the BYMV CP gene. The CPi (SMV-BYMV) fragment was then cloned in sense- and antisense-orientation into the vector pK7GWIWG2 (Karimi et al. 2002) by using Gateway technology (Invitrogen). The final construct is called pK7GW-CPi (SMV-BYMV), which we were able to transcribe in the cells to generate the hairpin RNA, thus inducing the RNAi mechanism to degrade the CP gene of SMV and BYMV. This was presented in our previous publication (Lo et al. 2014). In this study, we analyze resistance to SMV and BYMV in transgenic tobacco lines carrying the RNAi structure.

MATERIAL AND METHODS

Materials

Tobacco plants (*Nicotiana tabacum*) of the cultivar C9-1 were provided by the Vietnam Tobacco Economic Technical Institute. The SMV and BYMV viruses were provided by the Plant Protection Research Institute. The pK7GW-CPi (SMV-BYMV) vector was constructed based on the Gateway principles, which were described by Karimi et al. (2002) and transformed into the *Agrobacterium tumefaciens* strain CV58 PGV 2660.

The Polymerase Chain Reaction (PCR) primers SMV-CPi-Fi/BYMV-CPi-Ri were used to check for the presence of the CPi gene in transgenic plants, and the CP gene fragment was amplified by using primers SMVqF/SMVqR for real-time PCR (RT-PCR) (Table 1).

Methods

Agrobacterium-mediated transformation via leaf infection and regeneration of tobacco plants was carried out as previously described by Topping (1998). Total DNA was isolated from young leaves based on the method of Saghai et al. (1984).

Transgenic tobacco plants were grown in a greenhouse for 2-3 weeks to acquire 5 true leaves (each about 10 cm in length), which were then infected by SMV and BYMV by using artificial infection methods. Their viral resistance abilities were evaluated based on the method of Herbers et al. (1996).

The number of SMV copies in transgenic plants was analyzed by real time qRT-PCR (Real time Quantitative Reverse Transcription PCR) by using LightCycler® FastStart DNA MasterPLUS SYBR Green I (Roche) and the primers SMVqF and SMVqR (Table 1). The thermal cycle was as follows: denaturation at 95 °C for 10 min; amplification and binding for 45 cycles (95 °C for 10 s, 58 °C for 10 s, 72 °C for 20 s); and analysis of the flow temperature when the temperature increases from 65–95 °C for 1 min, continuing to collect fluorescent signals. Temperature Δ = 20°C sec^{-1}. We calculated and determined expression (R) by using the $R = 2^{-\Delta\Delta Ct}$ method of Livak and Schmittgen (2001).

Table 1. A list of oligonucleotide primers used in the study

Primers	DNA Sequence (5'- 3')
SMV-CPi-Fi	CACCGCAGCAGAAGCTTACA
BYMV-CPi-Ri	ATGTTCCGAACCCCAAGCAA
SMVqF	GCCTAGATATGGACTACTGAGGA
SMVqR	ATTCACATCCCTTGCAGTGT

RESULTS AND DISCUSSION

Transferring the RNAi construct [pK7GW-CPi (SMV-BYMV)] into the tobacco plant

To improve SMV and BYMV resistance in tobacco plants, the CPi (SMV-BYMV) fragment consisting of conserved regions originating from both SMV and BYMV CP genes was cloned into the vector pK7GWIWG2 (Karimi et al. 2002) in sense- and antisense-orientation. This RNAi-based construct, called pK7GW-CPi (SMV-BYMV) (Lo et al. 2014) (Figure 1), could be expressed as a hairpin RNA structure to trigger the RNAi mechanism once virus infection occurred.

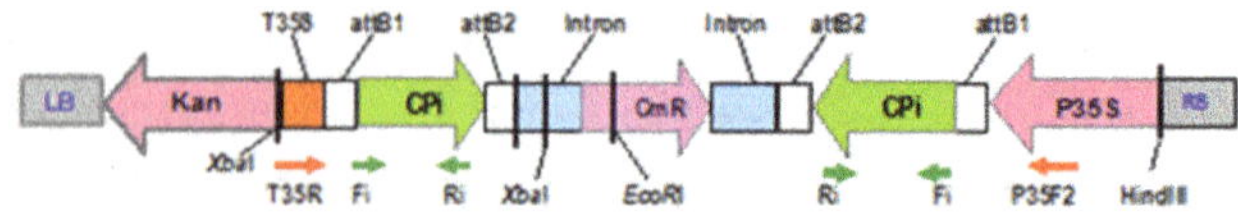

Figure 1. Diagram of pK7GW-CPi (SMV-BYMV). P35S: Promoter CaMV35S; attB1 and attB2: Recombinant positions for LB reaction; LB: left T-DNA border; RB: right T-DNA border; T35S: terminator 35S; Kan: kanamycin resistant; CmR: chloramphenicol resistant; CPi: CPi (SMV-BYMV) incorporation position; *Xba*I, *EcoR*I, *Hind*III: restriction sites; T35R, P35SF2, Fi, Ri: primers SMV-CPi-Fi/BYMV-CPi-Ri binding sites.

The pK7GW-CPi (SMV-BYMV) construct was transferred to *A. tumefaciens*. *Agrobacterium*-mediated transformation of this construct via leaf infection and regeneration of tobacco plants were performed as previously described by Topping (1998). Pieces of tobacco leaves were incubated with *A. tumefaciens* for 2 days in the dark, and then washed with antibiotics and transferred to shooting medium supplemented with antibiotics and BAP. After 2-3 weeks, the newly developed shoots appeared. Small buds were removed from the leaf explants and transferred to a fresh medium for rapid bud development. After 1-2 weeks, the shoots were transferred to a rooting medium for root development. The tobacco plants with developed roots growing in the antibiotic medium were transferred into pots.

The CPi (SMV-BYMV) structure was transferred to the tobacco leaf explants twice, using 30 leaf explants each time. The results presented in Table 2 show that from 60 initial explants, 38 survived and generated 82 shoots and 131 plants in vitro. Forty plants were transferred to pots and 26 developed properly. As a control, 30 leaf explants without the transgene were regenerated in a medium with or without antibiotics (Control-0 and Control-1, respectively). As expected, none of the Control-0 explants survived, whereas from 30 Control-1 explants, 105 plants were generated, and 10 were transferred to pots in a greenhouse.

The transgenic tobacco plants were grown in the greenhouse for 3 to 4 weeks, then their leaves were collected to check for the transgene presence by using the specific primers SMV-CPi-F and BYMV-CPi-R. The PCR results show specific bands of approximately 573 bp in size in all 26 lines (Figure 2), suggesting that the CPi (SMV-BYMV) structure has been successfully transferred into C9-1 tobacco plants.

Evaluation of BYMV and SMV resistance of the transgenic tobacco lines under artificial infection conditions

To test the virus resistance ability of transgenic tobacco plants, 26 transgenic tobacco lines and 10 controls were infected with SMV and BYMV three times at 15-day intervals. After each infection, virus development in the tobacco lines was observed. Figure 3 shows leaf morphology of transgenic and WT plants after infection. Most transgenic plants showed complete resistance, reflected in their normal leaf morphology and full growth (Figure 3A). In contrast, after being infected, all 10 control plants showed severe infection, with mosaic curly leaves and poor growth (Figure 3B). Virus resistance results from 26 transgenic tobacco lines and 10 control plants are presented in Table 3. The statistics from Table 3 show a high percentage of complete resistance of transgenic tobacco lines (19 resistant out of 26 infected, corresponding to 73.08%).

Figure 2. Presence of transgene in transgenic tobacco plants. (+) positive control pK7GW-CPi (SMV-BYMV) plasmid; (-) non-transgenic tobacco plants; M: Marker 1kb (Thermo Scientific); 1-26: 26 transgenic tobacco lines.

Table 2. Results of CPi (SMV-BYMV) transformation and tobacco plant regeneration

RNAi structure and controls	Transformed leaves	Surviving leaves	Shoots generated	Surviving plants	Plants transferred to pots	Transferred to greenhouse
pK7GW-CPi (SMV-BYMV)	2 x 30 = 60	38	82	131	40	26
*Control-0	30	0	-	-	-	-
*Control-1	30	30	79	105	10	10

*Control-0: negative control tobacco plants in the medium with antibiotics. *Control-1: tobacco plants with no transgene in the medium without antibiotics.

Table 3. Artificial infection and resistance ability to SMV and BYMV in transgenic and control plants

Infection times	Transgenic plants				Control plants			
	Infected	Complete resistance	No resistance	Resistance percentage	Infected	Complete resistance	No resistance	Resistance percentage
1	26	22	4++	84.62	10	3	7+++	30.33
2	22	22	0	100	3	0	3+++	0
3	22	19	3+	86.36	0	0		
Total	26	19	7	73.08	10		10	0

+, ++, +++ = mild, medium, and severe symptoms, respectively.

The analytic result from viral quantitative assays by real-time RT-PCR

The number of SMV copies from the transgenic tobacco plant structure was checked by using qRT- PCR. The leaves of three completely resistant lines, called $T_0$5 - resistant, $T_0$12 - resistant, and $T_0$19- resistant (Figure 3A), and three infected plants (wild type-WT), called WT1- infected, WT2 - infected, and WT3 – infected, from the artificial infection were used as materials for the reaction (Figure 3B).

Total RNA was extracted from 1 g of tobacco leaves, and 1 µg RNA was used for cDNA synthesis. The reaction curve was constructed from three reactions in which *E. coli* carrying the recombinant pBT-SMV was counted using a counting chamber. The DNA concentrations from 3 standard samples were 1000, 100, and 10 copies of DNA μL^{-1} respectively (1 cell corresponding to 1 copy). The standard curve equation was constructed using results from the qRT-PCR reaction with primers SMVqF and SMVqR; the linear curve to determine the number of DNA copies from the threshold cycle was: $Y = -168.15X + 4202.7$, in which Y is the number of copies and X is the threshold cycle values, and $R^2 = 0.898$ (R^2: squared correlation coefficient).

The number of threshold cycles to detect the number of SMV and SMV copies per 1 g of tobacco leaves are presented in Table 4. Table 4 shows that SMV viruses were found in all samples from transgenic and negative control lines after artificial virus infection. In the WT plants, there were 810 SMV copies on average, which is 1.94 times higher than the transgenic plants (average 416.66 copies). This result suggests that the pK7GW-CPi (SMV-BYMV) construct is activated when SMV infected, inhibiting SMV via the RNAi mechanism; thus, fewer SMV copies are found in these plants. However, in the transgenic plants, the number of copies of SMV is different (Table 4), demonstrating the different levels of disease in different transgenic lines. The qRT-PCR results show that

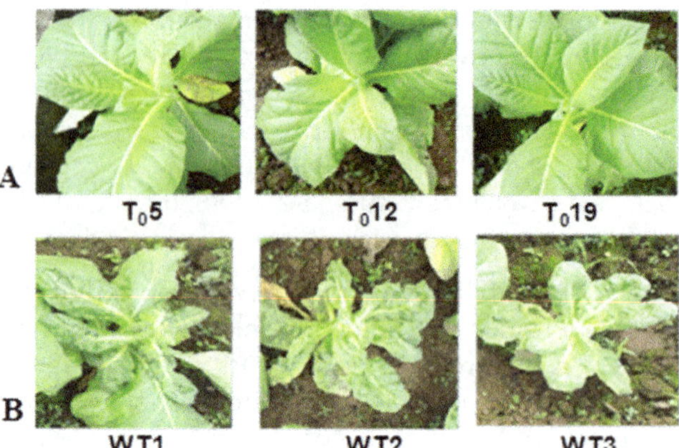

Figure 3. Leaf morphology of transgenic tobacco and WT plants after being infected three times with SMV and BYMV. A- The transgenic lines T_0-5, T_0-12, and T_0-19 are completely resistant; B- very severe infection in WT plants. WT1: WT1- infected, WT2: WT2 - infected, and WT3: WT3 - infected.

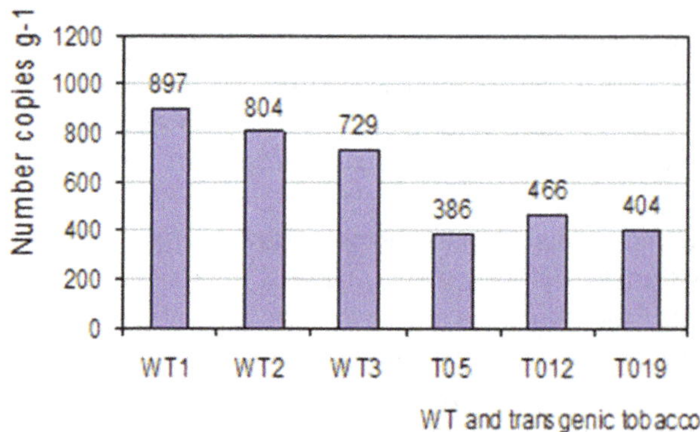

Figure 4. Comparative chart of the number of copies g^{-1} of tobacco leaf of transgenic tobacco lines and the WT plants after being infected 3 times with SMV and BYMV. WT1, WT2, WT3: the WT plants; T_0-5, T_0-12, T_0-19: the transgenic tobacco lines.

Table 4. Threshold cycle values for detecting SMV and the number of copies g^{-1} of tobacco leaf

Samples	Threshold cycle values	No. of copies g^{-1} of leaf
(-)	28.68	-
WT1 - infected	19.66	897
WT2 - infected	20.21	804
WT3 - infected	20.66	729
$T_0$5 - resistant	22.7	386
$T_0$12 - resistant	22.22	466
$T_0$19 - resistant	22.59	404
1x	19.52	1000
10x	23.21	100
100x	25.65	10

(-): negative control is H_2O, does not contain cDNA; WT1–infected, WT2–infected, and WT3–infected: three infected wild type plants; $T_0$5–resistant, $T_0$12–resistant, and $T_0$19–resistant: three resistant transgenic plants.

the expressed RNAi structure CPi (SMV-BYMV) improved SMV resistance in transgenic tobacco lines.

SMV and BYMV belong to the Potyvirus genus, the largest group of plant viruses (Dougherty and Carrington 1988). Their genomes consist of one single-stranded, linear, positive-sense RNA molecule (Dougherty and Carrington 1988). Mosaic disease caused by SMV and BYMV is one of the most common diseases in soybean, which severely reduces soybean yield and quality (Hartman et al. 1999). RNAi is considered to be a modern technique that can effectively protect plants against viruses. The effectiveness of RNAi in generating transgenic plants resistant to viruses has been demonstrated in a number of studies (Yan et al. 2007, Qiong et al. 2011, Pradeep et al. 2012, Zhang et al. 2012). In Vietnam, the first successful RNAi application in transgenic plants was the generation of transgenic tobacco plants resistant to Cucumber mosaic virus (CMV), to Tobacco mosaic virus (TMV), and to both of the viruses at the same time (Pham et al. 2008).

The approach of creating the RNAi structure from the viral CP gene was mentioned by Furutani et al. (2007) as an effective method to improve SMV resistance in soybean. However, in reality, soybean plants can be infected with multiple viruses simultaneously; thus, it is important to construct a transgene that can prevent both viruses causing mosaic disease. Based on the conserved regions of the CP coding sequences of SMV and BYMV, our team has developed an RNAi-based plant expression vector in order to suppress SMV and BYMV infection (Lo et al. 2014). In our study, using RNAi methods, we have achieved very promising results with 73.08% of transgenic plants showed complete resistance to SMV and BYMV. The analytic result of viral quantitative assays by real-time RT-PCR showed that the number of virus copies in transgenic tobacco plants was reduced on average by more than 51% compared to the control plants (Figure 4). The results obtained from this study reinforce the environmental-friendly approach to making virus-resistant crops. It also provides an important basis for transferring the CPi (SMV-BYMV) construct to soybean plants to generate soybean plants resistant to SMV and BYMV. Further studies include examining the effect of this method in generating soybean plants resistant to the SMV and BYMV viruses, as well as evaluating inheritance of this transgene in following generations in order to obtain a stable transgenic soybean line that is highly resistant to SMV and BYMV.

CONCLUSIONS

We have successfully generated new transgenic tobacco plants carrying RNAi [CPi (SMV-BYMV)] resistance to the two types of viruses and 73.08% of transgenic tobacco lines were fully resistant to SMV and BYMV. By using RT-PCR we have identified the number of virus copies in transgenic tobacco plants, which was reduced on average by more than 51% compared to the control plants. This promising result shows the potential of transferring the CPi (SMV-BYMV) structure to soybean, to increase resistance of soybean to SMV and BYMV and advance the aims of antiviral soybean breeding in Vietnam.

ACKNOWLEDGMENTS

The authors would like to express their gratefulness for the help of the Key Laboratory of Gene Technology, Institute of Biotechnology, Vietnam Academy of Science and Technology.

REFERENCES

Asad S, Haris WAA, Bashir A, Zafar Y, Malik KA, Malik NN and Lichtenstein CP (2003) Transgenic tobacco expressing geminiviral RNAs are resistant to the serious viral pathogen causing cotton leaf curl disease. **Archives of Virology 148**: 2341-2352.

Baulcombe D (2004) RNA silencing in plants. **Nature 431**: 356-363.

Bonfim K, Faria JC, Nogueira EO, Mendes EA and Aragão FJ (2007) RNAi-mediated resistance to bean golden mosaic virus in genetically engineered common bean (*Phaseolus vulgaris*). **Molecular Plant-Microbe Interactions 20**: 717-726.

Dougherty WG and Carrington JC (1988) Expression and function of potyviral gene products. **Annual Review of Phytopathology 26**: 123-143

Furutani N, Yamagishi N, Hidaka S, Shizukawa Y, Kanematsu S and Kosaka Y (2007) Soybean mosaic virus resistance in transgenic soybean caused by post-transcriptional gene silencing. **Breeding Science 57**: 123-128.

Hartman GL, Sinclair JB and Rupe JC (1999) **Compendium of soybean diseases**. 4th edn, The American Phytopathological Society Press, St. Paul, 128p.

Herbers K, Meuwly P, Frommer WB, Metraux JP and Sonnewald U (1996) Systemic acquired resistance mediated by the ectopic expression of invertase: Possible hexose sensing in the secretory pathway. **Plant Cell 8**: 793-803.

Karimi M, Inzé D and Depicker A (2002) GATEWAY™ vectors for *Agrobacterium*-mediated plant transformation. **Trends Plant Science 7**: 193-195.

Kasai M and Kanazawa A (2012) RNA silencing as a tool to uncover gene function and engineer novel traits in soybean. **Breeding Science 61**: 468-479.

Livak KJ and Schmittgen TD (2001) Analysis of relative gene expression data using real-time quantitative PCR and the 2(-Delta Delta C(T)) method. **Methods 25**: 402-408.

Lo TMT, Le VS, Chu HH and Chu HM (2014) Development of RNAi-Based vector aims at creating antiviral soybean plants in Vietnam. **International Journal of Bioscience, Biochemistry and Bioinformatics 4**: 208-211.

Pham TV, Nguyen VB, Le VS, Chu HH and Le TB (2008) Create tobacco plants resistant to cucumber mosaic virus disease by RNAi. **Vietnamese Journal of Biotechnology 6**: 679-687.

Pradeep K, Satya VK, Selvapriya M, Vijayasamundeeswari A, Ladhalakshmi D, Paranidharan V, Rabindran R, Samiyappan R, Balasubramanian P and Velazhahan R (2012) Engineering resistance against *Tobacco streak virus* (TSV) in sunflower and tobacco using RNA interference. **Biologia Plantarum 56**: 735-741.

Qiong H, Yanbing N, Kai Z, Yong L and Xueping Z (2011) Virus-derived transgenes expressing hairpin RNA give immunity to Tobacco mosaic virus and Cucumber mosaic virus. **Virology Journal 8**: 41.

Saghai MMA, Soliman KM, Jorgensen RA and Allard RW (1984) Ribosomal DNA spacer - length polymorphisms in barley: Mendelian inheritance, chromosomal location and population dymnamics. **Proceedings National Academic Science USA 81**: 8014-8018.

Topping JF (1998) Tobacco transformation. **Methods of Molecular Biology 81:** 365-372.

Yan PQ, Bai XQ, Wan XQ, Guo ZK, Li LJ, Gong HY and Chu CC (2007) Expression of TMV coat protein gene RNAi in transgenic tobacco plants confer immunity to tobacco mosaic virus infection. **Yi Chuan 29**: 1018-1022.

Zhang C, Song Y, Jiang F, Li G, Jiang Y, Zhu C and Wen FF (2012) Virus resistance obtained in transgenic tobacco and rice by RNA interference using promoters with distinct activity. **Biologia Plantarum 56**: 742-748.

2

Development and characterization of microsatellite loci for the Neotropical orchid Trichocentrum pumilum

Lia Maris Orth Ritter Antiqueira[1], Gabriel Dequigiovanni[2], Evandro Vagner Tambarussi[3], Jucelene Fernandes Rodrigues[2] and Elizabeth Ann Veasey[2*]

***Corresponding author:**
E-mail: eaveasey@usp.br

[1] Federal University of Technology - Paraná, Av Monteiro Lobato, km 4, 84.021-216, Ponta Grossa, PR, Brazil
[2] University of São Paulo, Luiz de Queiroz College of Agriculture, Av. Padua Dias, 11, PO Box 9, 13418-900, Piracicaba, SP, Brazil
[3] State University of Central West, PR 153, km 7, 84.500-000, Irati, PR, Brazil

Abstract: *Studies of genetic diversity and structure are key elements in designing effective in situ and ex situ management plans, especially for species experiencing forest fragmentation. To investigate the level of genetic diversity in populations of* Trichocentrum pumilum, *eight polymorphic microsatellite loci were developed and used for genotyping 96 specimens from four disturbed populations. Low genetic diversity within populations was found (average number of alleles per locus ranging from 3.75 to 4.25, observed and expected heterozygosities from 0.238 to 0.333 and from 0.450 to 0.482, respectively). The fixation index (F_{IS}) ranged from 0.35 to 0.47, with significant values for all populations. No genotypic disequilibrium was detected. A mixed breeding system was found through an apparent outcrossing rate estimate. Our results suggest that these microsatellite loci are suitable for genetic studies of this species, showing low within population genetic diversity and moderate structure for T. pumilum populations.*

Key words: *Conservation genetics, genetic diversity, Orchidaceae, population genetics, tropical orchid species.*

INTRODUCTION

The orchid family is one of the largest groups among vascular plants, corresponding to 10% of all flowering plants (Otero and Flanagan 2006). *Trichocentrum pumilum* (Lindl.) M. W. Chase & N.H. Williams is an epiphytic orchid with common occurrence in Brazil (Barros et al. 2013). This species is often found in gallery forest between the vegetation of semi-deciduous forest. It is considered to be self-incompatible and pollinator-limited, being exclusively visited and pollinated by two bee species (*Tetrapedia diversipes* and *Lophopedia nigrispinis*) (Pansarin and Pansarin 2011). The flowers are yellowish or greenish yellow with brown spots. In January, during fruit development, each plant produces a new pseudobulb, which sometimes produces a new lateral inflorescence for the next flowering season. The fruits come into dehiscence from June/July (Pansarin and Pansarin 2011).

Microsatellite markers are highly variable and, therefore, highly advantageous when compared to other markers (Kalia et al. 2011). This molecular marker has been used in population genetic studies of orchid species, including tropical epiphytic species, such as *Dendrobium loddigesii* (Cai et al. 2012), *Cattleya* spp.

(Almeida et al. 2013, Novello et al. 2013, Tambarussi et al. 2016), among others. However, there are no previous reports of genetic studies for *T. pumilum* and no information on the level of genetic diversity of its populations. Thus, this study aimed at the development of microsatellite markers and genetic characterization of four disturbed *T. pumilum* populations in order to generate useful information aiming at conservation strategies.

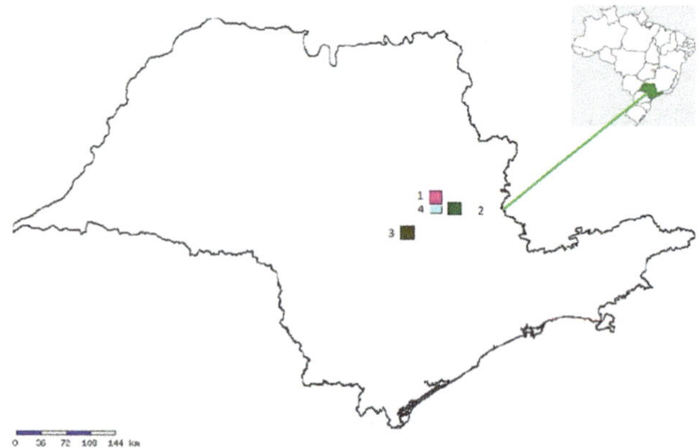

Figure 1. Collection sites of *Trichocentrum pumilum* in the State of São Paulo, Brazil. 1- Alto da Serra, São Pedro; 2- Iracemápolis; 3- Santa Maria da Serra; 4- São Pedro. Source: Atlas BIOTA FAPESP/ Center of Reference to Environmental Information (2015).

MATERIAL AND METHODS

Sampling procedures

Twenty-four individuals per population were sampled from four populations, at the municipalities of Iracemápolis (IRA), Santa Maria da Serra (SM), and São Pedro, in the State of São Paulo, Brazil (Figure 1). Populations sampled in São Pedro municipality were obtained in different areas of privately owned forest, one in São Pedro (SP) and the other in Alto da Serra (AS), at a higher altitude. All four populations were collected in forest fragments, next to agricultural landscape. The distance between populations varied from 1.6 (AS and SP) to 62.0 (IRA and SM) kilometers. Specimens of each population were deposited at the herbarium of the Escola Superior de Agricultura "Luiz de Queiroz", University of Sao Paulo, under the numbers ESA127076 and ESA127077.

DNA extraction, design, optimization of primers and sequencing/genotyping

DNA extraction from fresh leaves was based on Doyle and Doyle (1990), modified by Rodrigues et al. (2015). DNA quantification was performed on 1% agarose gels stained with GelRed (Biotium). Genomic library to obtain microsatellite markers was developed following Billote et al. (1999). DNA digestion, microsatellite enrichment, transformation of competent cells and sequencing of recombinant colonies were based on Tambarussi et al. (2013). We obtained 288 transformed clones, which were sequenced using universal T7 and SP6 primers with a BigDye v3.1 terminator kit on an ABI 3130*XL* Genetic Analyzer automated sequencer (Applied Biosystems, Foster City, USA). The sequences obtained were transformed in a FASTA format by BioEdit Sequence Alignment Editor program (Hall 1999) and vectors and adapters sequences were excluded. The program Websat (Martins et al. 2009) allowed the identification of microsatellite sequences. The parameters used for the identification of microsatellites were: dinucleotide length ≥5nt; tri-, tetra-, penta-, hexanucleotide with a repeat length ≥3 nucleotides. Primers were designed using Primer3 (Rozen and Skaletsky 2000) considering the following criteria: annealing temperature ranging from 54 ° to 62 °C, guanine and cytosine content between 40 and 60% and range between 100 and 300 base pairs. Gene Runner v.3.1 software (Spruyt and Buquicchio 1994) was used to confirm the values for each of the parameters and to indicate the formation of secondary structures which are undesirable. Finally, Chromas2 software (McCarthy 1996) was used to assess the quality of the sequencing primers and the regions of the microsatellites. All forward primers were labeled with M13 sequence (5'- TGTAAAACGACGGCCAGT-3') following a labeling protocol (Schuelke 2000).

The optimization of the microsatellites loci, conducted using six individuals per population, allowed the establishment of the optimal protocol and the appropriate temperature for each amplification primer. Microsatellite fragments were amplified using a MyCycler Thermal Cycler (Bio-Rad, Hercules, CA, USA) in a total reaction volume of 10 μL, containing 20 ng of genomic DNA template, 1 U *Taq* DNA polymerase (Fermentas, Vilnius, Lithuania), 1X polymerase chain reaction buffer (10 mM Tris-HCl (pH 8.3), 50 mM KCl, 1.5 mM $MgCl_2$, 0.08% Nonidet P40), 0.25 mM each dNTP, 1.5mM $MgCl_2$, 2.5 pmols of forward and M13 label primers (FAM, HEX or NED dyes) and 5 pmols of reverse primers. Polymerase chain reaction was carried out according to Schuelke (2000), consisting of 94 °C (5 min), then 30 cycles at 94 °C (30 s)/T_a °C (45 s) (T_a= annealing temperature (Table 1)/72 °C (45 s), followed by 8 cycles at 94 °C (30 s)/53 °C (45 s)/72 °C (45 s), and a final extension at 72 °C for 10 min. Allele sizes were determined using the ABI 3130xl Genetic Analyzer System (Applied Biosystems) at the Research Centre on the Human Genome and stem Cells of University of São Paulo. SSR patterns were scored based on the size standard ROX GSx500 (Life Technologies Inc.) using GENEMAPPER software v4.0 (Applied Biosystems). After selecting the polymorphic primers, 24 individuals from each population were genotyped using the same laboratory facilities and equipment cited above.

Data analysis

Fixation index (F_{IS}), genotypic disequilibrium, number of alleles per locus (*k*), observed (H_o) and expected heterozygosity (H_e) at Hardy-Weinberg equilibrium for each locus and as an average across all loci were estimated using FSTAT (Goudet 2002). To test whether F_{IS} and linkage disequilibrium between pairwise loci were significantly different from zero, we used Monte Carlo permutations of alleles between individuals and a Bonferroni correction (95%; α= 0.05). A global *F*-statistics analysis was obtained using GenAlex software (Peakall and Smouse 2012). The parameter apparent outcrossing rate ($\hat{t}_a$) was estimated considering the equation $\hat{t}_a$ = [(1–F)/(1+F)] (Vencovsky 1994) for each population.

RESULTS AND DISCUSSION

Two hundred and eighty-eight positive clones from the library were sequenced. A set of 30 microsatellite primers were developed (Table S1). From these 30 loci, eight were found to be polymorphic (Table 1). A low to moderate genetic diversity was found within *T. pumilum* populations. The number of alleles per loci ranged from one to eight, with averages of 4.00, 4.12, 4.25 and 3.75 for AS, IRA, SM and SP populations, respectively (Table 2). For the total sample considering all populations, we found 43 different alleles, with four private alleles within populations (data not shown). H_o and H_e ranged from 0.238 to 0.333 and from 0.450 to 0.482, respectively; F_{IS} was significantly higher than zero for all populations, varying from 0.35 (SP) to 0.47 (SM), suggesting excess of homozygotes (Table 2). After Bonferroni correction, no genotypic disequilibrium was detected in the studied populations (data not shown). *F*-statistics (F_{IS} = 0.337; F_{ST} = 0.064; F_{IT} = 0.417) confirmed moderate levels of genetic structure, showing that most of the diversity is within populations.

T. pumilum apparently is not under threat or in danger of extinction (IUCN 2002). However, the four populations considered in our study, collected from fragmented forest areas in the State of São Paulo, showed low within genetic variability, similar to the findings of Pandey et al. (2015) with *Cypripedium kentuckiense,* a terrestrial orchid native to North America *(A* = 4.00; H_o = 0.436; H_e = 0.448), which might be indicative of the occurrence of evolutionary processes, such as population fragmentation, reduction in gene flow and genetic drift, leading to a decrease of genetic diversity. Higher levels of genetic diversity were found by Trapnell et al. (2015) evaluating five populations of the epiphytic *Dendrobium*

Table 1. Microsatellite loci developed for *Thichocentrum pumilum*, including the primer name, forward and reverse sequences, repeat motif, size range, annealing temperatures (Ta) and the GenBank registration number

Locus	Sequences 5´- 3´	Repeat Motif	Size Range (pb)	Ta (°C)	GenBank
TpuB06	F: CTAACCAGAGCTTCCGCTGT R: AGTCATCTGTCCGCTACCTG	$(TTC)_5$	302 -314	56	KT271869
TpuC04	F: TGCATTTCTCCATTTTCAGCC R: AGCTGACCCCACATAGTGC	$(AAC)_8$	277 -299	56	KT271870
TpuB7	F: GAAATCATCGTCGTCTTCTGC R: GTGGCGATTCTTCCATTGTTG	$(CT)_{23}$	125 -161	56	KT271875
TpuG7	F: GTTGTCCCCATGAAACCATTG R: TGCCACCACTTCACAGAACT	$(GT)_7$	165 -169	56	KT271881
TpuC3_2	F: GGGTTGGGGTAGTTCTCGTG R: CCGTTTGTTTCAGCTTTTCC	$(GTTT)_3$	299 -303	56	KT271888
TpuC8_2	F: CGCCTTACCTAGCAGTGACA R: GTATCACTCAAAAGCCACTCA	$(AC)_{13}$	286 -304	56	KT271889
TpuD12_2	F: CCTGAGCTAAAATGGGTTGC R: GCGTGGACTAGCAAGAAAAC	$(CA)_4(GA)_5$	147 - 141	56	KT271892
TpuH4_2	F: GTCCTTACGGTGACCTTTCT R: GCAGATAACCAACGAAAATGC	$(TTTC)_3$	218 - 220	58	KT271896

Table 2. Genetic diversity parameters in 24 individuals of *Trichocentrum pumilum* for each population using eight microsatellite loci, including the number of alleles (k), the observed (H_o) and expected heterozygosities (H_e), and the fixation index within populations (F_{IS})

	Alto da Serra				Iracemápolis			
Locus	k	H_o	H_e	F_{IS}	k	H_o	H_e	F_{IS}
TpuB06	2	0.167	0.160	-0.07	2	0.208	0.190	-0.09
TpuC04	7	0.772	0.760	-0.02	8	0.834	0.792	-0.05
TpuB7	8	0.318	0.880	0.64*	8	0.042	0.839	0.95*
TpuG7	3	0.150	0.620	0.76*	3	0.143	0.606	0.76*
TpuC3_2	3	0.374	0.630	0.41	3	0.524	0.629	0.16
TpuC8_2	6	0.416	0.460	0.08	5	0.381	0.376	-0.01
TpuD12_2	1	NE**	0.000	NE	2	1.000	0.130	-0.05
TpuH4_2	2	0.095	0.260	0.63*	2	0.091	0.245	0.62*
Mean	4.00	0.333	0.460	0.39*	4.12	0.318	0.476	0.38*
SD***	2.62	0.231	0.308	0.345	2.59	0.284	0.277	0.43
Total	32	-	-	-	33		-	-
	Santa Maria				**São Pedro**			
Locus	k	H_o	H_e	F_{IS}	k	H_o	H_e	F_{IS}
TpuB06	3	0.174	0.170	-0.05	2	0.170	0.082	-0.02
TpuC04	6	0.400	0.590	0.32	6	0.539	0.732	0.09
TpuB7	8	0.273	0.840	0.68*	8	0.292	0.841	0.65*
TpuG7	3	0.227	0.680	0.66*	3	0.266	0.635	0.60*
TpuC3_2	3	0.250	0.230	-0.10*	3	0.195	0.632	0.14
TpuC8_2	7	0.458	0.780	0.41	4	0.574	0.732	0.26
TpuD12_2	2	0.000	0.080	1.00*	2	0.085	0.082	-0.02
TpuH4_2	2	0.125	0.260	0.51*	2	0.088	0.121	0.66*
Mean	4.25	0.238	0.450	0.47*	3.75	0.276	0.482	0.35*
SD	2.38	0.146	0.300	0.40	2.19	0.188	0.327	0.30
Total	34	-	-	-	30	-	-	-

*$P<0.05$ after a Bonferroni correction for multiple tests
**NE: not estimated
***SD is the standard deviation

calamiforme from Australia (A = 3.6; H_o = 0.489; H_e = 0.591), Mallet et al. (2014) with 10 populations of the epiphytic *Jumellea rossii* (A = 9.2; H_o = 0.463; H_e = 0.750), Kartzinel et al. (2013) studying 12 populations of *Epidendrum firmum* (A = 12.4; H_o = 0.785; H_e = 0.834), as well as Pinheiro et al. (2013) studying populations of *E. denticulatum*, including a population from a neighboring town in our study (Itirapina/SP), with H_o ranging from 0.331 to 0.516, and H_e from 0.395 to 0.529, all with nuclear SSR markers.

The apparent outcrossing rate ($\hat{t}_a$) showed values below 0.60 for each *T. pumilum* population, averaging 0.508, which is an indication that this species presents a mixed mating system, which may explain the higher levels of fixation index and lower levels of genetic diversity. Similar values of apparent outcrossing rate ($\hat{t}_a$ =0.43) was found for *Cattleya walkeriana* (Tambarussi et al. 2015). Besides the mixed breeding system found for *T. pumilum*, population fragmentation may also be a factor leading to lower genetic diversity values, since small fragmented populations may suffer severe loss of genetic diversity due to reduction in gene flow and genetic drift, and may experience extinction of the population in the long-term (Hamrick and Godt 1989). These are still speculative ideas, but future studies must also explore the putative role of fragmented and non-fragmented forest areas related to the loss of genetic diversity that could be tested with an increased number of populations.

In conclusion, the eight polymorphic loci reported in this study have proven to be useful for population genetics studies in *T. pumilum*. This is the first genetic study on this species, showing important data related to the genetic diversity of four populations and a preliminary indication of its mixed reproductive system, which should be of use in conservation programs and for further studies with other *T. pumilum* populations.

ACKNOWLEDGEMENTS

The authors wish to thank FAPESP for scholarships, especially the one attributed to the first author, which was also used to finance this study (Grant #2013/13918-6). EAV was supported by a CNPq research fellowship. The authors wish to thank Patricia Dias Santos for the contribution in the primers development.

REFERENCES

Almeida PRM, Roberts MCL, Vignada BBZ, Souza AP, Goes Neto AN and Van Den Berg C (2013) Microsatellite markers for the endangered orchids *Cattleya labiata* Lindl. and *C. warneri* T. Moore (Orchidaceae). **Conservation Genetics Resources 5**: 791-794.

Barros FD, Vinhos F, Rodrigues VT, Barberena FFVA, Fraga CN, Pessoa EM, Forster W and Menini Neto L (2013) Orchidaceae. In **Lista de espécies da flora do Brasil.** Jardim Botânico do Rio de Janeiro. Available at <http://floradobrasil.jbrj.gov.br/ jabot/floradobrasil/FB11942>. Accessed in Jan, 2016.

Billote N, Lagoda PJR, Risterucci AM and Baurens FC (1999) Microsatellite-enriched libraries: applied methodology for the development of SSR markers in tropical crops. **Fruits 54**: 277-288.

Cai X, Feng Z, Hou B, Xing W and Ding X (2012) Development of microsatellite markers for genetic diversity analysis of *Dendrobium loddigesii* Rolfe, an endangered orchid in China. **Biochemical Systematics and Ecology 43**: 42-47.

Doyle J and Doyle J (1990) Isolation of plant DNA from fresh tissue. **Focus 12**: 13-15.

Goudet J (2002) FSTAT, a program to estimate and test gene diversities and fixation indices (Version 2.9.3.2). Available at < http://www2.unil.ch/ popgen/softwares/ fstat.htm>. Accessed in April, 2016.

Hall TA (1999) BioEdit: a user-friendly biological sequence alignment editor and analysis program. **Nucleic Acids Symposium Series 41**: 95-98.

Hamrick JL and Godt MJW (1989) Allozyme diversity in plant species. In Brown AHD, Clegg MT, Kahler AL and Weir BS (eds) **Plant population genetics, breeding and genetic resources.** Sinauer Associates, Sunderland, p. 43-63.

IUCN - International Union for the Conservation of Nature (2002) Red list of threatened species. Available at < http://www.iucnredlist.org >. Accessed in Feb, 2016.

Kalia RK, Rai MK, Kalia S, Singh RA and Dhawan AK (2011) Microsatellite markers: an overview of the recent progress in plants. **Euphytica 177**: 309-334.

Kartzinel TR, Shefferson RP and Trapnell DW (2013) Relative importance of pollen and seed dispersal across a Neotropical mountain landscape for an epiphytic orchid. **Molecular Ecology 22:** 6048-6059.

McArthy M (2011) Chromas. v 2.01. School of Health Science, Griffith University, Australia. 1996-1998. Available at <http://technelysium. com.au/chromas.html>. Accessed in Nov, 2011.

Mallet B, Martos F, Blambert L, Pailler T and Humeau L (2014) Evidence for isolation-by-habitat among populations of an epiphytic orchid species on a small oceanic island. **Plos One 9**: e87469.

Martins WS, Lucas DCS, Neves KFS and Bertioli DJ (2009) WebSat - A Web software for microSatellite marker development. **Bioinformation 3**: 282-283.

Novello M, Rodrigues JF, Pinheiro F, Oliveira GCX, Veasey EA and Koehler S (2013) Simple-sequence repeat markers of *Cattleya coccinea* (Orchidaceae), an endangered species of the Brazilian Atlantic Forest. **Genetics and Molecular Research 12**: 3274-3278.

Otero JT and Flanagan NS (2006) Orchid diversity: Beyond deception. **Trends in Ecology and Evolution 21**: 64-65.

Pandey M, Richards M and Sharma J (2015) Microsatellite-based genetic diversity patterns in disjunct populations of a rare orchid. **Genetica 143**: 693-704.

Pansarin ER and Pansarin LM (2011) Reproductive biology of *Trichocentrum pumilum*: An orchid pollinated by oil-collecting bees. **Plant Biology 13**: 576-581.

Peakall R and Smouse PE (2012) GenAlEx 6.5: genetic analysis in Excel. Population genetic software for teaching and research—an update. **Bioinformatics 28**: 2537-2539.

Pinheiro F, Cozzolino S de BF, Gouveia, TMZM, Fay MF and Palma-Silva C (2013) Phylogeographic structure and outbreeding depression reveal early stages of reproductive isolation in the neotropical orchid *Epidendrum denticulatum*. **Evolution 67**: 2024-2039.

Rodrigues JF, Van Den Berg C, Abreu AG, Novello M, Veasey EA, Oliveira GCX and Koehler S (2015) Species delimitation of *Cattleya coccinea* and *C. mantiqueirae* (Orchidaceae): insights from phylogenetic and population genetics analyses. **Plant Systematics and Evolution 301**: 1345-1359.

Rozen S and Skaletsky H (2000) Primer3 on the WWW for General Users and for Biologist Programmers. In Krawetz S and Misener S (eds) **Bioinformatics methods and protocols: Methods in molecular biology**. Humana Press, Totowa, NJ, p. 365-386.

Schuelke M (2000) An economic method for the fluorescent labeling of PCR fragments. **Nature Biotechnology 18**: 233-234

Spruyt M and Buquicchio F (1994) Gene Runner. Available at <http:// www.generunner.net/>. Accessed in Jan, 2015.

Tambarussi EV, Sebbenn AM, Moreno MA, Ferraz EM, Kageyama PY and Vencovsky R (2013) Microsatellite markers for *Cariniana legalis* (Lecythidaceae) and their transferability to *C. estrellensis*. **Applications in Plant Sciences 1**: 1200493.

Tambarussi EV, Menezes LC and Ribeiro P (2015) Sistema de reprodução em *Cattleya walkeriana* Gardner. **Informativo da ACW 59**: 2-6.

Tambarussi EV, Menezes LC, Ibañes B, Antiqueira LMOR, Dequigiovanni G, Moreno MA, Ferraz EM, Zucchi MI, Veasey EA and Vencovsky R (2016) Microsatellite markers for *Cattleya walkeriana* Gardner, an endangered tropical orchid species. **Plant Genetic Resources 14**: 1-4.

Trapnell DW, Beasley RR, Lance SL, Field AR and Jones KL (2015) Characterization of microsatellite loci for an Australian epiphytic orchid, *Dendrobium calamiforme*, using Illumina sequencing. **Applications in Plant Sciences 3**: 1500016.

Vencovsky R (1994) Variance of an estimative of the outcrossing rate. **Brazilian Journal of Genetics 17**: 349-351.

3

Resistance to Fusarium wilt in common bean

Renata Oliveira Batista[1,2*], Ana Maria Cruz e Oliveira[2], Johnn Lennon Oliveira Silva[2], Alessandro Nicoli[1], Pedro Crescêncio Sousa Carneiro[3], José Eustáquio de Sousa Carneiro[2], Trazilbo José de Paula Júnior[4] and Marisa Vieira de Queiroz[5]

Abstract: *In breeding programs, understanding the potential of parents should be a way to spend significantly less time and costs to obtain new cultivars. For this, the objective of this study was to estimate the general and specific combining ability of parents aiming common bean breeding for resistance to Fusarium wilt (FW) based on disease severity and reduction in plant growth. Eight common bean genotypes were crossed in a 3 x 5 partial diallel mating scheme to obtain F_1 hybrids. The parents and their 15 F_1 hybrids were evaluated for severity of Fusarium wilt, area under the disease progress curve, percentage of plant height reduction and plant shoot fresh weight reduction and grain yield. The resistance of common bean to FW is controlled by a few dominant genes. The reduction in plant growth is controlled by a different set of genes that can increase the selection efficiency of parents for common bean breeding.*

Key words: *Genetic control, breeding for resistance, Fusarium oxysporum f. sp. phaseoli.*

***Corresponding author:**
E-mail: renata_agro@yahoo.com.br

[1] Universidade Federal Rural de Pernambuco (UFRPE), Departamento de Agronomia, 52.171-900, Recife, PE, Brazil
[2] Universidade Federal de Viçosa (UFV), Departamento de Fitotecnia, 36.570-900, Viçosa, MG, Brazil
[3] UFV, Departamento de Biologia Geral
[4] Empresa de Pesquisa Agropecuária de Minas Gerais (EPAMIG), 36.570-900, Viçosa, MG, Brazil
[5] UFV, Departamento de Microbiologia

INTRODUCTION

Fusarium oxysporum Schlecht., a widespread soil pathogen, causes vascular wilt in more than 100 plant species and is considered one of the 10 economically and scientifically most important pathogenic fungi in the world (Dean et al. 2012, Pantelides et al. 2013). In Brazil, *Fusarium oxysporum* f. sp. *phaseoli* (*Fop*) occurs in almost all common bean-producing areas (Toledo-Souza et al. 2012), reducing the yield significantly. According to Ramalho et al. (2012), this happens due to successively adoption of *Fop*-susceptible cultivars, which is hampering common bean cultivation under center pivot irrigation in several producing areas.

At *Fop*-free areas, the pathogen is generally introduced by infected seeds and contaminated agricultural implements. It survives either as a saprophyte or through the production of chlamydospores, which remain in the soil for many years. The integrated disease management of Fusarium Wilt (FW) is limited to crop rotation, seed treatment and use of resistant cultivars. Cross et al. (2000) and Pereira et al. (2009) stated that the use of resistant cultivars is the most effective method of controlling the economic losses caused by Fusarium wilt.

The severity of Fusarium wilt in common bean is usually assessed by a grade scale developed by Pastor-Corrales and Abawi (1987), based on the symptom of plant canopy wilting as main criterion. However, Pereira et al. (2013) reported that aside from wilting, the plant is affected by stunted growth in some *Fop*-infected bean cultivars. To assess this trait by quantifying the rate of stunted growth, plants of the same genotype are used as negative control (uninoculated), which restricts the evaluation to genotypes with high degree of uniformity (lines or hybrids).

Hybridization is a method used in bean breeding and an accurate selection of the parents for crossing is decisive to determine the success of the program (Ramalho et al. 2012). In diallel crosses, the general and specific combining ability can be estimated, and the predominant genetic control of the trait can be determined.

There are no reports of the use of stunted plant growth as a response of *Fop* susceptibility neither for choice of common bean parents to Fusarium wilt resistance. Thus, our objective was to estimate the general and specific combining ability of parents for common bean breeding for resistance to Fusarium wilt, based on disease severity and reduced plant growth.

MATERIAL AND METHODS

Plant material

Cultivars and elite lines of carioca bean with upright growth habit or high yield potential were chosen for group 1 of the partial diallel (BRS Estilo, VC 13 and VC 25). Group 2 included putative *Fop*-resistant genotypes BRSMG Talismã, CVIII 8511, Pérola, RC-I-8 and CNFC 11965. The diallel crosses were made by hand-pollinating the female parents without emasculating (Peternelli and Borém 1999). The crosses were made in a greenhouse, in a 3 x 5 partial diallel, in 2013, resulting in 15 F_1 hybrids. In stage R9, F_1 seeds were harvested and stored for later experiments.

Evaluated traits

Reaction to *Fop*

In the greenhouse, the eight parents and their 15 F_1 hybrids (totaling 23 treatments) were evaluated for *Fop* resistance. The experimental design was completely randomized with three replications in experimental units consisting of one pot with three plants.

The isolate FOP UFV 01 was collected from plants of the common bean cultivar Meia Noite with typical Fusarium wilt symptoms, in Coimbra (Minas Gerais, Brazil) (Pereira et al. 2013). Pereira et al. (2013) inoculated FOP UFV 01 in differential cultivars of beans, proposed by Woo et al. (1996) and Alves-Santos et al. (2002), and concluded that this isolate is a new race of the species. FOP UFV 01 inoculum was produced with PDA (potato - dextrose - agar) discs containing mycelia that were transferred to Petri dishes containing PDA. The plates were maintained for 14 days in a growth chamber at 25 ± 1 ºC, under a 12-h photoperiod. The spore suspension was prepared one hour before inoculation at a concentration of 1 x 10^6 conidia mL^{-1}, including macro and microconidia, as recommended by Pastor-Corrales and Abawi (1987).

To inoculate the genotypes, the roots were immersed in a conidia suspension, as proposed by Pastor-Corrales and Abawi (1987). In order to do so, seeds of the 23 genotypes were sown in 128 cell-trays containing Topstrato® vegetable substrate, and maintained in a greenhouse for germination and seedling growth. In stage V2 (fully expanded primary leaves), the seedlings were carefully removed from the trays, the roots were washed and 1/3 of their length was cut off. Immediately after cutting, the seedling roots were immersed in the macro and microconidia suspension of isolate FOP UFV 01 for 5 min. Thereafter, the seedlings were transplanted to plastic pots containing 2.5 L Topstrato® substrate and maintained in a greenhouse at 25 ± 3 ºC. The plants were irrigated daily and 10 days after inoculation (DAI), each pot was fertilized with 1.0 g urea as N source.

The reaction of genotypes to Fusarium wilt was evaluated in grades on a 1-9 disease severity scale, as described by Pastor-Corrales and Abawi (1987), based mainly on shoot wilting intensity, where: 1 = no visible symptoms; 3 = 1% to 10% of symptomatic leaves (leaves with mild chlorosis and wilting); 5 = 11% to 25% of symptomatic leaves (leaves with moderate chlorosis and wilting); 7 = 26% to 50% of symptomatic leaves (leaves with severe wilting and chlorosis) and 9 = dead or severely infected plant. The plants were assessed 15, 18 and 21 DAI, and the severity grade of Fusarium wilt in the last assessment (SFW) was used to classify genotypes for Fusarium wilt reaction. As suggested by Pastor-Corrales and Abawi (1987), the reaction of genotypes with a mean grade between 1.0 and 3.0 was considered resistant, from

3.1 to 6.0 intermediate and 6.1 to 9.0 susceptible.

The area under the disease progress curve (AUDPC) was calculated as proposed by Shaner and Finney (1977), in three assessments, as follows:

$$AUDPC = \sum_{i=1}^{n}\left[\left(\frac{Y_{i+1}+Y_i}{2}\right)(T_{i+1}-T_i)\right]$$ where:

Y_i = severity of Fusarium wilt at the ith observation,

T_i = time (days after inoculation) at the ith observation and

n = total number of evaluations.

Reduction in plant growth

The reduction in plant growth was assessed by reduction in plant height and fresh weight of the plant shoot. To this end, six seedlings of the same genotype were inoculated with FOP UFV 01 and 1/3 of their roots was cut off and immersed in distilled water for 5 min and transplanted into two pots, as a negative control. In this case, two replications and plots of one pot with three plants were used. After the last disease severity evaluation, one plant per pot of the FOP UFV 01-inoculated plots as well as of the negative control was cut and the height measured from the cotyledon node to the insertion of the last leaf, expressed in centimeters (cm). Then, each plant was wrapped separately in a paper bag and weighed to determine the shoot fresh weight (grams). The percentage of plant height reduction (PHR) and plant shoot fresh weight reduction (PSWR) in FOP UFV 01-inoculated plants, for each replication, was calculated from the mean height and fresh weight measured in the negative control plants, as follows:

$$PHR/PSWR = \frac{(Y_{ic} - Y_i) \times 100}{Y_{ic}}$$, where:

Y_{ic} = mean height or fresh weight of the shoot of genotype i plants inoculated in distilled water (negative control) and

Y_i = mean height or fresh weight of the shoot of genotype i plants inoculated with FOP UFV 01.

Grain yield

Grain yield (GY) was measured in a field experiment at the Experimental Station in Coimbra city (lat 20º 45’ S, long 42º 51’ W, alt 690 m asl). Seeds of the eight parents and their 15 F_1 hybrids (totaling 23 treatments) were sown in the 2014 dry season in a randomized block design with three replications. The plots consisted of two 1.5-m rows, spaced 0.50 m apart, with a planting density of 12 seeds per meter. At the time of harvest, all plants of the plot were cut by hand and the grains processed and weighed to determine grain yield in kg ha^{-1}. Note that no incidence of Fusarium wilt was detected in this season.

Genetic-statistical analyses

For the diallel analysis of SFW, AUDPC, PHR, PSWR and GY, the model of Griffing (1956) adapted to partial diallel was adopted, according to the following model:

$Y_{ij} = \mu + g_i + g'_j + s_{ij} + \bar{\varepsilon}_{ij}$, where

Y_{ij} = mean value of the hybrid combination between the ith parent of group I and the jth parent of group II; μ = overall mean of the diallel; g_i = effect of general combining ability of the ith group 1; g'_j = effect of general combining ability of the jth group of parent 2; s_{ij} = effect of specific combining ability between parents of order i and j, of groups 1 and 2, respectively, and $\bar{\varepsilon}_{ij}$ = mean experimental error. All genetic statistical analyses were performed with the software Genes (Cruz 2013).

RESULTS AND DISCUSSION

Performance of parents and F_1 hybrids

For the degree of SFW, the parents VC 13 and VC 25 from group 1, and BRSMG Talismã and CVIII 8511 of group 2 showed excellent performance, in terms of resistance to FOP UFV 01 (score = 1). Parent Pérola was also noteworthy (degree = 1.67) (Table 1). The other parents showed a reaction of susceptibility to *Fop* (scores > 7.7). Resistance is dominant to susceptibility based on the performance of the F1's. All families with one resistant parent produce resistant F1 progeny (R x R or R x S). All susceptible by susceptible crosses produced susceptible progeny (S x S). This is exactly what is expected from crosses between parents with dominant resistance genes. Other authors also observed dominance in the genetic control of Fusarium wilt resistance. Pereira et al. (2009) evaluated six crosses involving three resistant and four susceptible lines, and observed complete dominance in *Fop* resistance genetic control in common bean. The same result was reported by Cross et al. (2000) for two F_2 generations of crosses between *Fop*-resistant and *Fop*-susceptible Durango common bean parents.

The results of AUDPC for *Fop* reaction agreed with those of the SFW indicating that only the assessment at 21 days was sufficient to determine *Fop* severity and reaction (Resistant, Intermediate and Susceptible). *Fop* severity in common bean is usually evaluated only at 21 or 22 days (Pastor-Corrales and Abawi 1987, Cândida et al. 2009, Pereira et al. 2009). However, in evaluations performed in different years, at different locations and under different managements, as reported by Madden et al. (2007), the AUDPC proved effective to differentiate genotypes.

Considering the rate at which the growth of the parents is stunted, the percentage PHR and PSWR at *Fop*-susceptible plants was high, ranging from 43 to 63% (Table 1). Among the resistant parents, stunted growth comportment was different. For the parents VC 13, BRSMG Talismã and CVIII 8511, a PHR and PSWR of up to 11.11% was observed, while for VC 25 and Pérola, which are also resistant, the growth reduction rate was higher, ranging from 18 to 26%. The performance of the parents and hybrids for growth stunted measured by the PHR and PSWR in disagreement with *Fop*-resistance

Table 1. Means of parents and their F_1 hybrids for severity of Fusarium wilt (SFW), area under the disease progress curve (AUDPC), percentage of plant height reduction (PHR), percentage of shoot fresh weight reduction (PSWR), and grain yield (GY)

	Parents	**SFW**	**AUDPC**	**PHR** (%)	**PSWR** (%)	**GY** (kg ha^{-1})
	Group 1					
1	BRS Estilo	8.33	39.00	43.75	58.40	4195.51
2	VC 13	1.00	6.00	11.11	03.23	4150.62
3	VC 25	1.00	6.00	25.74	21.09	4944.86
	Group 2					
1	BRSMG Talismã	1.00	6.00	6.84	0	3745.32
2	RC-I-8	7.00	29.00	54.70	44.15	4256.23
3	CVIII 8511	1.00	6.00	4.76	02.52	4180.92
4	CNFC 11965	8.50	37.50	63.72	60.10	3830.17
5	Pérola	1.67	7.00	19.33	18.55	3950.71
Mean		3.69	17.06	28.00	26.00	4156.79
	Hybrids					
1	BRS Estilo x BRSMG Talismã	1.00	6.00	47.75	42.18	4377.88
2	BRS Estilo x RC-I-8	7.70	37.80	41.67	48.96	5301.02
3	BRS Estilo x CVIII 8511	1.00	6.00	37.84	39.45	4987.41
4	BRS Estilo x CNFC 11965	9.00	45.25	54.95	57.68	4011.69
5	BRS Estilo x Pérola	1.00	6.00	53.79	56.20	5627.19
6	VC 13 x BRSMG Talismã	1.00	6.00	14.73	20.16	4201.39
7	VC 13 x RC-I-8	1.00	6.00	30.13	43.48	4787.72
8	VC 13 x CVIII 8511	1.00	6.00	17.02	12.49	4532.40
9	VC 13 x CNFC 11965	1.00	6.00	37.31	41.05	4604.43
10	VC 13 x Pérola	1.00	6.00	19.22	19.06	4669.39
11	VC 25 x BRSMG Talismã	1.00	6.00	18.33	27.87	4689.64
12	VC 25 x RC-I-8	1.00	6.00	44.14	53.53	4773.31
13	VC 25 x CVIII 8511	1.00	6.00	09.43	06.04	4643.84
14	VC 25 x CNFC 11965	1.00	6.00	35.90	35.95	5521.64
15	VC 25 x Pérola	1.00	6.00	29.68	25.28	5522.47
Mean		1.98	10.74	32.66	35.29	4816.76

assessed by SFW or AUDPC indicates that the genetic control of stunted growth is more complex, being controlled by a set of dominant and recessive genes. Thus, the results of SFW compared with the PHR and PSWR indicate that the genes involved in Fusarium wilt resistance are probably not those involved in stunted growth. These results also show that the traits PHR and PSWR, in addition to SFW, can potentially be used in the selection and discrimination of parents for *Fop* resistance.

With regard to the growth reduction rate of the hybrids, in general, hybrids resulting from crosses involving at least one parent with high PSWR and PHR values also had a high reduction rate, e.g., hybrids 1, 2 3, 4, 5, 7, 10, 11, and 12 (Table 1). The hybrids 9 to 14, however, resulting from crosses of CNFC 11965 with VC 13 and VC 25, had stunted growth with values close to the mean growth reduction rate of their parents. Hybrids 6 and 8, derived from parents that are resistant and have low growth reduction, had higher reduction rates than their parents. Hybrid 15 resulting from a cross between VC 25 and Pérola, both of which are resistant and have high reduction rates, had a high reduction rate as well. However the growth reduction in hybrid 13, resulting from a cross between VC 25 (resistant with a high reduction rate) with CVIII 8511 (resistant with a low reduction rate) was lower. These divergent results evidence the complexity of genetic control of growth reduction rate of common bean in *Fop* infection.

The mean grain yield of the parents (4156.79 kg ha^{-1}) and their hybrids (4816.76 kg ha^{-1}) exceeded the national mean (1512 kg ha^{-1}) for this growing season. There was no incidence of Fusarium wilt in the evaluation experiment of grain yield of parents and their F_1 hybrids. This trait will be studied in more detail in the diallel analysis.

Diallel analysis

The treatments had a significant effect (p <0.01) on the traits SFW, AUDPC, PHR, PSWR, and GY (Table 2). Sum of squares of treatments was partitioned into effect of parents, hybrids and parents vs. contrast hybrids (Table 2). The effect of parents of groups 1 and 2 was also significant (p <0.05) on all traits, except for GY, for group 2 informing that, for the significant traits, there is variability between parents in the frequency of favorable alleles involved in the genetic control of the traits under study. The effect of G1 in comparison with G2 on the traits SFW, PHR, PSWR, and GY was significant (p <0.05) and non-significant on AUDPC.

The effect of hybrids was significant for all traits, indicating variability among the 15 hybrids (F_1 generation). This effect of hybrids was partitioned into effects of general combining ability (GCA_1 and GCA_2) and specific combining ability (SCA), which were significant for all traits (Table 2). The significance of the effects of general combining ability indicates that there is a difference in the frequency of favorable alleles between the parents of that group, while the significance of SCA indicates the presence of dominance deviations in the genetic control of the traits under study.

Table 2. Analysis of variance and diallel of the parents in groups 1 and 2 (G_1 and G_2) and their hybrids for severity of Fusarium wilt (SFW), area under the disease progress curve (AUDPC), percentage of plant height reduction (PHR), percentage of shoot fresh weight reduction (PSWR) and grain yield (GY)

Sources of Variation	df	Mean Square				
		SFW	AUDPC	PHR	PSWR	GY
Treatments	22	27.05**	553.13**	734.49**	980.17**	893893.77**
Parents (P)	7	38.01**	699.52**	1058.72**	1749.13**	407797.01*
Parents P_1	2	53.78**	1089**	630.09**	2433.26**	597176.41*
Parents P_2	4	39.42**	679.65**	1481.35**	1506.53**	145691.53
P_1 vs P_2	1	0.85*	0.57	225.44*	1351.29**	1077460.17*
Hybrids (H)	14	20.25**	474.70**	554.23**	480.50**	713829.40**
GCA_1	2	43.22**	1009.62**	1935.36**	1006.89**	854357.21*
GCA_2	4	16.42**	385.54**	389.09**	289.95**	904633.54**
SCA	8	16.42**	385.54**	291.52**	444.19**	583295.38**
P vs H	1	45.63**	626.34**	998.52**	2592.66**	6817472.19**
Error	46	0.17	4.44	57.74	44.34	189817.60
CV (%)		16.10	16.29	24.10	19.81	9.50
Mean		2.57	12.94	31.53	33.61	4587.21

df - degree of freedom. *,** Significant at 5%and 1% probability, respectively, by the F test.

The sum of squares of the GCA ($GCA_1 + GCA_2$) compared to the sum of squares of SCA was similar for the traits SFW, AUDPC, PSWR and GY in magnitude, indicating a predominance of genes with complete dominance in the control of these traits. For the trait PHR, the sum of squares of the GCA ($GCA_1 + GCA_2$) exceeded that of SCA, with percentages of 70% and 30%, respectively, indicating predominance of genes with additive effects in the genetic control of this trait.

General combining ability

The parents VC 13, VC 25, BRSMG Talismã, CVIII 8511, and Pérola stood out having the lowest GCA values for SFW and AUDPC indicating a higher frequency of *Fop* resistance alleles (Table 3). It is worth remembering that these parents were resistant to *Fop* and had the same GCA values. These results suggest that these parents have the same *Fop*-resistance alleles, and that *Fop*-resistance is controlled by few dominant genes. For stunted growth, among the resistant parents (Table 2), VC 13 of group 1 and CVIII 8511 of group 2 had the lowest GCA values for PHR and PSWR indicating that resistant-parents VC 13 and CVIII 8511 have lower frequency of alleles responsible for stunted growth caused by the pathogen. On the other hand, parent Pérola of group 2, also classified as resistant by SFW and AUDPC, had positive GCA for PHR and PSWR, indicating a higher frequency of alleles involved in stunted growth. These results indicate that different genes are involved in genetic control on traits of *Fop* severity (SFW and AUDPC) and stunted growth (PHR and PSWR) in common bean. Once traits SFW and AUDPC were evaluated on the scale of disease severity ratings developed by Pastor-Corrales and Abawi (1987), the shoot wilting symptom is the main criterion. However, others symptoms such as plant growth and reduction of shoot biomass are not considered on this scale, besides happening in screening essays. Andrade et al. (2009) and Araújo and Teixeira (2012) reported that genotypes with reduced biomass production have lower grain yield. Facing these divergent results related to the performance of resistant-parents, stunted growth must also be taken into account in the selection of parents for Fusarium wilt resistance. The GCA values of parents VC 25 of group 1 and BRSMG Talismã of group 2 were lower than those of Pérola, but higher than those of parents VC 13 and CVIII 8511.

The parents VC 25 in group 1 and Pérola from group 2, in addition to being *Fop*-resistant, had positive GCA and high GY (Table 4), indicating a higher frequency of favorable alleles for yield in these parents. BRS Estilo and RC-I-8 also had a positive but lower GCA. The lines of VC 13 of group 1 and CVIII 8511 of group 2, despite a negative GCA for GY, are *Fop*-resistant and have lower GCA values for stunted growth.

Specific combining ability

For the *Fop* severity traits (SFW and AUDPC), the SCA values were only negative (Table 4) in hybrids derived from crosses between contrasting parents for *Fop* resistance (R x S). All hybrids derived from crosses between resistant parents (R x R) had positive and equal SCA estimates. These results confirm the predominance of few genes with complete dominance in the genetic control of Fusarium wilt resistance. For the traits PHR and PSWR, the hybrid SCA estimates varied both in magnitude and in sign (Table 4), regardless of the *Fop*-reaction of their parents as evaluated by SFW and AUDPC, indicating greater complexity of the trait stunted growth as compared to *Fop*-resistance evaluated by SFW or AUDPC. Hallauer and Miranda (1988) reported that crosses involving lines with the same genetic basis have negative SCA estimates. However, for traits such as stunted growth, for which the lowest values are preferred, positive high-

Table 3. General capacity of the parents used in the partial diallel for severity of Fusarium wilt (SFW), area under the disease progress curve (AUDPC), percentage of plant height reduction (PHR), percentage of shoot fresh weight reduction (PSWR) and grain yield (GY)

	Parents	**SFW**	**AUDPC**	**PHR**	**PSWR**	**GY**
	Group 1					
1	BRS Estilo	1.96	9.47	12.90	9.21	44.27
2	VC 13	-0.98	-4.74	-8.50	-6.46	-257.69
3	VC 25	-0.98	-4.74	-4.40	-2.75	213.42
	Group 2					
1	BRSMG Talismã	-0.98	-4.74	-5.10	-4.68	-393.79
2	RC-I-8	1.25	5.86	0.01	3.25	137.25
3	CVIII 8511	-0.98	-4.74	-7.60	-6.91	-95.54
4	CNFC 11965	1.69	8.35	8.42	6.81	-104.18
5	Pérola	-0.98	-4.74	4.27	1.53	456.25

Table 4. Specific combining ability of parents used in the partial diallel for severity of Fusarium wilt (SFW), area under the disease progress curve (AUDPC), percentage of plant height reduction (PHR), percentage of shoot fresh weight reduction (PSWR), and grain yield (GY)

	Hybrid	SFW	AUDPC	PHR	PSWR	GY
1	BRS Estilo x BRSMG Talismã	-1.96	-9.47	5.65	-0.44	-89.37
2	BRS Estilo x RC-I-8	2.51	11.73	-5.54	-9.55	302.72
3	BRS Estilo x CVIII 8511	-1.96	-9.47	-1.76	-0.94	221.91
4	BRS Estilo x CNFC 11965	3.37	16.69	-0.67	3.57	-745.18
5	BRS Estilo x Pérola	-1.96	-9.47	2.32	7.36	309.90
6	VC 13 x BRSMG Talismã	0.98	4.74	0.81	-6.78	36.11
7	VC 13 x RC-I-8	-1.25	-5.86	-8.69	-5.40	91.40
8	VC 13 x CVIII 8511	0.98	4.74	12.63	17.33	68.89
9	VC 13 x CNFC 11965	-1.69	-8.35	3.09	2.62	149.54
10	VC 13 x Pérola	0.98	4.74	-7.84	-7.77	-345.93
11	VC 25 x BRSMG Talismã	0.98	4.74	-6.46	7.22	53.25
12	VC 25 x RC-I-8	-1.25	-5.87	14.24	14.95	-394.12
13	VC 25 x CVIII 8511	0.98	4.74	-10.87	-16.39	-290.80
14	VC 25 x CNFC 11965	-1.69	-8.35	-2.42	-6.19	595.64
15	VC 25 x Pérola	0.98	4.74	5.52	0.41	36.03

magnitude estimates indicate less divergence between the parents.

Among the hybrids obtained from crosses of resistant parents (R x R) with negative GCC values for stunted growth (parents with low frequency of alleles involved in stunted growth), hybrid 8 (VC 13 x CVIII 8511) had the highest SCA, and hybrid 13 (VC 25 x CVIII 8511) had the lowest SCA, indicating the low divergence of parent CVIII 8511 from parent VC 13 and high divergence from parent VC 25 for the traits PHR and PSWR.

The SCA estimate of hybrid 15 (VC 25 x Pérola) was positive for GY and its parents had the highest GCA values. This hybrid only performs well in grain yield, since no Fusarium wilt occurred in the field and the parents presented a moderate reduction in growth rate. The segregating population resulting from the cross between hybrids 13 and 15 is promising for simultaneous breeding of common bean for the traits grain yield, Fusarium wilt resistance and stunted growth.

The results of this study show that the use of the trait stunted growth as assessed by PHR and PSWR, associated with *Fop*-resistance by SFW and AUDPC, increases effectiveness on identification of lines that are sources of Fusarium wilt resistance as well as in choosing parents, based on diallel analysis. It is worth noting that these traits (PHR and PSWR) should be used in a diallel analysis with parents and F_1 generations, since a negative control is required to evaluate stunted growth, consisting of plants of the same genotype as those that are *Fop*-inoculated.

In common bean breeding, the step after choosing the segregating populations with the greatest potential for breeding of superior lines is the evaluation of inbred families, derived from these populations. These families are evaluated in two or three generations (Ramalho et al. 2001) for the traits with a more complex inheritance, including grain yield and plant architecture. At this stage, it would not be possible to evaluate stunted growth due to the lack of a negative control (plants of the same genotype). However, as there is strong evidence that severity of Fusarium wilt is controlled by a few dominant genes, the inoculation of F_2 plants with *Fop* and breeding of resistant lines only will ensure the presence of the resistance gene in plants within the lines in later inbred generations. When breeding lines are obtained, *Fop* inoculation could also be used again, considering only lines with resistant plants for evaluation in field trials in different years, seasons and locations. At this stage, the *Fop*-resistant lines can also be assessed for the trait stunted growth, increasing the chances of success in breeding common bean for Fusarium wilt resistance.

The resistance of common bean to Fusarium wilt (isolate FOP UFV 01) is controlled by a few dominant genes, while reduction in plant growth, a response of *Fop* susceptibility, is governed by another set with dominant and recessive genes. Stunted growth can potentially be used to select parents and inbred common bean lines for breeding for resistance to Fusarium wilt. The segregating population, derived from the cross VC 25 / CVIII 8511 // VC 25 / Pérola, is promising for common bean breeding for the traits grain yield, resistance to Fusarium wilt and stunted growth.

ACKNOWLEDGMENTS

This research was supported by the following Brazilian agencies: the Minas Gerais Science Foundation (FAPEMIG – Fundação de Amparo à Pesquisa do Estado de Minas Gerais) and, the National Council of Scientific and Technological Development (CNPq – Conselho Nacional de Desenvolvimento Científico e Tecnológico).

REFERENCES

Alves-Santos FM, Cordeiro-Lopez L, Sayagués JM, Martín-Domingues R, Garcia-Benavides P, Crespo MC, Días-Domingues JM and Eslava AP (2002) Pathogenicity and race characterization of *Fusarium oxysporum* f. sp. *phaseoli* isolates from Spain and Greece. **Plant Pathology 51**: 605-611.

Andrade CAB, Scapim CA, Braccini AL and Martorelli DT (2009) Produtividade, crescimento e partição de matéria seca em duas cultivares de feijão. **Acta Scientiarum, Agronomy 31**: 683-688.

Araújo AP and Teixeira MG (2012) Variabilidade dos índices de colheita de nutrientes em genótipos de feijoeiro e sua relação com a produção de grãos. **Revista Brasileira de Ciência do Solo 36**: 137-146.

Cândida DV, Costa JGC, Rava CA and Carneiro MS (2009) Controle genético da murcha do fusário (*Fusarium oxysporum*) em feijoeiro comum. **Tropical Plant Pathology 34**: 379-384.

Cross H, Brick MA, Schwartz HF, Panella LW and Byrne PF (2000) Inheritance of resistance to fusarium wilt in two common bean races. **Crop Science 40**: 954-958.

Cruz CD (2013) Genes - a software package for analysis in experimental statistics and quantitative genetics. **Acta Scientiarum 35**: 271-276.

Dean R, Van Kan JA, Pretorius ZA, Hammond-Kosack KE, Pietro AD, Spanu PD, Rudd JJ, Dickman M, Kahmann R, Ellis J and Foster GD (2012) The top 10 fungal pathogens in molecular plant pathology. **Molecular Plant Pathology 13**: 414-430.

Griffing B (1956) Concept of general and specific combining ability in relation to diallel crossing systems. **Australian Journal of Biological Sciences 9**: 463-493.

Hallauer AR and Miranda JB (1988) **Quantitative genetics in maize breeding.** 2nd edn, Iowa State University, Ames, 664p.

Madden, LV, Hughes G and Bosch FVD (2007) **The study of plant disease epidemics.** American Phytopathological Society, St. Paul, 421p.

Pantelides IS, Jamos SET, Pappa S, Kargakis M and Paplomatas EJ (2013) The ethylene receptor ETR1 is required for *Fusarium oxysporum* pathogenicity. **Plant Pathology 62**: 1302-1309.

Pastor-Corrales MA and Abawi GS (1987) Reactions of selected bean germplasms to infection by *Fusarium oxysporum* f. sp. *phaseoli*. **Plant Disease 71**: 990-993.

Pereira AC, Cruz MFA, Paula-Júnior TJ, Rodrigues FA, Carneiro JES, Vieira RF and Carneiro PCS (2013) Infection process of *Fusarium oxysporum* f. sp. *phaseoli* on resistant, intermediate and susceptible bean cultivars. **Tropical Plant Pathology 38**: 323-328.

Pereira MJZ, Ramalho MAP and Abreu AFB (2009) Inheritance of resistance to *Fusarium oxysporum* f. sp. *phaseoli* Brazilian race 2 in common bean. **Scientia Agricola 66**: 788-792.

Peternelli LA and Borém A (1999) Hibridação em feijão. In Borém A (ed) **Hibridação artificial de plantas**. UFV, Viçosa, p. 269-294.

Ramalho MAP, Abreu AFB and Santos JB (2001) Melhoramento de espécies autógamas. In Nass LL, Valois ACC, Melo IS and Valadares-Inglis MC (eds) **Recursos genéticos e melhoramento de plantas**. Fundação MT, Rondonópolis, p. 201-230.

Ramalho MAP, Abreu AFB, Carneiro JES, Wendland A, Paula Junior TJ, Vieira RF, Del Peloso MJ, Lobo Junior M and Pereira AC (2012) Murcha-de-fusário. In Paula Junior TJ and Wendland A (ed) **Melhoramento genético do feijoeiro-comum e prevenção de doenças.** Epamig, Viçosa, p. 127-138.

Shaner G and Finney RF (1977) The effects of nitrogen fertilization on the expression of show-mildwing in knox wheat. **Phytopathology 67**: 1051-1055.

Toledo-Souza ED, Silveira PM, Café-Filho AC and Lobo-Júnior M (2012) Fusarium wilt incidence and common bean yield according to the preceding crop and the soil tillage system. **Pesquisa Agropecuária Brasileira 47**: 1031-1037.

Woo SL, Zoina A, Del Sorbo G, Lorito M, Nanni B, Scala F and Noviello C (1996) Characterization of *Fusarium oxysporum* f. sp. *phaseoli* by pathogenic races, VCGs, RFLPs, and RAPD. **Phytopathology 86**: 966-973.

4

Application of microsatellite markers to confirm controlled crosses and assess genetic identity in common bean

Samara Rayane Pereira de Morais[1], Ariadna Faria Vieira[2], Laura Cristina da Silva Almeida[1], Luana Alves Rodrigues[3], Patrícia Guimarães Santos Melo[1], Luís Cláudio de Faria[3], Leonardo Cunha Melo[3], Helton Santos Pereira[3] and Thiago Lívio Pessoa Oliveira de Souza[3*]

***Corresponding author:**
E-mail: thiago.souza@embrapa.br

[1] Universidade Federal de Goiás (UFG), 74.690-900, Goiânia, GO, Brazil
[2] Universidade Federal de Minas Gerais (UFMG), 39.404-547, Montes Claros, MG, Brazil
[3] Embrapa Arroz e Feijão, 75.375-000, Santo Antônio de Goiás, GO, Brazil

Abstract: *This manuscript reports on the application of microsatellite markers to assess the effectiveness of controlled crosses and assess the genetic identity of seed samples from a same common bean cultivar (BRS Estilo). The DNA was extracted from leaf tissue collected from F_1 plants and their parents as well as from seed samples of the cv. BRS Estilo by the alkaline lysis method. In all cases, genotyping was carried out with 24 microsatellite markers distributed in four multiplex panels with six markers each. Of the 392 F_1 plants analyzed, obtained from 21 different biparental crosses, 325 (82.91%) was confirmed as real hybrids by the molecular analysis. The genetic analysis of the four BRS Estilo seed samples (two of certified seeds and two of saved seeds) detected 100.00% genetic identity with the samples used as control in only one certified seed sample.*

Key words: Phaseolus vulgaris, *molecular markers, genetic similarity, marker-assisted selection, plant breeding.*

INTRODUCTION

Common bean (*Phaseolus vulgaris* L.) is an agronomic legume species with key importance for the world, because of its wide use as staple food, especially in developing countries in Africa and Latin America (Broughton et al. 2003). Brazil is currently one of the leading producers and the world's largest consumer of common bean with a grain production of 2.7 million tons in 2014, on an acreage of 1.94 million hectares, resulting in an average yield of approximately 1,400 kg ha^{-1} (http://www.cnpaf.embrapa.br/socioeconomia/index.htm). Nevertheless, there are many factors that limit the crop performance in the country, related to the cropping system, distribution and marketing of the product, as well as associated with regional preferences. There are also other crop-related factors which can be addressed by breeding, such as grain yield and quality, resistance to pests and diseases, plant architecture and lodging, earliness, efficient biological nitrogen fixation and nutrient uptake, and tolerance to drought and heat (Barros and Souza 2012).

In the early 80s, with the introduction and more routine use of molecular genetic techniques, various types of molecular markers were tested, which proved useful as tools to assist and optimize the different stages of plant breeding programs. Among these markers are microsatellites, widely used in studies on genetic diversity and mapping, due to the simplicity, high repeatability and informative content of this technique, and for being codominant and multi-allelic (Muller et al. 2014).

The number of populations generated annually by crosses in the scope of a nationwide plant breeding programs is quite considerable. Thus, the use of molecular tools that can enhance or monitor this process is particularly important. Alzate-Marin et al. (1996) have already demonstrated the efficiency of RAPD (Random Amplified Polymorphic DNA) markers for the confirmation of biparental crosses between genetically close parents of common bean and soybean without phenotypic contrast in readily assessable morphological traits. Another important stage of plant breeding programs is the confirmation of the identity and genetic purity of the genotypes during the seed production process, which can also be monitored or improved with molecular markers.

The main goal of this study was to evaluate the efficiency of controlled biparental crosses in common bean and assess the genetic identity of BRS Estilo seed samples using microsatellite markers.

MATERIAL AND METHODS

Plant material

In the framework of the common bean breeding program conducted by Embrapa and partners, 392 F_1 plants resulting from 21 different biparental crosses performed under controlled conditions were evaluated. The purpose of these crosses was to develop segregating populations for inheritance studies and mapping of genes or QTLs (Quantitative Trait Loci) associated with disease resistance and slow darkening of grains of the Carioca market class.

Four seed samples of cv. BRS Estilo (Melo et al. 2010) were obtained from common bean growers of traditional producing areas in the State of Goiás. These samples represent four separate lots, two of certified seeds and two of saved seeds (grains used as seeds). All seed samples were analyzed regarding their genetic identity, using as reference two control samples consisting of genetic seed of BRS Estilo provided by Embrapa Arroz e Feijão, the developing institution and custodian of the cv. BRS Estilo. The tested samples and controls consisted of 1.0 kg of seeds each. It is noteworthy that during the breeding process for the development of BRS Estilo, plants were selected in the F_5 and again in the F_8 generation, as described by Melo et al. (2010). The common bean line 'LM 98202709', the original identification of BRS Estilo, was selected in the $F_{8:9}$ generation. This indicates that BRS Estilo is in fact a line of which all loci, or something very close to 100% of the loci, are expected to be homozygous.

DNA extraction and analysis with molecular markers

The DNA was extracted from leaf tissue samples of F_1 plants and their parents and from seed samples of cv. BRS Estilo using the alkaline lysis method, as described by Xin et al. (2003), but with the modifications proposed by Valdisser et al. (2013). For DNA extraction from seeds, subsamples of 5.0% of the total weight of the initial samples (50 g) were weighed and ground in the laboratory, i.e., one subsample of 50 g for each initial 1.0 kg sample, using a portable mill, model IKA A11 basic (IKA Werke GmbH & Co.).

In the molecular marker analyses, DNA samples from all parents were first analyzed using 24 microsatellite markers distributed in four multiplex panels with six markers each (Valdisser et al. 2013) (Table 1). For the genotyping of the resulting F_1 plants, at least one marker identified as polymorphic among the parents of each population was used to check the hybrid nature of those plants. The amplification reactions of the DNA from all tested plants were performed by multiplex PCR, with a final volume of 5.0 μL, as described by Valdisser et al. (2013). Capillary electrophoresis of the DNA amplified products was performed with an ABI 3500xL sequencer (Applied Biosystems), always using a marker with known molecular weight, the GeneScan 500 ROX Size Standard (Life Technologies). The final genotyping (allele calling) was performed with software GeneMapper (Applied Biosystems).

For the genetic identity analysis of four BRS Estilo seed samples, the genetic similarity of these samples was estimated in relation to the control samples consisting of genetic seeds of the cultivar. Of the 24 microsatellite markers used in the analyses those that generated clear and robust polymorphisms were considered. Subsequently, the alleles detected among the tested seed samples were coded in a data matrix which was used to calculate the estimates of genetic similarity (*SGij*), based on the simple matching coefficient using the software Genes (Cruz 2013).

Table 1. Microsatellite markers grouped in four panels used for molecular characterization of common bean, with the identification of the loci, their respective fluorescence, amplification range in base pairs (bp), primer sequences, and genome location

Panel	Marker	Fluores-cence	Amplification range (pb)	Primer F	Primer R	Chromo-some[a]
1	BM143	HEX	100-170	GGGAAATGAACAGAGGAAA	ATGTTGGGAACTTTTAGTGTG	Pv02
	PVBR25	6-FAM	140-180	GAGCTTCTCCGTCCTGTGT	CGAACTGAATCAGAAAGGAA	Pv09
	BM164	NED	130-190	CCACCACAAGGAGAAGCAAC	ACCATTCAGGCCGATACTCC	Pv02
	BM114	6-FAM	230-260	AGCCTGGTGAAATGCTCATAG	CATGCTTGTTGCCTAACTCTCT	Pv09
	BM138	NED	190-210	TGTCCCTAAGAACGAATATGGAATC	GAATCAAGCAACCTTGGATCATAAC	Pv05
	PVBR169	HEX	195-220	TGGAAAGTCGGAGGAGAAGA	AAAAGGGTCCCAACCAAAAC	Pv03
2	PVBR5	HEX	160-220	ATTAGACGCTGATGACAGAG	AGCAGAATCCTTTGAGTGTG	Pv06
	PVBR35	6-FAM	190-260	TCTACGCGTTCCCTCTGTCT	AGTGGATGTGTGGGAAAAGC	Pv04
	BM202	6-FAM	100-173	ATGCGAAAGAGGAACAATCG	CCTTTACCCACACGCCTTC	Pv11
	BM189	NED	80-120	CTCCCACTCTCACCCTCACT	GCGCCAAGTGAAACTAAGTAGA	Pv03
	BM210	NED	160-220	ACCACTGCAATCCTCATCTTTG	CCCTCATCCTCCATTCTTATCG	Pv07
	BM155	HEX	100-145	GTTCATGTTTGTTTGACAGTTCA	CAGAAGTTAGTGTTGGTTTGATACA	Pv05
3	BM187	HEX	120-220	TTTCTCCAACTCACTCCTTTCC	TGTGTTTGTGTTCCGAATTATGA	Pv06
	PVBR113	NED	60-110	TGCATTCTTCCTCCCATCTT	TTGATTTGATTTGATCAGTGGTG	Pv06
	PVBR87	NED	150-201	CTCATTGCGTCTACCAGTGC	CCTAGGTTCCGCAGCATGT	Pv05
	PVBR272	6-FAM	70-135	CAGAACAGAAGAAGAAACAGAAAATG	GCGTGTTCCTCTGTGTGTGT	Pv02
	BM154	6-FAM	205-317	TCTTGCGACCGAGCTTCTCC	CTGAATCTGAGGAACGATGACCAG	Pv09
	PVBR13	6-FAM	159-200	TGAGAAAGTTGATGGGATTG	ACGCTGTTGAAGGCTCTAC	Pv06
4	PVBR11	HEX	175-192	AAACTCAAAGTCGTTGTTCC	CCACTGACTCTAGCTCCTCC	Pv02
	BM181	NED	170-250	CAACAGTTAAAGGTCGTCAAATT	CCACTCTTAGCATCAACTGGA	Pv05
	BM183	6-FAM	130-170	CTCAAATCTATTCACTGGTCAGC	TCTTACAGCCTTGCAGACATC	Pv07
	PVBR163	6-FAM	180-350	TGAGAGTGGAGAAGGAGAGAGA	TGACAACACTGCAAACACCA	Pv06
	BM201	NED	90-120	TGGTGCTACAGACTTGATGG	TGTCACCTCTCTCCTCCAAT	Pv01
	PVBR251	HEX	193-220	TGAAGTTGCAGCTAGGTTGG	GGTTGTGCTTGTGTTGTT GG	Pv01

[a] Genome location based on mapping results (Grisi et al. 2007), the PhaseolusGenes database (http://phaseolusgenes.bioinformatics.ucdavis.edu/) or BLAST analysis using the reference genome of *Phaseolus vulgaris* (Schmutz et al. 2014).

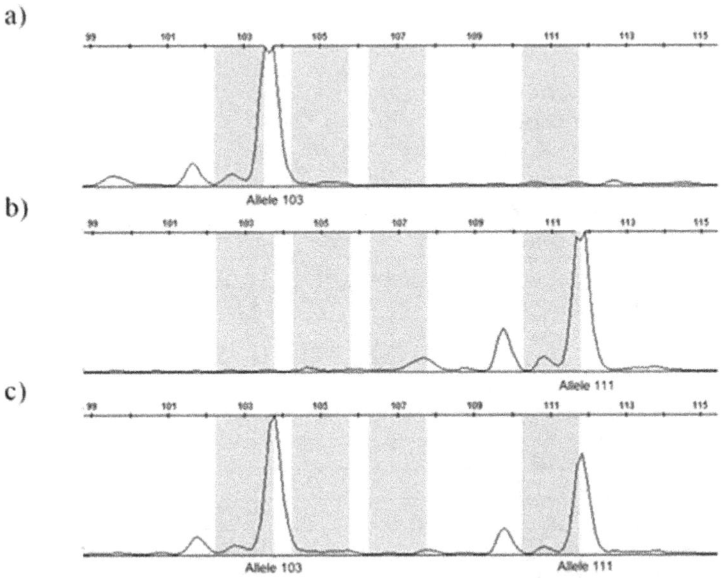

Figure 1. DNA amplification profile of common bean genotypes used as contrasting parents for anthracnose reaction with the microsatellite marker BM 189. a) Female parent presenting only allele 103 of BM 189 (homozygous); b) male parent presenting only allele 111 of BM 189 (homozygous); and c) F_1 plant presenting both alleles, 103 and 111 of BM 189 (heterozygous), confirming its hybrid nature.

RESULTS AND DISCUSSION

The hybrid nature of 325 (82.91%) of the 392 analyzed F_1 plants, derived from 21 different biparental crosses, was confirmed by the molecular analysis (Table 2). The microsatellite loci identified as polymorphic among their parents were heterozygous in these plants, that is, they contained two alleles, each specifically from one of the parents (Figure 1). The remaining 67 analyzed plants (17.09%) were originated by selfing, for containing only the same allele or homozygous genotype as the female parent. For the breeding program for anthracnose, the success rate of hybridization was 92.41% (207/224), for angular leaf spot 61.64% (45/73), for fusarium wilt 53.85% (21/39), and for slow darkening 92.86% (52/56). The variation in success rate of the four different programs was from 53.85 to 92.86%, and of the 21 tested biparental crosses from 33.33 to 100.00% (Table 2).

These results demonstrate the efficiency of the microsatellite markers used in this study to confirm common bean crosses. These markers allow the distinction between effective and ineffective crosses, avoiding a bias caused by plants originated by selfing. This would result in a loss of time and financial resources and produce inconsistent and misleading results about the genetic structure of the studied populations. The results further showed that this confirmation becomes essential when segregating populations are developed to be used in studies where the genetic structure of these populations must be preserved, e.g., in inheritance studies and gene/QTL mapping. In addition, based on the variation of the hybridization rate observed between the different crosses (Table 2), the effectiveness of the crosses did not only depend on the environmental conditions at the moment of the cross and on the experience and training of the cross operator, but also on the genetic suitability or compatibility among the parent genotypes.

In the analysis of the genetic identity of BRS Estilo seed samples, 22 of the 24 initially selected microsatellite loci were used, for having clearly evident and robust polymorphisms. Thus, these 22 loci were effectively used to estimate

Table 2. Common bean breeding programs, crosses, number of evaluated F_1 plants, number of obtained hybrids and hybridization rate between common bean genotypes that do not differ among each other based only on morphological traits, estimated using microsatellite markers

Program	Cross	No. of F_1 plants	No. of hybrids	Hybridization rate (%)
Anthracnose (ANT)	BRSMG Realce × CNFC15826	36	35	97.22
	BRS Pontal × Rosinha G2	35	35	100.00
	BRS Horizonte × Rosinha G2	17	14	82.35
	BRS Cometa × Rosinha G2	21	19	90.48
	BRS Horizonte × BRS Cometa	115	104	90.43
	Total ANT	224	207	92.41
Angular leaf spot (ALS)	BRS Sublime × BRS Horizonte	26	14	53.85
	BRS Sublime × BRS Cometa	13	10	76.92
	CNFC 15097 × BRS Horizonte	8	6	75.00
	CNFC 15097 × BRS Cometa	3	2	66.67
	CNFC 15082 × BRS Horizonte	22	12	54.55
	CNFC 15082 × BRS Cometa	1	1	100.00
	Total ALS	73	45	61.64
Fusarium wilt (FSW)	BRS Notável × BRS Horizonte	3	1	33.33
	BRS Ametista × BRS Horizonte	20	12	60.00
	BRS Esplendor × BRS Supremo	9	5	55.56
	CNFP 10794 × BRS Supremo	7	3	42.86
	Total FSW	39	21	53.85
Slow darkening (SDK)	BRSMG Madrepérola × CNFM 11940	5	3	60.00
	BRSMG Madrepérola × 1533-15	6	6	100.00
	BRSMG Madrepérola × AN512666-0	12	12	100.00
	CNFM 11940 × 1533-15	13	13	100.00
	CNFN 11940 × AN512666-0	10	8	80.00
	1533-15 × AN512666-0	10	10	100.00
	Total SDK	56	52	92.86
Total		392	325	82.91

the genetic similarity of the four seed sources tested in the present work in relation to the control samples formed by genetic seeds of the cv. BRS Estilo.

Of the four seed sources tested, in only one 100.00% genetic identity with the two control samples was observed, both of which also presented 100.00% relative genetic similarity to each other (Figure 2). This source consisted of a lot of certified seed. However, the genetic identity of the other certified seed sample was estimated at 93.00%, indicating a potential mixture with other genotypes with similar phenotypic patterns of plants and seeds, possibly undetectable by conventional genetic purity tests. Neither sample of saved seed was totally similar to the control samples, with estimates of genetic similarity of 87.00 and 91.00% (Figure 2). This result demonstrates that the use of uncertified seeds by the common bean growers represents a real risk, since the genetic identity of a considerable part of the plantation will be unknown, enabling potential discrepancies between plants in terms of yield, environmental adaptation, cycle, and reaction to pests and diseases. Thus, these results also reinforce the importance of using certified seed from a reliable origin.

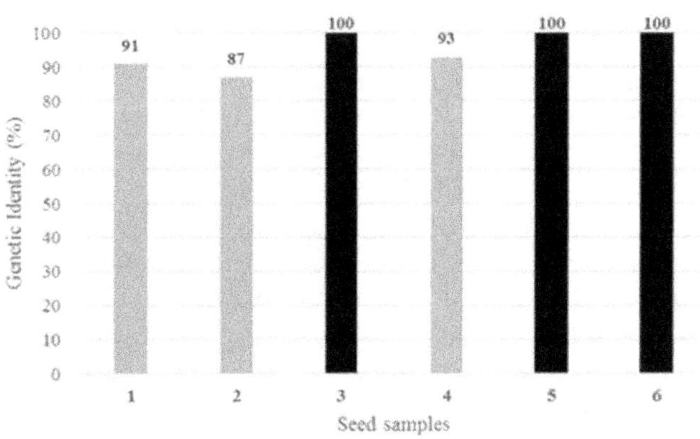

Figure 2. Genetic identity between seed samples of the Carioca common bean cultivar BRS Estilo, estimated based on the relative genetic similarity using 22 microsatellite loci. Samples 1 to 4 represent the different seed sources tested, i.e., samples 1 and 2 consisted of saved seeds (grains used as seeds) and samples 3 and 4 of certified seed. Samples 5 and 6 are the controls, formed by genetic seeds of BRS Estilo, provided by Embrapa Arroz e Feijão.

The advantages of using molecular markers in plant breeding programs over the use of only phenotypic or morphological markers lie in the fact that molecular markers are unlimited in number, usually easily detectable, generate a high level of polymorphism between tested genotypes, are neutral in relation to environmental effects, with little or no effect of epistasis or pleiotropy, and generally behave as traits with simple and predictable inheritance (Barros and Souza 2012).

This study demonstrates the efficiency of microsatellite markers as useful tools to support key decision-making phases in common bean breeding programs. The panel of 24 microsatellite loci used in this study is being routinely used at different stages of the common bean breeding program conducted by Embrapa and partners. In addition to the confirmation of controlled crosses and the detection of genetic identity of genotypes, other microsatellite marker applications include studies of genetic diversity among parents and elite lines, monitoring of allelic diversity and genetic representation of parents in breeding populations obtained by backcrosses and recurrent selection, aside from the control of genetic purity during the seed production process (Batista et al. 2014, Cardoso et al. 2014). It is worth mentioning that this molecular tool is available for immediate adoption and use by other common bean breeding programs in Brazil and worldwide.

In Brazil, marker-assisted selection is only routinely used in breeding programs of plant species of high commercial value, such as soybean, maize and cotton, in particular those led by the private sector. For common bean, although several markers and examples of application in the academic context are available, the routine use is only just being implemented. Thus, this study is an interesting example of the effective application of molecular markers in routine activities of a national common bean breeding program, led by a public research institution. Thousands of SNP (Single Nucletide Polymorphism) markers are currently available for common bean. This number increased greatly due to the recent publication of the genome sequence of the species (*Phaseolus vulgaris* v1.0; http://www.phytozome.net) (Schmutz et al. 2014). Therefore, efforts are currently under way at Embrapa Arroz e Feijão to develop DNA chips and genotype panels using SNP markers for genetic analysis and marker-assisted selection.

ACKNOWLEDGEMENTS

The authors would like to thank the Universidade Federal de Goiás and Embrapa Arroz e Feijão for the opportunity to develop this work, and are indebted to the entire research team of the common bean breeding program conducted by Embrapa and partners for the constant support. The authors are also grateful to the Conselho Nacional de Desenvolvimento Científico e Tecnológico (CNPq), for scholarships for undergraduate students and for technological development and innovative extension.

REFERENCES

Alzate-Marin AL, Baía GS, Martins-Filho S, Paula-Júnior TJ, Sediyama CS, Barros EG and Moreira MA (1996) Use of RAPD-PCR to identify true hybrid plants from crosses between closely related progenitors. **Brazilian Journal of Genetics 19**: 621-623.

Barros EG and Souza TLPO (2012) Biotecnologia na cultura do feijoeiro. In Cançado GMA and Londe LN (eds) **Biotecnologia aplicada à agropecuária**. Editora Epamig, Caldas, p. 351-370.

Batista CEA, Bueno LG, Resende TN, Lima VR, Wendland A, Pereira HS, Faria LC, Melo LC Souza TLPO (2014) Parental genetic representativeness in black seeded common bean progenies from the Embrapa recurrent selection program for tolerance to BGMV. **Annual Report of the Bean Improvement Cooperative 57**: 221-222.

Broughton WJ, Hernández G, Blair M, Beeb S, Gepts P and Vanderleyden J (2003) Beans (*Phaseolus* spp.) – model food legumes. **Plant and Soil 252**: 55-128.

Cardoso PCB, Brondani C, Menezes IPP, Valdisser PAMR, Borba TCO, Del Peloso MJ and Vianello RP (2014) Discrimination of common bean cultivars using multiplexed microsatellite markers. **Genetics and Molecular Research 13**: 1964-1978.

Cruz CD (2013) GENES – a software package for analysis in experimental statistics and quantitative genetics. **Acta Scientiarum Agronomy 35**: 271-276.

Grisi MCM, Blair MW, Gepts P, Brondani C, Pereira PAA and Brondani RPV (2007) Genetic mapping of a new set of microsatellite markers in a reference common bean (*Phaseolus vulgaris*) population BAT93 × Jalo EEP558. **Genetics and Molecular Research 6**: 691-706.

Melo LC, Del Peloso MJ, Pereira HS, Pereira HS, Faria LC, Costa JGC, Diaz JLC, Rava CA, Wendland A and Abreu AFB (2010) BRS Estilo – common bean cultivar with carioca grain upright growth and high yield potential. **Crop Breeding and Applied Biotechnology 10**: 377-379.

Muller BSF, Sakamoto T, Menezes IPP, Prado GS, Martins WS, Brondani C, Barros EG and Vianello RP (2014) Analysis of BAC-end sequences in common bean (*Phaseolus vulgaris* L.) towards the development and characterization of long motifs SSRs. **Plant Molecular Biology 86**: 455-470.

Schmutz J, McClean PE, Mamidi S, Wu GA, Cannon SB, Grimwood J, Jenkins J, Shu S, Song Q, Chavarro C, Torres-Torres M, Geffroy V, Moghaddam SM, Gao D, Abernathy B, Barry K, Blair M, Brick MA, Chovatia M, Gepts P, Goodstein DM, Gonzales M, Hellsten U, Hyten DL, Jia G, Kelly JD, Kudrna D, Lee R, Richard MMS, Miklas PN, Osorno JM, Rodrigues J, Thareau V, Urrea CA, Wang M, Yu Y, Zhang M, Wing RA, Cregan PB, Rokhsar DS and Jackson SA (2014) A reference genome for common bean and genome-wide analysis of dual domestications. **Nature Genetics 46**: 707-713.

Valdisser PAMR, Mota APS, Bueno LG, Menezes IPP, Coelho GRC, Magalhães FOC and Vianello (2013) **Protocolo de extração de DNA e genotipagem de SSRs em larga escala para uso no melhoramento do feijoeiro comum (*Phaseolus vulgaris* L.)**. Editora Embrapa, Santo Antônio de Goiás, 6p. (Comunicado Técnico, 208).

Xin Z, Velten J, Oliver MJ and Burke JJ (2003) High-throughput DNA extraction method suitable for PCR. **Biotechniques 34**: 820-826.

5

Be-Breeder - Learning: a new tool for teaching and learning plant breeding principles

Roberto Fritsche-Neto[1*] and Filipe Inácio Matias[1]

Abstract: *The Be-Breeder application is an on-line tool constructed through the R software for the purpose of assisting in some of the main genetic and statistical analyses related to the area of plant breeding. In addition, Be-Breeder provides a section called "Learning", which in a simple click-point manner allows explanation of theories related to the effect of inbreeding, population structure, qualitative and quantitative traits, heterosis, population size, effect of selection, and composition of hybrids. Be-Breeder is available for network use on the website of the Allogamous Plant Breeding Laboratory (Laboratório de Melhoramento de Plantas Alógamas) of ESALQ-USP through the link: http://www.genetica.esalq.usp.br/alogamas/R.html.*

Key words: *Heterosis, quantitative genetics, inbreeding, population structure, selection.*

***Corresponding author:**
E-mail: roberto.neto@usp.br

[1] USP/ESALQ – Genetics, Av. Pádua Dias, 11, CP 83, 13418-900, Piracicaba, SP, Brazil

INTRODUCTION

Most Agronomy courses have mandatory subjects related to plant breeding. This determination of mandatory subjects by CONFEA (CONFEA 1973) and the Ministry of Education (MEC 2006) in the course curriculum is based on the importance of this topic for the education/training of the agronomist and for agriculture. For example, it is estimated that half of the increase in yield of the main agricultural crops has come about through plant breeding processes (Kuhr et al. 1985, Byerlee and Moya 1993). Facts of this nature have also led to the creation of numerous graduate study programs in this line of research. The aim of these programs is to train human resources in understanding the mechanisms and concepts in regard to this theme, as well as possible applications of the principles that guide plant breeding.

In this context, seeking to assist teaching and learning, different tools, such as computer applications or software, can be used in clarification of more complex theories and equations. For this purpose, the Be-Breeder application, a compilation of on-line routines visualized through the Shiny package of the R software (Chang et al. 2015), has a section called *Learning*, with interactive learning strategies. They are an easily applied alternative tool for explaining theories related to the effect of inbreeding, population structure, qualitative and quantitative traits, heterosis, population size, effect of selection, and composition of hybrids.

BE-BREEDER APPLICATION

The Be-Breeder application is teaching tool freely available for on-line use at http://www.genetica.esalq.usp.br/alogamas/R.html. In the *Learning* section, the user has interactive tabs that deal with themes related to quantitative genetics, population genetics, and breeding methods. The tabs available and suggestions on how to use them are provided below.

Inbreeding Effect

Inbreeding is the process of raising the frequency of gene loci in homozygosity by means of successive processes of crosses between individuals with kinship, with maximum expression through self-pollination in plants (Nass 2001, Bos and Caligari 2007). This allows the formation of pure lines that have various applications (Shull 1909, Johannsen 2014).

In dealing with the inbreeding process in plants and reporting reduction in the percentage of loci in heterosis by half in each self-pollination cycle, the user can use visualization of histograms for a single locus with two alleles: "A" and "a". Upon choosing generation of self-pollination from F_1 to F_9 and the F.Inf (∞), it can be observed that at each cycle the percentage of the genotype "Aa" is cut in half, beginning at 100% in F_1 to 0% in F.Inf; in contrast, the percentage of homozygous genotypes "AA" and "aa" increases from zero in generation F_1 to 50% in generation F.Inf.

In this tab, the user can also observe the fluctuation in additive (σ_a^2) and dominant (σ_d^2) variance between and within populations as a function of the inbreeding coefficient (*F* of Wright). Additive variance between is estimated by the expression $\sigma_{aA}^2 = 2F\sigma_a^2$, additive variance within by $\sigma_{aW}^2(1 - F)\sigma_a^2$, dominant variance between by $\sigma_{dA}^2(1 - F)\sigma_d^2$, and dominant variance within by the expression $\sigma_{dW}^2(1 - F)\sigma_d^2$ (Nass 2001).

Qualitative vs. Quantitative

The number of genes and their different forms of intra-allelic interaction influence the number of phenotypic classes and also the frequency of each class in the population. Consequently, in the total number of genes that control the trait, as well as the types of interactions, concepts arise and transition of qualitative traits to quantitative ones (Ramalho et al. 2008, Borém and Miranda 2013). In this respect, a simple algorithm was developed in Be-Breeder in which the user can construct a genetic structure of a trait, choosing the number of genes with dominance effect, the number of genes with partial dominance effect, and the number of genes with additive effect. As a response, a histogram relating frequency and number of classes can be visualized in the output box. It should be emphasized that, just as in natural biological systems, the randomness factor was embedded in the algorithm in such a way that, although the number of genes chosen is the same, it will not indicate that the number of classes will necessarily be the same for polygenic effects, allowing inferences to be made regarding segregation.

Progeny Size

Observation of a determined genotype is dependent on the number of genes acting on a trait, such that the greater the number of genes, the more individuals are necessary in population sampling to verify all the possible genotypes or the genotype desired. Inbreeding, promoted by self-pollination, has a direct influence on this observation through the fact of increasing loci in homozygosity in the population, reducing genotypic variability. The size of the population to be evaluated is given by the expression $n^o = \frac{log(1-p)}{log(1-IH)}$, in which *p* represents the probability of observing a determined genotype, and *IH* is estimated by $IH = \left(\frac{2^{m-1}}{2^m}\right)^n$, in which *m* is the number of generations of self-pollination and *n* is the number of genes that control the trait. This number of individuals can easily be obtained in this tab of Be-Breeder, in which the user can simulate numerous scenarios and observe fluctuation in the population size as a function of generation of self-pollination, the number of genes, and the probability of observation.

Effect of Selection (HWE)

In population genetics, it is relevant to check if a determined gene locus is in Hardy-Weinberg equilibrium (HWE) as a function of the frequency of the alleles *A*(*p*) and *a*(*q*). From this information, it is possible to determine the number of individuals expected for each genotype using the expressions $n^o\ AA = (p^2 + pqF)N$, $n^o\ Aa = (2pq - 2pqF)N$, and $n^o\ aa = (q^2 + pqF)N$, in which *N* is the total number of individuals of the population and *F* is the inbreeding coefficient of Wright. Upon comparing the number expected with the number observed for each genotype, it is possible to perform a chi-square test (χ^2) to verify significance through the expression $\chi^2 = \sum_i \frac{(O_i - E_i)^2}{E_i}$ for $i = (AA, Aa, aa)$, in which O_i is the number of individuals observed for genotype i and E_i is the number of individuals expected for genotype i.

Nevertheless, the effect of selection is the main piece of information the breeder uses to conduct a breeding program. Thus, upon applying a selection intensity in a population with an original mean value (μ_o), a selected population is obtained with a mean value (μ_s), in which the difference between these mean values is equivalent to the differential of selection (*DS*). By multiplying *DS* by the heritability of the trait (h^2), gain from selection is obtained ($GS = DS^* h^2$), which upon being added to μ_o will give rise to the predicted mean of the improved population in the next evaluation cycle (Falconer et al. 1996, Bernardo 2010).

In the application, the user can simulate numerous scenarios, modifying the allelic frequencies as a function of the phenotypic observations, thus allowing speculations in regard to the effect of the composition of the population, heritability of the trait, and the intensity of selection in a breeding population. The output is a table with the following information: allelic frequency, genotypic frequencies, the number of individuals observed, the number of individuals expected, and the χ^2 test to verify Hardy-Weinberg equilibrium. In addition, the user can simulate selection by identifying the number of individuals selected in each genotypic class and the respective genetic values, obtaining the parameters μ_o, μ_s, *DS*, and *GS* as output.

Components of Genetic Variance

Genetic variance is composed of variance of additive and non-additive effects that fluctuate as a function of allelic frequencies (Bernardo 2010). So as to deal with these concepts, a function was developed that allows the user to indicate the frequency of the allele "A" (p), which is obtained from the difference with allele "a" (q). The user can also modify the additive effect (a) and dominance effect (d) and observe the effect of these factors (p, a, and d) on the magnitude and relationship between total genetic variance and its additive and dominance components. The expressions used to estimate the variances are $\sigma^2_A = 2pq[a + d(q - p)]^2$ for the additive, $\sigma^2_D = (2pqd)^2$ for the dominance, and $\sigma^2_G = \sigma^2_A + \sigma^2_D$ for the total of a determined gene locus (Falconer et al. 1996).

Constructing Synthetic Populations

The number of alleles of a gene in a breeding population indicates variability, such that the greater the number of alleles present in the population, the greater the variability of the heterozygotes will be, according to the expression $\sigma^2_{gH} = \sum_i (p_i)^2 - \left(\sum_i (p_i)^2\right)^2$. However, this does not indicate that the number of heterozygotes in the population will increase indefinitely; that is, retention of heterozygosity (RH) reaches a plateau with a determined number of alleles. Although this value increases toward 1.00, the number of heterozygotes in the population will remain constant, just as indicated by the expression $RH = 1 - \sum_i (p_i)^2$(Nietlisbach et al. 2016). It is known that the objective is to maintain heterozygosity at high levels in an allogamous population, so as to exploit heterosis and avoid inbreeding depression (Falconer et al. 1996). Nevertheless, an excessive increase in the number of alleles can increase genetic variability to critical levels, which can impede selection and standardization of the population for important agronomic traits or descriptors.

In this regard, in this function the user provides the number of alleles of a gene (maximum of 10 alleles) and their respective initial frequencies in the composition of the population. The sum total should be equal to one. Retention of heterozygosity and the genetic variance of the heterozygotes can be observed in the output window as a result, representing that, although genetic variance increases the number of heterozygotes in the population, this reaches a plateau and stabilizes. Thus, it is possible to identify the ideal composition of parents for formation of a population for numerous situations.

Recurrent Selection

Intrapopulational (IRS)

Intrapopulational recurrent selection is a breeding procedure that leads to an increase in the frequencies of alleles of interest in the population without, however, drastically reducing its variability, improving the performance per se of the population in each selective and recombination cycle (Bernardo 2010). In this context, Be-Breeder allows estimation of gain from selection (*GS*), effective size of the population (*Ne*), and inbreeding coefficient (*F* of Wright) for the *IRS* breeding arrangement in different selection scenarios, number of progenies evaluated, selection intensity, and heritability.

The expression of response to intrapopulational recurrent selection, according to Falconer et al. (1996), is given by $GS = \left[i.c.\frac{(\sigma^2_a + F.D_1)}{\sigma_P} - \frac{ID}{2Ne} \right]$, in which *GS* is gain from selection, *i* is the standardized selection differential, *c* and D_1 are values that depend on the selection arrangement by parental control (Table 1), σ^2_a is additive genetic variance, σ_p is the phenotypic standard deviation from the unit of selection, and *ID* refers to inbreeding depression given in percentage. The *Ne* parameter refers to the effective size of the population given the expression $Ne = Ne_{tab} {*} N {*} i$, in which Ne_{tab} is the value dependent on the selection arrangement (Table 1), *N* is the total size of the population, and *i* is selection intensity. The inbreeding coefficient *F* of Wright is estimated by $F = \frac{1}{2Ne}$.

Table 1. Selection arrangement in regard to intrapopulational recurrent selection for the population of evaluation, population of recombination (HS – half sibs, FS – full sibs, and S1 – self-pollination), *c* index, effective size (Ne_{tab}), and coefficient between additive and dominance effects of the homozygotes (D_1)

Evaluation	Recombination	*c*	Ne_{tab}	D_1
HS	HS	¼	4	0
HS	S_1	½	1	0
FS	FS	½	2	0
FS	S_1	½	1	0
S_1	S_1	1	1	0.5
S_2	S_2	3/2	2/3	5/4

b) Reciprocal (RRS)

Reciprocal or intrapopulational recurrent selection is a breeding arrangement that leads to an increase in complementarity between two heterotic groups or populations from crosses. The *RRS* brings about superior hybrids by crossing these groups in each selection cycle, in which intragroup selection and recombination of the most complementary parents increases the frequency of favorable alleles within each group (Bernardo 2010). Thus, as for IRS, in Be-Breeder it is also possible to simulate different selection arrangements, number of progenies evaluated, selection intensity, and heritability for *RRS*.

For that purpose, the response of reciprocal recurrent selection was estimated by the expression $GS = i_1.c.\frac{\sigma^2_{a1}}{\sigma^2_{p1}} + i_2.c.\frac{\sigma^2_{a2}}{\sigma^2_{p2}}$, in which *GS* is gain from selection, *i* is the standardized selection differential for each group (1 and 2), *c* is a value that depends on the selective arrangement of parental control (Table 2), σ^2_a is additive genetic variance, and σ_p is the phenotypic standard deviation for each group (1 and 2) (Falconer et al. 1996). The effective size of the population (*Ne*) and inbreeding coefficient *F* of Wright are also provided for each heterotic group (1 and 2) in the output window; they are estimated in a manner similar to that described in the item *IRS*.

Hybrids (Jenkins)

Prediction of Hybrids

Among the main products coming from plant breeding, hybrids stand out for the important role they exercise in the world economy (USDA 2016). Among them, single-cross hybrids (*SH*), three-way hybrids (*TH*), and double-cross hybrids (*DH*) (Shull 1910, Jones 1918) are the most representative products on the market. In this section, the user of Be-Breeder can find the predicted genotypic value of three-way hybrids (*TH*) and double-cross hybrids (*DH*) through the input of a *.txt* document containing the mean phenotypic dataset of the single-cross hybrids (*SH*) coming from the lines of interest. For this purpose, the expressions of Jenkins (1934) are used as a basis, in which $HT_{(AB)C} = \frac{HS_{(Ac)} + HS_{(Bc)}}{2}$ and $HD_{(AB)(CD)} = \frac{HS_{(Ac)} + HS_{(Ac)} + HS_{(Bc)} + HS_{(BD)}}{4}$. The sequence of the columns in the *txt* file must follow the example indicated in Table 3.

Table 2. Selective arrangement in reference to reciprocal recurrent selection for the population of evaluation, the population of recombination (TC – Testcross, HS – half sibs, FS – full sibs, and S1 – self-pollination), *c* index, tabulated effective size (Ne_{tab})

Evaluation	Recombination	c	Ne_{tab}
HS	S	¼	1
FS	S^1	¼	1
HS	HS	1/8	4
TC	HS	1/16	4

Table 3. Example of *.txt* file for input in the Be-Breeder application in the Learning section, "Hybrid Effect" tab so as to predict the phenotypic value expected of three-way and double-cross hybrids from the mean phenotypic value observed from the single-cross hybrids between the lines of interest

Lines	A	B	C	D
A	0	15	14	11
B	15	0	18	12
C	14	18	0	10
D	11	12	10	0

b) Number of Hybrids

In this tab of the application, the user can obtain the possible number of single-cross hybrids (*SH*), three-way hybrids (*TH*), and double-cross hybrids (*DH*), given the number of lines (*n*) of a breeding population. This information can be obtained for a single population through the expressions (Vencovsky and Barriga 1992): $n° \ HS = \frac{n(n-1)}{2}$, $n° \ HT = \frac{n(n-1)(n-2)}{2}$, and $n° \ HD = \frac{n(n-1)(n-2)(n-3)}{8}$. In dealing with two heterotic groups (*1* and *2*) or two populations (*1* and *2*), the number of *SH*, *TH,* and *DH* depends on the number of lines belonging to group *1* (*a*) and the number of lines belonging to group *2* (*b*), according to the expressions: $n° \ HS = a^*b$, $n° \ HT = a^*(a-1)^*b + b^*(b-1)^*a$, and $n° \ HD = a^*(a-1)^*b^*(b-1)$.

Genotype x Environment

In plant breeding, the statistical significance of the component of the genotype × environment interaction defines the selection strategy and commercial recommendation. The absence of interaction indicates that the environments of evaluation do not have a different influence on the behavior and on the ordering of the genotypes, and so it is sufficient to choose a single environment because the crop recommendation will be the same. Simple interaction indicates that the genotypes respond differently to environmental influences, but not enough to for there to be change in ordering, maintaining the same commercial recommendation among them. In contrast, in complex interaction, there are changes in ordering of genotypes among the environments, and a crop recommendation per location is necessary (Borém and Miranda 2013). In this context, Be-Breeder provides a tab for simulation of interactions among three genotypes in two environments, and it is possible to construct various scenarios, simulate recommendations, and make inferences regarding the implications of the G × A effect in the selection process and in data analysis. The user has columns that range from zero to five for each genotype/environment combination (genotype value of individual *i* in environment *j*), obtaining the mean values of genotypes and of environments separately as output, as well as graph visualization of each scenario.

Heterosis

According to the Dominance and Repulsion Hypothesis (Borém and Miranda 2013), the mean performance of hybrids (F1) in relation to the mean of the parents is related to allelic complementarity between the parents (*P1* and *P2*), the genetic divergence between them, and the magnitude of the dominance deviations (Nass 2001). By being observed mainly in quantitative traits, the number of genes they control also has a certain influence on the magnitude of the heterosis observed. In this context, hybrid vigor in F1, also called biological heterosis (*H*), is estimated by the expression $H = F1 - \frac{P1 + P2}{2}$ or $H = (p - r)^2 d$ (Falconer et al. 1996). Hybrid performance in relation to superior performance (Best Parent - *B.P*), for its part, receives the name heterobeltiosis (*Hb*) or agronomic heterosis, and is estimated by $Hb = F1 -$

B.P. In light of the foregoing, in this tab of the application, it is possible to simulate different scenarios as a function of the number of genes, of deviation of dominance, and of divergence among the parents, making it possible to observe the fluctuation of heterosis and of heterobeltiosis for the simulated hybrids.

FINAL CONSIDERATION

Through use of the Be-Breeder - Learning application, it can be observed that it provides a teaching interface of the R software, which can be easily handled and which offers an alternative for understanding concepts and expressions of quantitative and population genetics applied in plant breeding. This tool can assist researchers, professors, and students in experiments, predictions, and academic studies, among other applications, both in learning and in teaching the principles that guide plant breeding.

REFERENCES

Bernardo R (2010) **Breeding for quantitative traits in plants**. Stemma Press, Woodbury, 400p.

Borém A and Miranda GV (2013) **Melhoramento de plantas**. UFV, Viçosa, 523p.

Bos I and Caligari P (2007) **Selection methods in plant breeding**. Springer Science & Business Media, Dordrecht, 461p.

Byerlee D and Moya P (1993) **Impacts of international wheat breeding research in the developing world, 1966-1990**. CIMMYT, Texcoco, 87p.

Chang W, Cheng J, Allaire J, Xie Y and McPherson J (2015) Shiny: web application framework for R. **R package version 0.11 1**.

CONFEA (1973) Resolução Nº 218, de 29 de Junho de 1973. **Diário Oficial da União (DOU)**: 0218-73.

Falconer DS, Mackay TF and Frankham R (1996) Introduction to quantitative genetics (4th edn). **Trends in Genetics 12**: 280.

Jenkins MT (1934) Methods of estimating the performance of double crosses in corn. **Agronomy Journal 26**: 199-204.

Johannsen W (2014) The genotype conception of heredity. **International Journal of Epidemiology 43**: 989-1000.

Jones DF (1918) The effect of inbreeding and crossbreeding upon development. **Proceedings of the National Academy of Sciences 4**: 246-250.

Kuhr S, Johnson V, Peterson C and Mattern P (1985) Trends in winter wheat performance as measured in international trials. **Crop Science 25**: 1045-1049.

MEC - Ministério da Educação e Cultura (2006) Resolução Nº 1, de 2 de fevereiro de 2006. **Diário Oficial da União (DOU)** de 03/02/2006 Seção I: 31-32.

Nass L (2001) **Recursos genéticos e melhoramento-Plantas**. Fundação MT, Rondonópolis, 1183p.

Nietlisbach P, Keller L and Postma E (2016) Genetic variance components and heritability of multiallelic heterozygosity under inbreeding. **Heredity 116**: 1-11.

Ramalho M, Santos JB and Pinto CB (2008) **Genética na agropecuária**. UFLA, Lavras, 463p.

Shull GH (1909) A pure-line method in corn breeding. **Journal of Heredity 5**: 51-58.

Shull GH (1910) Hybridization methods in corn breeding. **Journal of Heredity 6**: 98-107.

USDA - United States Department of Agriculture (2016) Gain reports. Available at <http://gain.fas.usda.gov/Pages/Default.aspx>. Access on July 28, 2016.

Vencovsky R and Barriga P (1992) **Genética biométrica no fitomelhoramento**. Revista Brasileira de Genética, Ribeirão Preto, 486p.

6

CD 1104 - extra strong wheat with high yield potential

Volmir Sergio Marchioro[1*], Francisco de Assis Franco[1], Ivan Schuster[1], Tatiane Dalla Nora Montecelli[1], Mateus Polo[1], Fábio Junior Alcântara de Lima[1], Adriel Evangelista[1] and Diego Augusto dos Santos[1]

Abstract: *CD 1104 is a cultivar indicated for the wheat-producing regions 1, 2, 3, 4 (irrigated) and 4 (rainfed) of the states of RS, SC, PR, SP, MS, MG, DF, GO, and MT. Its suitability for the industrial segment of strong flours and yield potential (mean of 4.427 kg ha*$^{-1}$*) are high.*

Key words*:* Triticum aestivum *L., bread baking quality, tolerance to soil aluminum.*

***Corresponding author:**
E-mail: volmir@marchioro.eng.br

[1] COODETEC - Desenvolvimento, Produção e Comercialização Agrícola Ltda, BR 467, km 98, 85.813-450, Cascavel, PR, Brazil

INTRODUCTION

Wheat (*Triticum aestivum* L.) is highly important in the agricultural economy of the world, ranking third in global grain production. Therefore, the search for greater productivity is one of the main goals of breeding programs worldwide (Carvalho et al. 2008). On the other hand, it is very important that at the time of harvest, the wheat grains have the desired technical properties that meet the demands of the processing industry and consequently of consumers.

According to Pomeranz (1987), the wheat grain quality can be influenced by a number of factors such as soil, climate, pests, diseases, management, harvesting, drying, storage, and milling. In view of these different factors, the use of cultivars with genetic potential for a specific industrial purpose is the best way of ensuring an end product with higher quality. Wheat with high gluten strength can be used in flour blends to improve the baking quality of other wheat varieties with low gluten strength (Gutkoski et al. 2007). With a view to the establishment of cultivars with high yield potential, along with the technical property high gluten strength, which is currently particularly valued by the food industry, COODETEC developed wheat cultivar CD 1104.

BREEDING METHODS

Cultivar CD 1104 was obtained from the cross between cultivars CD 108 and BRS 220, by COODETEC in 2002, in Palotina. The F1 seeds were sown in the same year in a greenhouse, in Cascavel. The plants were harvested at maturity and all ears bulk-threshed, resulting in the F2 population. The F2 population was grown in a field in Palotina, in 2003, applying the modified mass method. This procedure consists of selecting the best plants within a population, threshing all ears of all selected plants together, of which a significant sample of seeds is sown on a plot with individual plants to obtain the next generation. The F3 population was conducted by the above method in Cascavel, in 2004. In 2005 and 2006, respectively, the populations F4 and F5 were grown in Palotina by the genealogical method. This procedure consists of selecting plants, the ears of each selected plant are threshed together and the seeds of each plant sown in the next generation in a plot with individual plants. The F6 population was also selected by the genealogical method in the field in Cascavel, in 2007, when the traits for various sibling lines were fixed, of which one line was selected (line CD 1034), from which CD 1104 was derived. The pedigree of this line is CC16440-00P-00T-7T-1Q-0T.

TRAITS AND PERFORMANCE

Line CD 1034 was included in the HC (Hot/Cold) genotype collection in 2008, and was tested in 2009 in Preliminary tests in Cascavel/PR, Palotina/PR, Guarapuava/PR, Não me Toque/RS, and in São Gotardo/MG, where yields exceeded those of the controls. The VCU trials of 2010-2014 were distributed at different locations of the wheat-producing regions and in different seasons (Table 1). The experimental design was in randomized blocks, with three replications in plots with six 5-m long rows, spaced 0.17 m apart. Fertilization and disease, pest and weed control were applied according to official technical recommendations (Comissão 2011). Prior to sowing, the seeds were treated with Triadimenol + Imidacloprid.

Table 1. Locations and seasons of experiments of Value for Cultivation and Use (VCU) with cultivar CD 1104, in the wheat-growing regions 1, 2, 3, 4 (irrigated) and 4 (rainfed) including the states of RS, SC, PR, MS, MS, GO, and MG

Region/Location	2010	2011	2012	2013	2014
Wheat-growing region 1	-	-	-	9	9
Cruz Alta/RS	-	-	-	1	1
Passo Fundo/RS	-	-	-	1	1
Não Me Toque/RS	-	-	-	2	2
Vacaria/RS	-	-	-	1	1
Guarapuava/PR	-	-	-	2	2
Ponta Grossa/PR	-	-	-	1	1
Campos Novos/SC	-	-	-	1	1
Wheat-growing region 2	4	4	6	11	11
Santo Augusto/RS	-	-	-	2	2
Santa Rosa/RS	-	-	-	1	-
São Luiz Gonzaga/RS	-	-	-	1	1
Abelardo Luz/SC	-	-	-	2	2
Campo Mourão/PR	-	2	2	1	2
Cascavel/PR	3	2	3	3	3
Itaberá/SP	1	-	1	-	-
Itapeva/SP	-	-	-	1	1
Wheat-growing region 3	7	5	8	7	7
Arapongas/PR	-	1	1	1	-
Palotina/PR	4	4	4	3	3
Rolândia/PR	1	-	-	-	1
Dourados/MS	1	-	2	2	2
Manduri/SP	1	-	1	1	-
Santa Cruz do Rio Pardo/SP	-	-	-	-	1
Wheat-growing region 4	3	1	3	4	4
Catalão/GO	1	-	1	2	2
São Gotardo/MG	2	1	2	2	2

* In the years in which two trials were carried out in Catalão/GO and São Gotardo/MG, one was conducted in a dryland and the other in an irrigated cropping system. For the other locations, more than one test per year indicate different sowing times within the season recommended for cultivation specifically for that location.

The variables measured in the VCU tests were grain yield, days from emergence to heading, days from emergence to maturity, plant height, lodging, hectoliter weight, 1000-grain weight, pre-harvest sprouting, and tolerance to soil aluminum. The grains of three replications per treatment were mixed, generating composite samples, including quality analyses, thereby obtaining the variables: falling number, gluten strength and alveograph tenacity/extensibility ratio; farinograph stability; and flour color. The latter was determined by the L, a, b system by which the values of L (lightness) vary from 0 (black) to 100 (white) and the a and b values (chromaticity coordinates) range from -a (green) to +a (red) and from -b (blue) to +b (yellow). At strategic locations, the genotypes included in the VCU tests were grown together without disease control, and the diseases leaf rust, leaf spot, powdery mildew, fusarium head blight, blast and mosaic virus were evaluated, among others.

The plant height of cultivar CD 1104 is medium (81 cm), ranging from 60 to 95 cm. The cycle is medium (52-86 d from emergence to heading; 105-133 d from emergence to maturity). In the mean, these traits were, respectively, 68 and 120 d. CD 1104 was classified as moderately susceptible to lodging, moderately resistant to moderately susceptible to pre-harvest sprouting and moderately tolerant to soil aluminum. The mean hectoliter weight was 79 kg hL^{-1} and 1000-grain weight 35 grams.

In collections comprising the genotypes of the VCU tests at specific locations from 2010 to 2014 without disease control, information was obtained for the classification of cultivar CD 1104 with regard to the reaction to the main diseases. The cultivar was classified as moderately resistant to powdery mildew (*Blumeria graminis* f.sp. *tritici*), blast (*Pyricularia grisea*) and wheat mosaic virus (soil-borne wheat mosaic virus), and moderately susceptible to powdery mildew (*Blumeria graminis* f.sp. *tritici*), scab (*Fusarium graminearum*), leaf spots (*Septoria tritici* and *Bipolar sorokiniana*),

Table 2. Means of the general gluten strength (W), yield stability (YST), falling number (FN), tenacity/extensibility ratio (P/L), flour color (COL L, COL a and COL b) per wheat-growing region of samples of tests conducted in the states of RS, SC, PR, MS, MS, GO, and MG

Wheat-growing region	No. of samples	W ($x10^{-4}$ J.)	YST (min.)	FN (min.)	P/L (relation)	COL L (89 to 96)	COL a (-1.0 to +1.0)	COL b (6 to 10)
1	4	392	16.2	392	1.7	92.2	-0.02	10.4
2	11	422	17.7	395	1.6	92.2	-0.02	11.2
3	13	438	17.5	364	1.5	92.3	0.01	11.7
4	4	447	18.4	368	1.7	92.3	0.10	11.2
Mean	**32**	**429**	**17.5**	**377**	**1.6**	92.3	0.01	11.3

Table 3. Grain yield means (kg ha^{-1}) of cultivar CD 1104 and the controls in the tests carried out in the wheat-growing regions 1, 2, 3, 4 (irrigated) and 4 (rainfed) including the states of RS, SC, PR, MS, MS, GO, and MG

Wheat-producing region	Cultivar	2010	2011	2012	2013	2014	Mean	%
1	CD 1104	-	-	-	5690	5047	**5369**	**102**
	C_1	-	-	-	5593	4922	**5258**	**99**
	C_2	-	-	-	5604	5025	**5315**	**101**
	C_M	-	-	-	5599	4973	**5286**	**100**
2	CD 1104	5106	3817	3866	5144	4573	**4501**	**109**
	C_1	4154	3280	3698	4198	4377	**3941**	**96**
	C_2	4529	3413	3758	4845	4896	**4288**	**104**
	C_M	4342	3347	3728	4521	4636	**4115**	**100**
3	CD 1104	3292	3513	3353	2110	3407	**3135**	**114**
	C_1	3119	3305	2866	2031	3131	**2890**	**105**
	C_2	2610	3199	2411	1887	2974	**2616**	**95**
	C_M	2865	3252	2638	1959	3053	**2753**	**100**
4 (irrigated)	CD 1104	7413	7700	5642	6653	6706	**6823**	**111**
	C_1	6285	7115	5407	6162	6343	**6262**	**102**
	C_2	6634	6785	5196	5526	5982	**6025**	**98**
	C_M	6460	6950	5302	5844	6162	**6144**	**100**
4 (rainfed)	CD 1104	-	-	2358	1967	2748	**2358**	**104**
	C_1	-	-	2292	1849	2734	**2292**	**101**
	C_2	-	-	2245	1782	2709	**2245**	**99**
	C_M	-	-	2268	1816	2721	**2269**	**100**

C_M = Control means. In the wheat-growing regions 1 and 2, controls C_1 and C_2 were, respectively, BRS Guamirim and QUARCZO in all years; in wheat-growing region 3, the controls C_1 and C_2 were, respectively, CD 150 and IPR 85, in 2010 and CD 150 and BRS PARDELA in 2011, 2012, 2013, and 2014; in wheat-growing region 4 (irrigated), controls C_1 and C_2 were, respectively, CD 150 and BRS 264, in 2011 and 2013 and CD 150 and CD 116 in 2012 and 2014; and in the wheat-growing region 4 (rainfed), controls C_1 and C_2 were, respectively, CD 116 and BRS 264 in 2012, 2013 and 2014.

glume blotch (*Septoria nodorum*), and to leaf rust (*Puccinia triticina).*

In the analysis of processing quality of 32 samples of the experiments conducted from 2009 to 2014 in the wheat-growing regions 1, 2, 3 and 4, mean of 429 $x10^{-4}$ joules of gluten strength (W) and stability of 17.5 minutes were found, classifying CD 1104 in the group of strong wheat cultivars (Table 2). The high mean W value of cultivar CD 1104 indicates that the flour obtained from grain of this cultivar can be used in blends with weak wheat flour, to improve the flour quality for bread-baking. In terms of gluten strength and stability in different environments, cultivar CD 1104 had a medium performance and was thus classified as strong wheat in the wheat-growing region 1 as well. In this way, it became part of the choice group of high-W cultivars released on the market by COODETEC (Franco et al. 2009, Franco et al. 2011, Franco et al. 2013).

The grain yield means of cultivar CD 1104 in the wheat-growing regions 1, 2, 3, 4 (irrigated) and 4 (rainfed), were 2%, 9% 14%, 11%, and 4% higher than those of the control mean, respectively (Table 3). The overall grain yield mean was 4427 kg ha^{-1}, exceeding the controls by 8%. In view of the grain yield performance of cultivar CD 1104, it was indicated for cultivation in the wheat-growing regions listed above, for the states of RS, SC, PR, SP, MS, MG, DF, GO, and MT. The cultivar was registered by the Registro Nacional de Cultivares (no 32282). Cultivar CD 1104 has a high yield potential, strong wheat quality, high gluten strength and stability in different environments, representing a promising option for farmers of the segment of strong wheat production.

BASE SEED PRODUCTION

The Cooperativa Central de Pesquisa Agrícola - COODETEC (BR 467 - km 98 - PO Box 89 - CEP. 85813-450, Cascavel, Paraná, Brazil), is authorized to license seed companies for the production of protected varieties (law nº 9456/97), to multiply and sell seed to grain producers. Cultivar CD 1440 was released on the market in 2013, with an availability of 5000 bags 40 kg seeds.

REFERENCES

Carvalho FIF, Lorencetti C, Marchioro VS and Silva AS (2008) **Condução de populações no melhoramento genético de plantas.** UFPel, Pelotas. 288p.

Comissão Brasileira de Pesquisa de Trigo e Triticale (2011) **Informações técnicas para trigo e triticale - safra 2011.** Coodetec, Cascavel, 170p.

Franco AF, Marchioro VS, Dalla Nora T, Schuster I, Oliveira EF, Vieira ESN and Lima FJA (2009) CD 116: A healthy wheat cultivar with industrial quality. **Crop Breeding and Applied Biotechnology 9**: 166-198.

Franco AF, Marchioro VS, Dalla Nora T, Oliveira EF, Schuster I, Vieira ESN, Sobrinho AA and Evangelista A (2011) CD 150 - short wheat cultivar with high quality and high yield. **Crop Breeding and Applied Biotechnology 11**: 186-188.

Franco AF, Marchioro VS, Dalla Nora T, Schuster I, Evangelista A and Lima FJA (2013) CD 151: Cultivar de trigo de ampla adaptação. **Bioscience Journal 29**: 101-103.

Gutkoski LC, Klein B, Agnussatt FA and Pedó I (2007) Características tecnológicas de genótipos de trigo (*Triticum aestivum* L.) cultivados no cerrado. **Ciência e Agrotecnologia 31**: 786-792.

Pomeranz Y (1987) **Modern cereal science and technology**. VHC, New York, 486p.

7

UENF Rio Dourado: a new passion fruit cultivar with high yield potential

Alexandre Pio Viana[1], Fernando Higino de Lima e Silva[1], Gus-tavo Menezes Gonçalves[2], Marcelo Geraldo de Morais Silva[3], Rulfe Tavares Ferreira[1], Telma Nair Santana Pereira[1], Messias Gonzaga Pereira[1], Antonio Teixeira do Amaral Júnior[1] and Geraldo Franscisco de Carvalho[1]

Abstract: *This work aimed at introducing the characteristics of the passion fruit cultivar UENF Rio Dourado, developed from three cycles of recurrent selection, to the scientific community. The cultivar presents yield of 25 tons ha^{-1}, mean values of 250 for number of fruits and 175g for fruit weight, providing a percentage increase in yield, number of fruits and fruit weight of 36%, 25% and 3%, respectively, when compared to the standard cultivar tested.*

Key words: Passiflora edulis *Sims., plant breeding, fruits crops, VCU.*

***Corresponding author:**
E-mail: pirapora.alexandre@gmail.com

[1] Universidade Estadual do Norte Fluminense Darcy Ribeiro (UENF), Laboratório de Melhoramento Genético Vegetal, Av. Alberto Lamego, 2000, Parque Califórnia, 28.013-602, Campos dos Goytacazes, RJ, Brazil
[2] Petrobras-Biofuel, Av. República do Chile, 65, 20.031-912, Rio de Janeiro, RJ, Brazil
[3] Instituto Federal Fluminense (IFF), Campus Avançado Cambuci, Estrada Cambuci, km 05, s\n, Três Irmãos, 28430-000, Cambuci, RJ, Brazil

INTRODUCTION

Passion fruit belongs to the Passifloraceae family, and the main cultivated species is *Passiflora edulis* Sims, also known as sour passion fruit. This species represents about 90% of the Brazilian orchards, mainly due to its quality, yield, and consumer preference.

Brazil is the largest world producer of passion fruit. In 2013, yield was of about 780 million tons (IBGE 2014). However, the low national mean yield in 2013 (15 tons ha^{-1}) in Brazil (IBGE 2014) was due to many factors, such as the low number of cultivars and hybrids available for the producers in the market, besides viral and fungal diseases (Santos et al. 2015). Thus, these problems are key barrier to continued expansion for this important Brazilian fruit crop.

There are currently no more than 25 sour passion fruit cultivars available in the seed market in Brazil (MAPA 2015). This number of cultivars is low, considering the importance of Brazil in the global scenario as the biggest passion fruit producer. Thus, due to the economic and social importance of this crop, it is very important to improve breeding programs, which will enable faster development of new cultivars.

Passion fruit breeding is directly related to the fruit, which focuses on three main points: meet the market demands (quality), increase yield, and develop cultivars which are resistant to diseases (Gonçalves et al. 2009).

With this objective, the Universidade Estadual do Norte Fluminense Darcy Ribeiro (UENF) tested a population with potential to develop a commercial cultivar for the north and northwest regions of the state of Rio de Janeiro. In this context, the purpose of this study is to inform the scientific community on the characteristics of this new cultivar.

CULTIVAR ORIGIN AND DEVELOPMENT

Passion fruit breeding program developed by the Universidade Estadual do Norte Fluminense Darcy Ribeiro started in 1998 with the early collection of various genotypes in three different producing regions of the state of Rio de Janeiro (Viana et al. 2003). Based on these first studies, a wider sampling was carried out in commercial areas of the Northern Rio de Janeiro region.

Figure 1 shows a schema of the passion fruit breeding program of UENF. In 2002, a larger number of progenies was obtained with the use of appropriate genetic designs, such as the Design I proposed, by Comstock and Robinson (1948). Thus, based on the results of these previous analyses, an intrapopulation recurrent selection program was prepared. Currently, passion fruit breeding program carried out by UENF is in the fourth cycle of recurrent selection.

Data from 81 full-sib progenies from the third cycle were evaluated between 2011-2013 in Campos dos Goytacazes (lat 21º 45' S, long 41°20' W, alt 11m asl), in the northern state of Rio de Janeiro, and in Itaocara (lat 21º 40' S, long 42° 04' W, alt 76m asl), in the northwestern state of Rio de Janeiro.

The experiments were arranged in randomized complete block design with two replications and five plants per plot. Weed, pest, fertilization and disease management were carried out according to standard recommendations. Daily drip irrigation was used during the dry season.

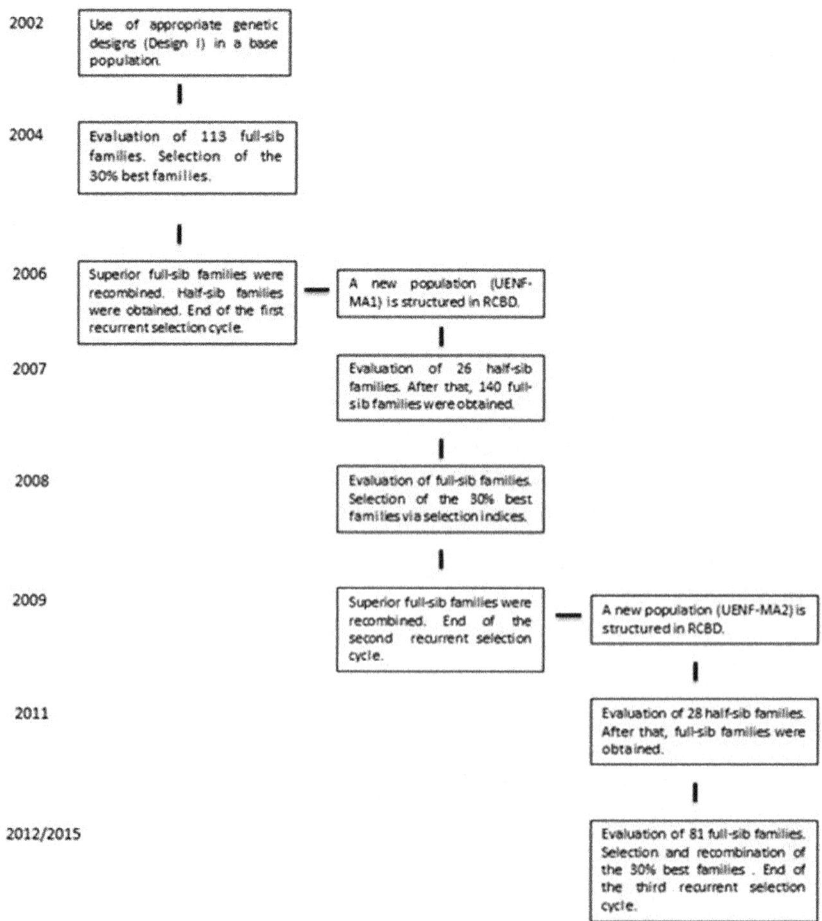

Figure 1. Schema of passion fruit breeding program of UENF.

The 22 best progenies were selected regarding fruit characteristics via BLUP (best linear unbiased prediction). The selected progenies were planted in separate field in the municipalities (previously specified) of Campos dos Goytacazes and Itaocara.

Fruits seeds resulting from the open-pollination crossing between the best selected progenies composed the released cultivar UENF Rio Dourado. For comparison, it was used the open-pollinated cultivar 'FB 200' in the VCU tests.

CULTIVAR CHARACTERISTICS

UENF Rio Dourado cultivar was evaluated based on agronomic and morphological characteristics (leaves, flowers and fruits) proposed by the National Cultivar Protection Service (SNPC) of the Ministry of Agriculture, Livestock and Supply (MAPA).

For the cultivar characterization, the used characteristics were: branch color; sinus depth; nectary position; skin color; blade length (mm): blade width (mm); petiole length (mm); flower diameter (mm); corona diameter (mm); sepal length (mm); sepal width (mm); sepal length (mm); petal width, (mm); bracts length (mm); peduncle length (mm); androgynophore length (mm). For the above characteristics, it was used the mean value of a random sample of five observations per plot.

For the characterization and selection of 22 best progenies, the used characteristics were: a) number of fruit (NF) – measured by counting all the fruits in each plot; b) total yield (Yield, tons ha^{-1}) from the first harvest in August 2012 until the end of April 2013; c) mean fruit weight (FW, g $fruit^{-1}$); d) fruit length (FL, mm) – obtained by longitudinal measurements of the fruits with a digital caliper; e) fruit width (FD, mm) – obtained by transversal measurements of the fruits with a digital caliper; f) pulp percentage (PP) – obtained by the ratio between the pulp weight and the total fruit weight (seeds, aril and juice); f) skin thickness (ST, mm) – the mean value after measuring the skin thickness in a transversal cut; g) soluble solids content of the juice (SSC, °Brix) – measured by a hand-held digital refractometer (Atago USA, Inc., Belleview, WA). For the characteristics FW, FL, FD, PP, ST and SSC, the mean value was obtained from a random sample of 15 fruits per plot.

PERFORMANCE

UENF Rio Dourado cultivar presented better performance than FB 200 cultivar for the three main evaluated characteristics (Table 1).

It is important to emphasize that UENF Rio Dourado cultivar originates from open-pollination. If crossing system via manual pollination had been used, yield would have been better than a when using the open-pollination system (Krause et al. 2012). Thus, these results show the genetic potential of UENF Rio Dourado cultivar as alternative to retake passion fruit cultivation in the regions in question. Due to its high yield potential and favorable fruit characteristics, UENF Rio Dourado cultivar is a new option for passion fruit growers in the north and northwestern region of the state of Rio de Janeiro. Moreover, after more performance tests,

Table 1. Mean values for number of fruits, yield (tons ha^{-1}) and fruit weight (grams) of UENF Rio Dourado cultivar and one control, in two locations (Campos dos Goytacazes and Itaocara), in the state of Rio de Janeiro

Traits	UENF Rio Dourado	FB 200
Number of fruits	250	186
Yield	25	16
Fruit weight	175	171

Table 2. Qualitative and quantitative characteristics of UENF Rio Dourado cultivar measured in leaves, flowers and fruits

Traits	UENF Rio Dourado
Branch color	Green-Purple
Sinus depth	Deep
Nectary position	Adjacent to the limbus
Skin color	Yellow
Blade length	166.65 mm
Blade width	214.70 mm
Petiole length	61.98 mm
Flower diameter	79.87 mm
Corona diameter	68.06 mm
Sepal length	33.79 mm
Sepal width	12.69 mm
Petal length	33.81 mm
Petal width	10.35 mm
Bracts length	26.91 mm
Peduncle length	51.93 mm
Androgynophore length	11.68 mm
Fruit length	84.50 mm
Fruit width	76.00 mm
Percent pulp	41.00 %
Skin thickness	7.10 mm
Soluble solids content	13.50 °Brix
pH	3.20

there is the possibility of recommending this cultivar for other Brazilian regions.

OTHER CHARACTERISTICS

The cultivar has the following characteristics (Table 2).

There is no specified form to register passion fruit cultivars at MAPA. *Passiflora* breeders need to fill the "other species" form (annex IX). Thus, besides the descriptors used by MAPA, the authors aimed to contribute by describing UENF Rio Dourado cultivar and by proposing descriptors for the register of *Passiflora edulis Sims* species.

BASIC SEED PRODUCTION

UENF Rio Dourado was registered (nº 34328) by MAPA. The genetic seed stock is maintained by UENF.

ACKNOWLEDGEMENTS

The authors thank FAPERJ, CNPq and CAPES for the financial support for the passion fruit breeding program of UENF.

REFERENCES

Comstock RE and Robinson HF (1948) The components of genetic variance in populations of biparental progenies and their use in estimating the average degree of dominance. **Biometrics 4**: 254-266.

Gonçalves GM, Viana AP, Bezerra Neto FV, Amaral Junior AT and Pereira MG (2009) Genetic parameter estimates in yellow passion fruit based on design I. **Brazilian Archives of Biology and Technology 52**: 523-530.

IBGE (2014) **Sistema IBGE de recuperação automática – Sidra. Produção agrícola municipal: produção de maracujá**. Available at <http://www.sidra.ibge.gov.br/>. Accessed on August 27, 2014.

Krause W, Neves LG, Viana AP, Araújo CAT and Faleiro FG (2012) Produtividade e qualidade de frutos de cultivares de maracujazeiro-amarelo com ou sem polinização artificial. **Pesquisa Agropecuária Brasileira 47**: 1737-1742.

MAPA (2015) **Ministério da Agricultura Pecuária e Abastecimento**. Available at <http://extranet.agricultura.gov.br/php/snpc/cultivarweb/cultivares_registradas.php>. Acessed on May 12, 2015.

Santos EA, Viana AP, Freitas JCO, Rodrigues DL, Tavares RF, Paiva CL and Souza MM (2015) Genotype selection by REML/BLUP methodology in a segregating population from an interspecific *Passiflora spp.* crossing. **Euphytica 204**: 1-11.

Viana AP, Pereira TNS, Pereira MG, Souza MM, Maldonado JFM and Amaral Júnior AT (2003) Simple and canonic correlation between agronomical and fruit quality traits in yellow passion fruit (*Passiflora edulis* f. flavicarpa) populations. **Crop Breeding and Applied Biotechnology 3**: 133-140.

8

URS Altiva – a new oat cultivar with high agro-nomic performance

Itamar C. Nava[1], Marcelo T. Pacheco[1] and Luiz C. Federizzi[1*]

***Corresponding author:**
E-mail: federizi@ufrgs.br

[1] Universidade Federal do Rio Grande do Sul (UFRGS), Faculdade de Agronomia, Departamento de Plantas de Lavoura, Av. Bento Gonçalves, 7712, 91501-970, Porto Alegre, RS, Brazil

Abstract: *The oat cultivar URS Altiva, developed from the simple cross 'UFRGS 995090-2 x URS 21', and released by the Oat Breeding Program of the Federal University of Rio Grande do Sul (UFRGS) in 2015, presents high grain yield, high grain quality, desirable agronomic performance, and partial resistance to crown rust.*

Key words: *Avena sativa L., oat breeding program, crown rust.*

INTRODUCTION

The cultivated hexaploid oat (*Avena sativa* L., 2n = 6x =42, AACCDD) is an important cereal crop used for food, feed, and forage worldwide. Oat has several nutritional properties suitable for human consumption, which are associated with health benefits. Oat grains contain high amounts of valuable nutrients, such as proteins, soluble fibers (β-glucans), unsaturated fatty acids, vitamins, minerals, and antioxidants. These attributes make oatmeal a functional food with beneficial effects on lowering cholesterol and reducing glycemic response in humans (Ames et al. 2014).

Oat presents wide adaptation and is cultivated predominantly in temperate regions or in winter seasons. In the Southern Hemisphere, oat cultivation extends from the latitude of 21° (Brazil) to 45° S (New Zealand), while in the Northern Hemisphere, oat is grown from 19° (Mexico) to 65° N (Finland). However, oat production is mainly concentrated between the latitudes 23° and 38° S, and between 35° and 55° N. In subtropical environments, such as Southern Brazil, oat plays an important role for grain production during the winter/spring seasons in no-tillage crop system rotation, usually with soybeans (Locatelli et al. 2007). The area cultivated with hexaploid oat in Brazil was approximately 189.500 hectares with mean grain yield of 1.853 kg ha^{-1} in the growing season of 2015. The states of Rio Grande do Sul (RS) and Paraná (PR) were the main oat producers (CONAB 2016).

Oat breeding in Brazil is still recent; it started in the 1970s. The identification of agronomic traits that meet the demand of farmers, industry and final consumers, and their incorporation into elite oat germplasm have been crucial to the development of successful new oat cultivars. Some of the most important traits selected by the UFRGS Oat Breeding Program include wide adaptation, high grain yield and grain quality, short plant cycle, reduced plant height, lodging resistance, frost tolerance, aluminum tolerance, and genetic resistance to the main diseases, such as crown and stem rust, leaf spot, fusarium, and BYDV (barley yellow dwarf virus). The objective of this work is to present the pedigree, breeding method, and agronomic performance of the new released oat cultivar URS Altiva.

PEDIGREE AND BREEDING METHOD

URS Altiva is a hulled F_7-derived line developed from the simple cross 'UFRGS 995090-2 x URS 21'. Both parents were developed in Brazil by the UFRGS Oat Breeding Program. The genealogy of the parent UFRGS 995090-2 is 'UFRGS 881971 // Pc68/*5 Starter F_4', while the genealogy of the parent URS 21 is 'UFRGS 10 / CTC 84B993'. URS 21 was released as a cultivar in 2000, and has still been cultivated, mainly due to its partial resistance to crown rust, a disease caused by the fungus *Puccinia coronata* Corda f. sp. *avenae*. This genetic resistance enables URS 21 to yield well even in the years which are favorable for the disease development. Crown rust has historically been the major problem in most oat-growing areas in Southern Brazil, causing devastating grain yield losses when using susceptible cultivars, or when chemical control with fungicides is not adequately applied.

The cross between UFRGS 995090-2 and URS 21 was carried out in 2004 at the UFRGS Agronomy Experimental Station, located in Eldorado do Sul, RS, Brazil. The line UFRGS 995090-2 was employed as the female parent and the cultivar URS 21 was employed as the male parent. Artificial hybridization between UFRGS 995090-2 and URS 21 was carried out following the open flower technique, as described by Bertagnolli and Federizzi (1994). Two seeds from the first filial generation (F_1) were obtained from this cross, which were sown in the field at the growing season of 2005 under the identification code 'F_1-67/05'. The number 67 represented the original number of this cross, carried out in 2004. Panicles from the two F_1 plants were harvested and bulk-threshed, giving rise to seeds of the second filial generation (F_2).

The segregating population developed from the cross UFRGS 995090-2 x URS 21 was advanced and selected from F_2 to F_6 by means of a modified pedigree method, carried out at the UFRGS Agronomy Experimental Station. The main characteristics and modifications of the breeding method consisted of: i) plant density varied from 25 to 30 seeds per linear meter in the F_2 generation; ii) selection of the best plants started in F_2, and only one panicle from each selected plant was harvested; iii) selection of the best families started in $F_{2:3}$; iv) each panicle harvested in the previous generation composed a two-meter long double-row in the next generation; v) plant density was determined by the number of seeds present in the previously selected panicle; and vi) all generation was carried out in a no-tillage system, with soybeans as the preceding crop. The main selection criteria used in the field included: vigor and biomass, when plants had 6 to 7 leaves (approximately 40 days after emergence); plant height; number of days to flowering and to maturation; disease resistance; lodging resistance; and panicle fertility at the maturity stage. In the laboratory, traits related to visual grain quality were selected, such as size, shape, uniformity, grain filling, and health.

A population of approximately 600 F_2 plants was cultivated in 2006. The population was identified by the number 062062, in which '06' represented the year of 2006, '2' represented the F_2 generation, and '062' represented the evaluated population. From this population, a total of 24 panicles were selected. In the growing season of 2007, these panicles were sown in the field, originating 24 $F_{2:3}$ families, identified with the numbers 073031-1 - 073031-24. From the family 073031-1, six panicles were selected. These panicles were sown in 2008, originating six $F_{3:4}$ families, and were identified with the numbers 084128-1 - 084128-6. From the family 084128-6, six panicles were selected to compose the next generation.

The $F_{4:5}$ families were grown in 2009 and identified with the numbers 095113-1 - 095113-6. From the family 095113-5, four individual panicles with partial resistance to crown rust were selected. Partial resistance was characterized by abundant chlorosis and necrosis around small pustules on the leaf laminae, and early telia formation on green leaf tissue of the plants. Similar symptoms have been identified in the cultivar URS 21 and other oat genotypes that presented partial resistance to crow rust (Graichen et al. 2011, Zambonato et al. 2012), and more recently, in the cultivar URS Brava (Federizzi et al. 2015). The resistance mechanism in the oat-crown rust system is not yet fully understood, even though reactive oxygen species seem not to be important in this pathosystem. It seems that in the partial resistance mechanism, observed in URS 21, the death of the fungus occurs first than the death of the cells, differing from the hypersensitive response, in which the cell is killed first (Graichen et al. 2011). Another particularity observed in URS 21 is that the production of phenolic compounds was more pronounced than in a cultivar that presented hypersensitive response (Figueiró et al. 2015).

During the growing season of 2010, the four $F_{5:6}$ lines were sown under the identification numbers 106088-1 - 106088-4. In that year, the line which gave origin to the cultivar 'URS Altiva' presented high phenotypic uniformity (no visual segregation) and was harvested in bulk, threshed, and coded as 'UFRGS 106088-1'. The new line was first tested in a preliminary trial (2011) and then evaluated during three consecutive years in a net of cooperative trials, including the Regional Trial (2012), the National Trial of first (2013) and second-year (2014). In the cooperative trials, the line 'UFRGS 106088-1' was tested in 30 experiments, carried out in from 9 to 11 locations each year, distributed in the states of Paraná, Rio Grande do Sul, and São Paulo. In all trials, UFRGS 106088-1 was compared with three check cultivars.

Taking together the results obtained in the preliminary trial and in the three years of cooperative tests, the line UFRGS 106088-1 reached satisfactory agronomic performance (data presented below) to be released as a new cultivar. In all trials, the new line achieved grain yield close to or greater than 5% of the best check, which is the main standard established by the Brazilian Oat Research Committee (Comissão Brasileira de Pesquisa de Aveia – CBPA), in order to approve the release of a new oat cultivar using their cooperative trials. The new oat cultivar was released in 2015 and was denominated URS Altiva. The name 'Altiva' (proud, noble) was selected for this cultivar to highlight its robustness, which results in high grain yield potential, grain quality and resistance to crown rust and lodging, when compared with other oat cultivars currently available in Brazil.

AGRONOMIC PERFORMANCE

Approximately 400 g of F_7-seeds of the oat line UFRGS 106088-1 were harvested in 2010 and used in the preliminary trial, which was carried out at the UFRGS Agronomy Experimental Station in 2011. The new line was compared with the check cultivars URS 21, Barbarasul, and URS Taura for the agronomic traits grain yield, test weight, number of days from emergence to heading, plant height and lodging. Among the check cultivars, URS Taura had the highest mean grain yield of 4504 kg ha^{-1}, and the line UFRGS 106088-1 had mean grain yield of 5277 kg ha^{-1}, corresponding to 117.2% of the best check. For test weight, an important measure of physical grain quality, the line UFRGS 106088-1 showed weight of 62.8 kg hL^{-1}, which was higher than 56.2 kg hL^{-1} presented by the best check URS 21. The line UFRGS 106088-1 presented 82 days from the emergence to heading, which was inferior to that observed for the three check cultivars. Considering plant height, the new line was taller, reaching mean of 125 cm, compared with 119, 121 and 109 cm observed for the check cultivars URS 21, Barbarasul and URS Taura, respectively. However, the greatest height did not result in increased plant lodging in the new line, which presented lodging of 40%, while the check cultivars URS 21 and Barbarasul presented 50 and 80%, respectively, and the cultivar URS Taura did not present lodging in this trial.

During the cooperative tests, the oat line UFRGS 106088-1 was evaluated in 11 locations across the states of Rio Grande do Sul (Augusto Pestana, Eldorado do Sul, Passo Fundo and Pelotas), Paraná (Guarapuava, Londrina, Mauá da Serra, Pato Branco, Ponta Grossa and Santa Tereza do Oeste), and São Paulo (Capão Bonito). These trials were carried out under the coordination of the Brazilian Oat Research Committee. Table 1 shows the results for grain yield, test weight, thousand kernel weight, plant height, and lodging obtained for the line UFRGS 106088-1 and the check cultivars, evaluated in the Regional and National Trials of Oat Lines.

In the Regional Trial carried out in 2012, the line UFRGS 106088-1 was compared with the check cultivars URS 21, Barbarasul, and URS Taura. These check cultivars were among the best oat cultivars available in Brazil. The cultivar URS 21 presented the highest agronomic performance for grain yield among the check cultivars, i.e., mean grain yield of 2821 kg ha^{-1}, whereas the line UFRGS 106088-1 presented mean grain yield of 3082 kg ha^{-1}, corresponding to 109.3% of the best check. When the test weight was evaluated, the cultivar URS 21 was the best check, showing mean test weight of 45.2 kg hL^{-1}; on the other hand, the line UFRGS 106088-1 had 51.8 kg hL^{-1}, corresponding to 114.5% of the best check. For the trait thousand kernel weight, the cultivar URS Taura was the best check with mean thousand kernel weight of 28.6 g, whereas the line UFRGS 106088-1 had mean thousand kernel weight of 34.9 g, corresponding to 122.1% of the best check. For the traits plant height and lodging, the line UFRGS 106088-1 was taller than all the check cultivars, and exhibited a higher level of lodging resistance when compared with the cultivars URS 21 and Barbarasul. In the Regional Trial of 2012, results for grain yield, test weight, thousand kernel weight, plant height, and lodging were available in 11, 11, 9, 10 and 7 locations, respectively (Table 1).

In the first year of the National Trial of Oat Lines, carried out in 2013, the line UFRGS 106088-1 was compared with the check cultivars URS 21, Barbarasul, and URS Taura. Results demonstrated that the cultivar URS 21 was the best check for grain yield, test weight, and thousand kernel weight, with mean of 3195 kg ha^{-1}, 48.0 kg hL^{-1}, and 27.7 g, respectively. The experimental line UFRGS 106088-1 presented mean grain yield of 3570 kg ha^{-1}, mean test weight of 53.4 kg hL^{-1}, and mean thousand kernel weight of 33.4 g, equivalent to 111.8%, 111.2%, and 120.6% of the best check, respectively. Considering the traits plant height and lodging, the line UFRGS 106088-1 was taller than all the check cultivars, but had the highest level of lodging resistance, even when compared with the check cultivar URS Taura. In 2013, data for grain yield, test weight, thousand kernel weight, plant height, and lodging were available in 10, 10, 7, 8 and 5 locations, respectively (Table 1).

In the National Trial of Oat Lines of second-year, carried out in 2014, the line UFRGS 106088-1 was compared with the check cultivars URS 21, Barbarasul, and URS Corona. The cultivar URS Corona was the best check for grain yield and thousand kernel weight, showing mean of 3488 kg ha^{-1} and 31.2 g, respectively. However, the cultivar URS 21 was the

Table 1. Grain yield, test weight, thousand kernel weight, plant heigth and lodging of the oat line UFRGS 106088-1 and the check cultivars evaluated in the Regional Trial of Oat Lines (2012), National Trial of Oat Lines of first-year (2013), and National Trial of Oat Lines of second-year (2014)

Cultivar	**Grain yield** (kg ha^{-1})				
	2012	2013	2014	BC annual†	$BC_{URS\ 21}$‡
URS 21 (C)*	2821	3195	3282	98.0	100
Barbarasul (C)	2650	2589	3093	87.9	89.6
URS Taura (C)	2635	2191	-	81.0	80.2
URS Corona (C)	-	-	3488	100	106.3
UFRGS 106088-1	3082 (**109.3**)§	3570 (**111.8**)	3656 (**104.8**)	**108.6**	110.9
Number of locations	11	10	8	29	29
	Test weight (kg hL^{-1})				
URS 21 (C)	45.2	48.0	46.8	100	100
Barbarasul (C)	42.6	43.2	44.2	92.9	92.8
URS Taura (C)	44.2	43.0	-	93.6	93.5
URS Corona (C)	-	-	46.3	98.8	98.8
UFRGS 106088-1	51.8 (**114.5**)	53.4 (**111.2**)	53.2 (**113.6**)	**113.1**	**113.1**
Number of locations	11	10	9	30	30
	Thousand kernel weight (g)				
URS 21 (C)	28.2	27.7	27.8	95.9	100
Barbarasul (C)	26.5	24.6	26.2	88.4	92.3
URS Taura (C)	28.6	26.0	-	96.9	97.6
URS Corona (C)	-	-	31.2	100	112.4
UFRGS 106088-1	34.9 (**122.1**)	33.4 (**120.6**)	33.7 (**108.0**)	**116.9**	**121.9**
Number of locations	9	7	8	24	24
	Plant height (cm)				
URS 21 (C)	107.4	115.3	119.1	115.7	100
Barbarasul (C)	103.7	108.9	108.6	108.9	94.0
URS Taura (C)	93.6	94.0	-	100	84.2
URS Corona (C)	-	-	111.0	102.2	93.2
UFRGS 106088-1	112.4 (**120.1**)	120.9 (**128.5**)	119.6 (**110.1**)	**119.6**	**103.2**
Number of locations	10	8	8	26	26
	Lodging (%)				
URS 21 (C)	50.9	44.5	56.2	135.7	100
Barbarasul (C)	47.6	35.8	48.8	118.8	87.2
URS Taura (C)	30.5	35.5	-	100	69.3
URS Corona (C)	-	-	59.3	121.3	105.4
UFRGS 106088-1	35.4 (**116.1**)	29.8 (**83.8**)	20.5 (**42.0**)	**80.5**	**56.5**
Number of locations	7	5	6	18	18

*Check cultivar.
†Mean performance relative to the best check cultivar, within each year of test, as a percentage.
‡Mean performance relative to the check cultivar URS 21, as a percentage.
§Values shown in brackets for the line UFRGS 106088-1 demonstrate its performance when compared with the best check within the year of evaluation, in percentage.

best check for test weight, with mean of 46.8 kg hL^{-1}. The line UFRGS 106088-1 presented mean grain yield of 3656 kg ha^{-1}, mean test weight of 53.2 kg hL^{-1}, and mean thousand kernel weight of 33.7 g, corresponding to 104.8, 113.6, and 108.0% of the best check for each agronomic trait, respectively. For plant height and lodging, the line UFRGS 106088-1 was taller and more resistant to lodging than all the check cultivars. In 2014, data for for grain yield, test weight, thousand kernel weight, plant height, and lodging were recorded in 8, 9, 8, 8 and 6 locations, respectively (Table 1).

Over the three-year test for grain yield, the line UFRGS 106088-1 was evaluated in 29 experiments, and presented mean grain yield of 3436 kg ha^{-1}, corresponding to 108.6% of the best check cultivar within each year. Additionally, the line UFRGS 106088-1 presented grain yield performance equal to 110.9% of that exhibited by the check cultivar URS 21 over time (Table 1). These results clearly indicate the high adaptability and stability of the line UFRGS 106088-1 for grain yield over years and locations of evaluation.

Table 2 shows the number of days from emergence to flowering, days from flowering to maturation, and days from emergence to maturation, reflecting the cycle of the line UFRGS 106088-1 during the three-year test. Considering the number of days from emergence to flowering, the line UFRGS 106088-1 showed mean vegetative cycle of 71.1, 78.5, and 72.2 days in the years of evaluation 2012, 2013, and 2014, respectively. These results demonstrate the earliness of line UFRGS 106088-1, when compared with the check cultivars in the three years of evaluation. However, the line UFRGS 106088-1 had opposite performance for the number of days from flowering to maturation, showing mean of 43.2, 42.9, and 43.1 days, corresponding to 111.3, 114.4, and 107.6% of the earlier check cultivar Barbarasul in 2012 (38.8 days), URS Taura in 2013 (37.5 days), and URS Corona in 2014 (43.1 days), respectively. When the full cycle was measured by the number of days from emergence to maturation, small differences were observed between UFRGS 106088-1 and the check cultivars (Table 2). The longer period from flowering to maturation may be associated with the high grain yield obtained for the line UFRGS 106088-1, as it would allow the plants to accumulate more photoassimilates and translocate them to the grains.

Table 2. Days from emergence to flowering, days from flowering to maturation and days from emergence to maturation of the oat line UFRGS 106088-1 and the check cultivars evaluated in the Regional Trial of Oat Lines (2012), National Trial of Oat Lines of first-year (2013), and National Trial of Oat Lines of second-year (2014)

Cultivar	Days from emergence to flowering				
	2012	2013	2014	BC annual[†]	$BC_{URS\ 21}$[‡]
URS 21 (C)*	77.3	81.8	76.2	102.0	100
Barbarasul (C)	77.7	85.9	79.2	105.2	103.2
URS Taura (C)	72.9	82.3	-	100.3	97.6
URS Corona (C)	-	-	76.6	100.6	100.6
UFRGS 106088-1	71.1 (**97.4**)[§]	78.5 (**96.1**)	72.2 (**94.8**)	**96.1**	94.3
Number of locations	10	9	8	27	27
	Days from flowering to maturation				
URS 21 (C)	39.3	40.1	44.0	103.0	100
Barbarasul (C)	38.8	39.5	43.7	101.8	98.9
URS Taura (C)	40.0	37.5	-	101.0	97.6
URS Corona (C)	-	-	43.1	100	98.1
UFRGS 106088-1	43.2 (**111.3**)	42.9 (**114.4**)	46.4 (**107.6**)	**110.7**	**107.5**
Number of locations	7	8	7	22	22
	Days from emergence to maturation				
URS 21 (C)	116.8	121.2	119.1	101.4	100
Barbarasul (C)	117.4	125.4	121.6	103.4	102.0
URS Taura (C)	114.1	119.6	-	100	98.2
URS Corona (C)	-	-	118.7	100	99.6
UFRGS 106088-1	115.3 (**101.0**)	120.9 (**101.1**)	117.6 (**99.1**)	**100.4**	**99.0**
Number of locations	7	8	7	22	22

*Check cultivar.
[†]Mean performance relative to the best check cultivar, within each year of test, in percentage.
[‡]Mean performance relative to the check cultivar URS 21, in percentage.
[§]Values shown in brackets for the line UFRGS 106088-1 demonstrate its performance when compared with the best check within the year of evaluation, in percentage.

Table 3. Crown rust severity, stem rust severity and leaf spot severity of the oat line UFRGS 106088-1 and the check cultivars evaluated in the Regional Trial of Oat Lines (2012), National Trial of Oat Lines of first-year (2013), and National Trial of Oat Lines of second-year (2014)

Cultivar	**Crown rust severity** (%)				
	2012	2013	2014	BC annual[†]	$BC_{URS\ 21}$[‡]
URS 21 (C)*	25.1	24.5	21.2	158.4	100
Barbarasul (C)	32.9	30.5	27.6	204.8	128.7
URS Taura (C)	31.9	36.1	-	137.2	137.1
URS Corona (C)	-	-	7.7	100	36.3
UFRGS 106088-1	6.6 (**26.1**)[§]	8.5 (**34.9**)	3.6 (**46.8**)	35.9	26.4
Number of locations	10	7	7	24	24
	Stem rust severity (%)				
URS 21 (C)	3.9	3.6	6.2	107.8	100
Barbarasul (C)	5.2	6.0	7.0	146.3	132.5
URS Taura (C)	3.6	10.1	-	188.8	181.7
URS Corona (C)	-	-	5.3	100	86.5
UFRGS 106088-1	3.0 (**82.5**)	3.4 (**93.1**)	4.4 (**82.8**)	**86.1**	78.7
Number of locations	9	4	3	16	16
	Leaf spot severity (%)				
URS 21 (C)	17.7	28.8	28.9	106.6	100
Barbarasul (C)	18.6	31.6	29.8	113.0	106.2
URS Taura (C)	22.5	26.8	-	113.6	106.1
URS Corona (C)	-	-	25.7	100	89.0
UFRGS 106088-1	12.8 (**72.4**)	23.9 (**89.3**)	15.0 (**58.5**)	**73.4**	**68.7**
Number of locations	8	7	8	23	23

*Check cultivar.

[†]Mean performance relative to the best check cultivar, within each year of test, in percentage.

[‡]Mean performance relative to the check cultivar URS 21, in percentage.

[§]Values shown in brackets for the line UFRGS 106088-1 demonstrate its performance when compared with the best check within the year of evaluation, in percentage.

The high grain yield observed for the line UFRGS 106088-1 is also associated with its genetic resistance against the main oat diseases. Table 3 shows the results of crown rust severity, stem rust severity, and leaf spot severity. The line UFRGS 106088-1 presented very low severities for crown rust, during the three years of cooperative tests. Mean crown rust severity for the line UFRGS 106088-1 was 6.6, 8.5, and 3.6% in 2012, 2013, and 2014, respectively. These results corresponded to 26.1, 34.9, and 46.8% of the best check cultivar in each year of evaluation. URS 21 was the best check cultivar for crown rust in 2012 and 2013, whereas URS Corona was the best check cultivar in 2014. The line UFRGS 106088-1 also presented desirable levels of resistance against stem rust caused by the fungus *Puccinia graminis* f. sp. *avenae*. In the three-year test, the line had mean stem rust severity of 3.0, 3.4, and 4.4%, corresponding to 82.5, 93.1, and 82.8% of the best check URS Taura, in 2012, URS 21, in 2013, and URS Corona, in 2014, respectively. The severity of leaf spot, mainly caused by the fungus *Pyrenophora chaetomioides* Speg., was relatively low in the line UFRGS 106088-1, when compared with the check cultivars. In the three-year test, the line showed mean leaf spot severity of 12.8, 23.9, and 15.0%, corresponding to 72.4, 89.3, and 58.5 of the best check in each year. The cultivar URS 21 was the best check in the Regional Trial, carried out in 2012; the cultivar URS Taura was the best check in the National Trial of first-year, carried out in 2013; and the cultivar URS Corona was the best check in the National Trial of second-year, carried out in 2014 (Table 3).

The cultivar URS Altiva, after its release, was registered in the National Registry of Cultivars ('Registro Nacional de Cultivares'), of the Ministry of Agriculture, Livestock and Supply ('Ministério da Agricultura, Pecuária e Abastecimento - MAPA'), under the number 34272. The new cultivar was protected by the National Service for Cultivar Protection ('Serviço Nacional de Proteção de Cultivares'), under the certificate number 20160023. The cultivation of URS Altiva is recommended for the states of Rio Grande do Sul, Santa Catarina, Paraná, and São Paulo. A volume of approximately 30 tons of breeder seeds harvested in 2015 will be used for the production of foundation seeds.

REFERENCES

Ames N, Rhymer C and Storsley J (2014) Food oat quality throughout the value chain. In Chu Y (ed) **Oats Nutrition and Technology.** Wiley, Oxford, p. 33-70.

Bertagnolli PF and Federizzi LC (1994) Cruzamentos artificiais em aveia. **Pesquisa Agropecuária Brasileira 29**: 601-606.

CONAB - Companhia Nacional de Abastecimento (2016) Série histórica de área, produtividade e produção de aveia. Available at < http:// www. conab.gov.br >. Accessed in March 2016.

Federizzi LC, Pacheco MT and Nava IC (2015) URS Brava a new oat cultivar with partial resistance to crown rust. **Crop Breeding and Applied Biotechnology 15**: 197-202.

Figueiró AA, Reese N, Hernandez JLG, Pacheco MT, Martinelli JA, Federizzi LC and Delatorre CA (2015) Reactive oxygen species are not increased in resistant oat genotypes challenged by crown rust isolates. **Journal of Phytopathology 163**: 795-806.

Graichen FAS, Martinelli JA, Wesp CL, Federizzi LC and Chaves MS (2011) Epidemiological and histological components of crown resistance in oat genotypes. **European Journal of Plant Pathology 131**: 497-510.

Locatelli AB, Federizzi LC, Milach SCK and McElroy AR (2007) Flowering time in oat: genotype characterization for photoperiod and vernalization response. **Field Crops Research 106**: 242-247.

Zambonato F, Federizzi LC, Pacheco MT, Arruda MP and Martinelli JA (2012) Phenotypic and genetic characterization of partial resistance to crown rust in *Avena sativa* L. **Crop Breeding and Applied Biotechnology 12**: 261-268.

9

Identification of heterotic patterns between expired proprietary, NDSU, and industry short-season maize inbred lines

Mohammed A. A. Bari[1], Marcelo J. Carena[1*] and Messias G. Pereira[2]

***Corresponding author:**
E-mail: marcelo.carena@ndsu.edu

[1] North Dakota State University, Department of Plant Sciences, Dep. #7670, Fargo, ND 58108-6050, United States of America
[2] Universidade Estadual do Norte Fluminense, Departamento de Genética e Melhoramento de Plantas, Av. Alberto Lamego, 2000, 28.013-602, Campos, RJ, Brazil

Abstract: *Maize (*Zea mays *L.) inbred lines are under restricted use, protected by Patent and Plant Variety Protection (PVP) laws. Research objectives were i) to identify and validate heterotic groups and patterns, and ii) to determine if ex-PVP lines are useful for continued genetic progress in short-season environments. Three groups of crosses were made following North Carolina Mating Design II (NCII) including 12 NDSU, 24 ex-PVP lines, and seven current industry testers. Hybrids were planted in four different experiments at six ND environments following partially balanced lattice experimental designs in 2011 and 2012. Top heterotic patterns were selected based upon grain yield and other agronomic traits. Our research indicates most ex-PVP lines are genetically narrow and may not be immediately useful. Less protection (5-yr vs. 20-yr) might increase usefulness of ex-PVP lines. This change in intellectual property will allow public breeders to develop better versions of industry lines carrying known weaknesses.*

Key words: *Zea mays L., ex-PVP, heterotic groups, SCA.*

INTRODUCTION

Maize breeding for hybrid production is a confidential and highly profitable business. In order to protect this business, The USA decided to protect inbred parents and hybrids by the U.S. Patent/or U.S. Plant Variety Protection Act (PVPA). Expired-PVP (ex-PVP) inbred lines, after being protected for 20 years, are maintained at the North Central Regional Plant Introduction Station (NCRPIS) at Ames, IA (Mikel 2006, Bari and Carena 2014). These lines have become annually available after protection ended. They potentially represent new germplasm sources for many public and private breeding programs for study and use (Nelson et al. 2008). However, many breeders doubt their usefulness due to their original development date, which is over 20 years old.

Maize breeding programs are dependent on the identification and utilization of heterotic groups and heterotic patterns (Melani and Carena 2005). Assigning and validating ex-PVP inbred lines to heterotic groups will be useful to exploit desirable heterotic patterns. Heterotic groups represent groups of germplasm sources that when crossed with each other produce consistently better crosses than when crosses are made within those groups (Hallauer and Carena 2009). Identifying heterotic patterns, which are crosses between known genotypes (from different heterotic groups) expressing a high level of heterosis (Carena and Hallauer 2001, Troyer 2006, Mendes et al. 2015), is key to the development of successful maize hybrids (Eyherabide and Hallauer 1991, Barata and Carena 2006, Carena and Wicks III 2006). The North American dent maize germplasm is composed of multiple heterotic groups that when crossed to each other can optimize hybrid performance (Mikel and Dubley 2006). The identification of genotypes in these groups helps exploit suitable heterotic patterns. Dubreuil et al. (1996) emphasized that the accurate assignment of inbred lines to heterotic groups is a prerequisite for efficient utilization of germplasm. Heterotic groups in dent maize have been subdivided into Iowa Stiff Stalk Synthetic (BSSS) and non-BSSS (Lu and Bernardo 2001). A similar grouping consists of Reid Yellow Dent (related to BSSS), Lancaster, Iodent, and miscellaneous heterotic groups (Gethi et al. 2002). Troyer (1999) divided maize into five genetic backgrounds: Reid Yellow Dent (Iodent Reid and BSSS), Minnesota 13 (W153R and SD105), Northwestern Dent (A48, A509, and A78), Lancaster Sure Crop (Mo17 and Oh43), and Leaming (Oh07). The Reid Yellow Dent is the largest group, and has made significant contributions to commercial hybrids.

There are several methods to classify maize inbreds into heterotic groups. Two major classification methods are widely used across the world (Fan et al. 2009). The traditional method uses specific combining ability with line-pedigree information, and/or field hybrid-yield information, to assign maize lines to heterotic groups. A more challenging method is to use different molecular markers to compute genetic similarity (GS) or genetic distance (GD) estimates to assign maize lines to a particular heterotic group, which is not always accurate. Fan et al. (2009) executed a third approach, by using heterotic group's specific and general combining ability (HSGCA) to classify inbreds into heterotic groups. They claimed their way is efficient compared to SSR markers and yield-based specific combining abilities. Menkir et al. (2004) classified inbred lines into heterotic groups by yield-based specific combining ability and molecular marker based approaches. They reported that yield-based combining ability derived heterotic groups did not match with groups established using molecular markers. Melchinger (1999) extensively discussed the potentials of DNA markers in assigning inbreds of unknown genetic origin to established heterotic groups. However, he concluded that if a large number of genotypes are available and proven testers exist, the testcross performances should be the main criteria for classifying materials into heterotic groups. In addition, Barata and Carena (2006) observed large inconsistencies between molecular marker-based classification and field trial based classification (e.g., testcross and diallel data) of a diverse set of inbreds. They concluded that groups of similar germplasm and heterosis properties could not be identified accurately and reliably with molecular markers. Consequently, they recommended extensive field evaluation across environments to classify inbred lines into heterotic groups. Mating designs (Hallauer et al. 2010) can be used to test a large set of progenies, extensively over locations and years to classify inbreds to heterotic groups. Many ex-PVP lines do not have assigned heterotic groups yet; an approximation can be deduced based on PVP documents and their genesis. Moreover, reported heterotic groups may not be stable in different situations. The objectives of the study were i) to identify and validate heterotic groups and patterns of ex-PVP inbreds, industry testers, and NDSU inbred lines, and ii) to determine if ex-PVP lines are useful for continued genetic progress in short-season environments.

MATERIAL AND METHODS

For full details on plant materials, crossing procedures, and field trials see Bari and Carena (2014). Ex-PVP lines were selected because they had the fewest number of silking days and growing degree-days in their public descriptions. Along with current testers and NDSU lines, they represent earliness pools for northern U.S. maize breeding. Ex-PVP materials were requested and obtained from the North Central Regional Plant Introduction Station (NCRPIS) in Ames, IA. Additional information on the the 12 NDSU lines and industry testers utilized can be found in Carena and Wanner (2003, 2009), Carena et al. (2003, 2010), and Carena (2013). Industry testers represent known heterotic groups available in the northern U.S. Corn Belt. T1 is an Iodent line, T2 is B14-derived SS line, T3 is another Iodent line, T4 is a B14/B73 derived SS line, T5 is an LH82 derived non-SS line, T6 is B14-derived SS line, and T7 is a B14/B73 derived SS line.

Twelve NDSU lines were crossed with 12 ex-PVP lines in the 2010 NDSU Fargo summer nursery following the North Carolina Mating Design II (NCII) (Comstock and Robinson 1948). The same 12 NDSU lines were crossed with another set of 12 ex-PVP lines, following a second NCII mating design at the 2010-2011 northern New Zealand winter nursery. All 24 ex-PVP lines (i.e. the lines used in the first and second sets of crosses) were also crossed in the winter with seven current industry inbred testers following a third NCII design. Hybrids were planted in four different partially balance lattice design trials along with five industry hybrids as checks across six North Dakota (ND) environments. Experimental checks represented a wide maturity range for ND (83RM to 100RM). Experimental plots were planted and harvested using machines that had been modified for small experimental plots. Plant density was approximately 80,000 plants ha^{-1}. Fertilization and field management practices were as recommended for ensuring optimum maize production. While harvesting, approximately 500 g seed samples were kept from each plot for grain quality assessment.

Data were recorded on an individual plot basis. Grain weight, grain moisture, and test weight were measured electronically on the combine while harvesting. Grain yield (Mg ha^{-1}) was adjusted to 155 g kg^{-1} grain moisture. Root lodging was measured as percentage of plants in a plot leaning at an angle greater than 30° from vertical while stalk lodging was measured as a percentage of plants in a plot with stalks broken at or below top ear. Lodging notes were taken just before harvest and analyzed as percentages to total stands per plot. Grain protein (g kg^{-1}), grain starch (g kg^{-1}), and grain oil (g kg^{-1}) data were collected with near infrared technology (OmegAnalyzer G, Bruins Instruments).

Plot means of all phenotypes were used for statistical analyses. Analyses of variance were performed for all traits at each location, as well as combined across locations and years using SAS 9.3 software (SAS 2010), for the four experiments. Data were collected and summarized in Excel files and then exported to SAS for analyses. Homogeneity of error variances was tested before combining data across environments. Mean comparisons among genotypes were assessed by Fisher's protected least significant difference (FLSD) at <0.05 level of significance, which has been shown to be an appropriate test for detecting differences (Carmer and Swanson 1971). A combined ANOVA was computed for grain yield (Mg ha^{-1}), grain moisture at harvest (g kg^{-1}), test weight (kg hL^{-1}), root lodging (%), stalk lodging (%), grain protein (g kg^{-1}), grain starch (g kg^{-1}), and grain oil (g kg^{-1}). General and specific combining ability effects were estimated considering year by location combination as environments. Combining abilities were further partitioned into male, female, and interactions of male and female (Scott et al. 2009). SCA effects for yield, along with mean grain yield were utilized to estimate and validate heterotic groups of short season ex-PVP and NDSU lines from the known heterotic groups of testers. Pedigrees of particular inbred lines were also used to determine heterotic groups when inbreds combined well with contrasting heterotic testers.

RESULTS AND DISCUSSION

SCA effects and means for grain yield have been widely used to classify maize heterotic groups (Menkir et al. 2004, Melani and Carena 2005, Fan et al. 2008). Therefore, we used SCA effects and mean grain yield of research trial III to validate heterotic groups of our first set of 12 ex-PVP inbred lines. In this trial we used hybrids between ex-PVP lines and known industry testers. We arranged SCA effects in descending order and selected the lines with top SCA effects (Table 3). Inbreds that were combining well with two contrasting testers (belonging to different heterotic groups) are also presented in Table 1.

Table 1. SCA effects for grain yield and mean grain yield per se, utilized in trial III to determine heterotic groups of first set of 12 ex-PVP maize inbred lines

Hybrids	SCA Effects (Mg ha^{-1})	Yield (Mg ha^{-1})	Testers	Testers HG†	Ex-PVP inbreds	Ex-PVP inbreds HG
T4 x PH207	1.31	9.22	T4	SS	PH207	Iodent?
T4 x Q381	1.20	9.36	T4	SS	Q381	UR?
T1 x Lp5	1.03	8.75	T1	Iodent	Lp5	SS?
T1 x CR1Ht	0.96	9.29	T1	Iodent	CR1Ht	Lancaster
T1 x NK794	0.77	8.75	T1	Iodent	NK794	SS
T1 x LH52	0.71	8.61	T1	Iodent	LH52	Lancaster
T1 x DK78010	0.56	7.68	T1	Iodent	DK78010	SS?
T7 x DK78010	0.47	7.38	T7	SS	DK78010	SS?
T1 x LH54	0.52	8.50	T1	Iodent	LH54	Lancaster
T3 x DKFAPW	0.44	7.17	T3	Iodent	DKFAPW	SS
T2 x DJ7	0.41	6.79	T2	SS	DJ7	SS?
T4 x NK779	0.34	6.16	T4	SS	NK779	Non-SS
T5 x NK807	0.21	6.01	T5	Non-SS	NK807	SS

†known heterotic group (HG) of testers were used to assign the HG of ex-PVP inbred lines; 'SS' refers to Stiff Stalk, 'non-SS' refers non-Stiff Stalk

Heterotic group determination for ex-PVP lines (first set of 12 ex-PVP maize inbred lines)

Mikel (2006) and GRIN have proposed heterotic groups of ex-PVP inbred lines based on the disclosure made by industry. Our study wanted to validate the classification.

SS inbreds

Lp5 combined well with Iodent tester T1, resulting in a grain yield average value of 8.75 Mg ha^{-1} and SCA effect of 1.03 Mg ha^{-1}. Lp5, therefore, seems to be in agreement with the PVP documents as presented in Table 1 (Mikel 2006). DK78010 combined well with both Iodent type tester T1 and SS tester T7. This inbred had above average combining ability with testers of the same and opposite heterotic groups. Even though this line would require more testing efforts, combining well across heterotic groups is a positive and desirable attribute of an inbred. Barata and Carena (2006) also found ND278 and ND282 combined very well across testers. So, the reported heterotic group, SS of DK78010 may not be an accurate classification. Inbred NK807 combined well with non-SS tester T5, therefore,it could belong to the SS group. However, NK807 has Lancaster background so further testing might be needed (Mikel 2006). DKFAPW combined well with the Iodent tester T3. Therefore, the SS background of DKFAPW could be correct. Inbred NK794 combined well with Iodent tester T1 and seems to belong to the SS group. The combining ability of DJ7 was above average with SS tester T2. Therefore, it showed the inbred may not belong to the SS group as published by Mikel (2006). The unrelated proportion of DJ7 could combine well with SS testers.

Non-SS inbreds

Q381 combined very well with SS tester T4 (with a yield of 9.36 Mg ha^{-1} and SCA effect of 1.20 Mg ha^{-1}) (Table 1). The heterotic background of Q381 was proposed as unrelated. In our extensive evaluation, Q381 seems to belong to a non-SS group. PH207 combined well with SS tester T4 (with a yield of 9.22 Mg ha^{-1} and SCA effect of 1.31 Mg ha^{-1}). Our evaluation infers that PH207 has a non-SS background, in addition to Iodent. NK779 combined well with the SS tester T4.NK779 evaluation seems to agree with the non-SS heterotic group assignation, which supports the line's MN13 and unrelated composition (Mikel 2006).

Lancaster inbred

CR1Ht combined very well with Iodent tester T1 (with a yield of 9.29 Mg ha^{-1}and SCA effect of 0.96 Mg ha^{-1}) (Table 1). CR1Ht was reported to have Lancaster and MN13 genetic backgrounds (Table 1), which agrees with our findings. LH52 and LH54 also combined well with Iodent tester, which agrees with past research suggestingthat both lines belongto the Lancaster group.

Heterotic group determination for ex-PVP lines (second set of 12 ex-PVP maize inbred lines)

SS inbreds

PHJ40 combined well with Iodent tester T1 (Table 2). The results for PHJ40 were in agreement with the PVP documented report for this line. Similar results confirmed data for NKS8324, CR14, and LH205, as expected. The combining ability of NKS8324 with Iodent tester T1 was above average. CR14, LH205 (with a yield of 7.62 Mg ha^{-1}and SCA effect of 0.77 Mg ha^{-1}), and RS710 (with a yield of 6.72 Mg ha^{-1}and SCA effect of 0.72 Mg ha^{-1}) also combined well with Iodent tester T1.

Non-SS inbreds

PHK05 and PHK76 combined well with SS tester T4 (Table 2). Based on our results, PHK05 and PHK76 belong to the non-SS group, as expected. Similarly, OQ603, PHP02 (with a yield of 7.47 Mg ha^{-1}and SCA effect of 0.54 Mg ha^{-1}), and L127 combined well with SS testers. PHR25 combined well with SS tester T4, so it could be considered Iodent but further evaluation would be needed to confirm this. PHT77 combined well with SS tester T7. The heterotic classification of PHT77 somewhat agrees with Mikel (2006), who reported that the line belongs to Lancaster and unrelated groups.

Heterotic group determination for NDSU lines

The first set of 12 ex-PVP lines was crossed to 12 NDSU lines and the resulting hybrids were evaluated in trial I.

SS inbreds

ND2004 combined well with non-SS inbred NK779 and SS inbred NK807 (Table 3). The testcross results from Carena and Wanner (2009) infer that ND2004 has a SS heterotic group (Table 2). However, ND2004 has unique combining ability with both heterotic groups based on these trials, which has been useful due to its excellent properties as male and female. ND2002 combined well with the Lp5 line, which was defined as mostly SS. The unrelated proportion of Lp5 might have contributed to combine well with the SS part of the line ND2002 (Table 3). Both ND2002 and ND2004 have been licensed exclusively. ND2003 and ND08-343 combined well with Lancaster line LH54.Testcross data from Carena and Wanner (2009) supports the SS background for ND2003 validating previous information. Also, data confirmed ND08-343 belongs to the SS heterotic group. ND2006 had an above average combining ability with SS and unrelated inbred Lp5. However, ND2006 was reported to have an SS heterotic background so more testing is needed. ND2010 combined well with the Lancaster inbred PH207 and SS inbred NK807. Further testing is, therefore, needed to confirm heterotic grouping of ND2010. ND2007 combined well with Iodent inbred PH207. ND2007 could carry an SS background based on the heterotic combinations tested and its pedigree.

Table 2. SCA effects for yield and mean grain yield per se, utilized in trial IV to determine heterotic groups of second set of 12 ex-PVP maize inbred lines

Hybrids	SCA effects (Mg ha^{-1})	Yield (Mg ha^{-1})	Testers	Testers HG†	Ex-PVP	Ex-PVP inbreds HG
T1 x LH205	0.77*	7.62	T1	Iodent	LH205	SS
T1 x RS710	0.72	6.72	T1	Iodent	RS710	SS
T6 x PHP02	0.54	7.47	T6	SS	PHP02	Non-SS
T1 x CR14	0.39	6.53	T1	Iodent	CR14	SS
T7 x OQ603	0.35	6.75	T7	SS	OQ603	Non-SS
T6xPHK76	0.35	6.64	T6	SS	PHK76	Non-SS
T1 x NKS8324	0.33	7.01	T1	Iodent	NKS8324	SS
T4 x L127	0.33	6.88	T4	SS	L127	Non-SS
T4 x PHR25	0.31	6.43	T4	SS	PHR25	Iodent?
T7 x PHT77	0.24	7.08	T7	SS	PHT77	Lancaster
T1 x PHJ40	0.12	6.58	T1	Iodent	PHJ40	SS
T4 x PHK05	0.17	5.57	T4	SS	PHK05	Non-SS

* Significance at $P<0.05$; †testers' heterotic group (HG) were used to derive ex-PVP inbreds' HG; 'SS' refers Stiff Stalk, 'non-SS' refers non-Stiff Stalk

Table 3. SCA effects for yield and mean grain yield per se, utilized in trial I to determine heterotic groups of NDSU maize lines

Hybrids	SCA effects (Mg ha^{-1})	Yield (Mg ha^{-1})	Ex-PVP Inbreds	Ex-PVP's HG†	NDSU Inbreds	NDSU's HG
ND2005 x NK779	0.75*	5.42	NK779	Non-SS	ND2005	SS/Non-SS?
ND2004 x NK779	0.52	5.45	NK779	Non-SS	ND2004	SS?
ND2004 x NK807	0.39	5.46	NK807	SS	ND2004	SS?
Lp5 x ND2002	0.46	6.34	Lp5	SS & UR	ND2002	SS?
DJ7 x ND291	0.45	5.54	DJ7	SS & UR	ND291	SS/non-SS
ND2003 x LH54	0.41	5.54	LH54	Lancaster	ND2003	SS
ND2007 x PH207	0.39	5.20	PH207	Iodent	ND2007	SS
ND2011 x DJ7	0.37	5.25	DJ7	SS & UR	ND2011	Non-SS
ND08-343 x LH54	0.32	5.06	LH54	Lancaster	ND08-343	Non-SS
ND2010 x PH207	0.31	5.78	PH207	Iodent	ND2010	SS?
ND2010 x NK807	0.32	5.45	NK807	SS	ND2010	SS?
ND2006 x Lp5	0.27	4.92	Lp5	SS & UR	ND2006	SS?
DK78010 x ND2001	0.25	4.90	DK78010	SS	ND2001	SS/non-SS
ND2000 x Q381	0.17	4.96	Q381	Non-SS	ND2000	SS
ND2000 x 794	0.16	4.68	NK794	SS	ND2000	SS

* Significance at $P<0.05$; †Ex-PVP inbreds' heterotic groups (HG) were used to derive NDSU inbreds' HG; 'SS' refers Stiff Stalk, 'non-SS' refers non-Stiff Stalk;

Non-SS inbreds

The highest SCA effect was observed in hybrids between ND2005 and NK779 (with a yield of 5.42 Mg ha^{-1} and SCA effect of 0.75* Mg ha^{-1}) (Table 3). The heterotic group of ND2005 was presented as non-SS (Bari and Carena 2014). However, ND2005 combined well with testers from two heterotic groups (Carena and Wanner 2009). Barata and Carena (2006) reported that certain inbred lines, especially if genetically diverse, could have the advantage to combine well across heterotic groups. ND2011 combined well with DJ7 confirming to belong to a non-SS heterotic group. We observed good combining ability between ND2001 and DK78010, leaving room to an alternative heterotic group. ND2001 was derived from MN13 (Carena et al. 2010).

SS/non-SS lines

Inbred ND2000 has been largely used in the short-season market and has shown to have unique capabilities to combine with both SS and non-SS inbred lines (Table 3). However, Carena and Wanner (2003) reported, through molecular marker analysis and yield data, that it belonged to the SS group. Further investigation is needed. ND291 showed good combining ability with DJ7.ND291 showed a broad heterotic base of SS/non-SS groups, which is in agreement with the previous molecular marker analysis and yield trial data (Carena et al. 2003).

Heterotic pattern detection

The most used heterotic pattern utilized in the U.S. is the one represented by BSSS x Lancaster, normally used with genetically narrow germplasm. Table 4 shows two heterotic patterns for ex-PVP and current industry lines. The SS x non-SS pattern was represented by hybrid T4 x Q381. The Iodent x Lancaster heterotic pattern was represented in the hybrids T1 x CR1Ht and T4 x PH207. The three hybrids had significantly ($P \leq 0.05$) higher yield than the average of industry checks while similar agronomic characteristics. In addition, these hybrids also had significantly higher ($P \leq 0.05$) protein and oil content, but statistically lower starch than the top check. Hybrids were comparable to the top check. Table 5 showed a few unique combinations. The top yielding hybrid was T1 x LH205, representing the Iodent x SS heterotic pattern. It had similar agronomic properties to the top check, but showed statistically higher grain moisture at harvest and more

Table 4. Heterotic patterns of selected maize hybrids between current industry testers and first set of ex-PVP inbred lines, from the combined analysis across six environments, trial III

Hybrids	**Yield** (Mg ha^{-1})	**MSTR**[†] (g kg^{-1})	**TWT**[‡] (kg hL^{-1})	**PRL**[§] (%)	**PSL**[¶] (%)	**Protein** (g kg^{-1})	**Oil** (g kg^{-1})	**Starch** (g kg^{-1})	**Heterotic Pattern**[#]
T4 x Q381	9.36	151	68.39	1.32	1.41	113	41.67	694	SS x non-SS
T1 x CR1Ht	9.29	150	66.28	1.18	4.18	114	44.01	684	Iodent x Lan
T4 x PH207	9.22	142	69.81	1.85	1.74	112	41.73	695	SS x non-SS
T1 x Lp5	8.75	173	65.55	6.64	5.19	110	41.80	694	Iodent x SS
T1 x NK794	8.75	167	66.83	0.00	3.36	112	41.31	693	Iodent x SS
T6 x Q381	8.72	146	73.44	4.39	0.28	113	39.63	695	SS x non-SS
T1 x LH52	8.61	151	67.94	3.03	1.72	114	42.95	691	Iodent x Lan
T1 x LH54	8.50	169	65.13	0.00	0.81	116	41.78	690	Iodent x Lan
T6 x CR1Ht	8.49	150	73.19	0.71	0.18	116	42.35	690	SS x Lan
T7 x Q381	7.95	156	70.70	1.47	3.19	107	37.86	702	SS x non-SS
T3 x LH52	7.87	138	69.78	0.73	2.28	120	42.93	686	Iodent x Lan
T1 x DJ7	7.80	185	65.45	1.85	3.58	112	43.17	692	Iodent x Lan
T6 x DKFAPW	7.70	143	71.28	0.67	1.95	113	39.68	694	SS x SS?
T1 x DK78010	7.68	151	65.98	1.85	3.49	111	40.51	694	Iodent x SS
Pioneer 38N88[††]	9.41	144	70.86	0.33	0.33	103	39.58	704	
Mean of Checks[‡‡]	8.05	147	70.09	0.75	2.80	101	39.58	706	
Exp. Mean[§§]	6.78	153	69.92	3.18	3.78	111	41.58	694	
LSD (0.05)	1.15	7	1.39	5.04	4.25	5	1.24	5	
CV (%)	21.15	5	2.47	197.11	140.08	5	3.69	1	

[†] Grain moisture at harvest,[‡] test weight,[§] percent root lodging,[¶] percent stalk lodging,[#]'Lan' refers to Lancaster, 'SS' refers to Stiff stalk
[††] best performing check of the trial, [‡‡] mean of five industry checks, [§§]Experimental mean of 72 entries; hybrids were selected based on an index combining higher yield, lower grain moisture, higher test weight, lower per cent root and stalk lodging, and higher grain quality traits.

Table 5. Heterotic patterns of selected maize hybrids between industry testers and second set of ex-PVP inbreds, from combined analysis across six environments, trial IV

Hybrids	Yield (Mg ha⁻¹)	MSTR† (g kg⁻¹)	TWT‡ (kg hL⁻¹)	PRL§ (%)	PSL¶ (%)	Protein (g kg⁻¹)	Oil (g kg⁻¹)	**Starch** (g kg⁻¹)	**Heterotic Pattern#**
T1 x LH205	7.62	198	66.97	2.24	1.51	111	45.11	690	Iodent x SS
T7 x PHP02	7.55	189	70.56	0.07	1.10	102	38.04	707	SS x non-SS
T6 x PHP02	7.47	179	69.87	4.57	1.95	106	38.75	704	SS x non-SS
T7 x PHT77	7.08	185	68.45	0.30	1.45	106	36.63	705	SS x Lan
T1 x S8324	7.01	187	67.56	1.13	2.11	111	47.33	689	Iodent x SS
T4 x PHP02	6.93	182	66.82	1.42	3.23	108	41.20	699	SS x non-SS
T1 x PHT77	6.90	178	67.52	0.01	5.51	104	39.80	703	Iodent x Lan
T4 x L127	6.88	182	69.41	0.43	8.11	107	44.16	697	SS x non-SS
T7 x OQ603	6.75	195	69.45	0.63	0.74	103	38.85	702	SS x non-SS
T1 x RS710	6.72	158	69.68	0.55	4.34	105	44.75	697	Iodent x SS
T6 xPHK76	6.64	190	72.79	1.41	2.77	107	39.13	704	Non-SS x SS
T1 x PHJ40	6.58	157	71.96	1.61	3.94	105	42.57	700	Iodent x SS
T4 x PHR25	6.43	158	69.07	2.19	0.02	108	41.88	698	SS x Iodent
T6 x L127	6.24	184	73.15	0.41	1.54	113	41.33	695	SS x non-SS
T6 x PHR25	6.16	154	71.64	0.12	2.09	110	39.01	699	SS x Iodent
Pioneer 38N88††	7.22	164	68.96	1.38	1.82	103	41.97	703	
Checks mean‡‡	6.49	165	69.79	0.50	3.89	104	40.77	703	
Exp. Mean§§	5.93	174	69.91	2.36	4.74	109	42.00	697	
LSD (0.05)	1.24	11	1.72	3.85	6.54	4	1.44	4	
CV (%)	25.73	8	3.05	201.30	170.25	5	4.25	0.73	

†Grain moisture at harvest,‡ test weight,§ percent root lodging,¶ percent stalk lodging,#'Lan' refers to Lancaster, 'SS' refers to Stiff stalk
†† best performing check of the trial, ‡‡ mean of five checks, §§ Experimental mean of 64 entries; hybrids were selected based on an index combining higher yield, lower grain moisture, higher test weight, lower per cent root and stalk lodging, and higher grain quality traits.

grain protein and oil content. Hybrid T7 x PHP02 represented the typical SS x non-SS heterotic pattern with higher test weight, stalk lodging resistance, and starch content than the average of checks.

Most NDSU lines are genetically broad-based and have unique capabilities to combine well with a wide range of inbred lines, which is reflected in the more diverse heterotic patterns observed. The universal combination of SS x non-SS/ Lancaster is prevalent in trial II. However, in trial I, we found inbred lines with combining ability across heterotic groups. This is often an advantage to develop outstanding hybrids with lines serving as male and/or female parents and further testing is encouraged. Hybrids ND2002 x CR1Ht and ND2002 x PHP02 are recommended as they showed statistically similar grain yield and agronomic traits when compared with the best performing check.

Short- season maize lines were categorized into SS, non-SS, Lancaster, and Iodent heterotic groups. Some of them belonged to both SS and non-SS because same inbreds showed above average combining ability with testers derived from different backgrounds. Heterotic groups are conceptual (Hallauer and Carena 2009) and often are not confined to a particular group. A specific line from one heterotic group may fall into another heterotic group based on performance in a particular hybrid combination. Pedigree information is still a good reference in order to classify inbred lines into heterotic groups and patterns, and these have demonstrated to be very useful to exploit. They represent a cost efficient way to group inbred parents to identify top and unique hybrid combinations. We found the SS x non-SS heterotic pattern to be frequent in our trials representing short-season ex-PVP and current industry genetic materials, in agreement with Mikel and Dubley (2006). Further sampling for larger genetic diversity (Carena et al. 2009, Carena 2011, Sharma and Carena 2012) is needed to develop alternative and productive heterotic patterns. Integration of exotic germplasm improvement with inbred line development has produced NDSU inbred lines associated with different heterotic groups. Ex-PVP lines do not seem to represent a useful source of genetic variation for continued genetic progress in short-season environments. Less protection will encourage the use ex-PVP lines as a complement to genetically broad-based maize. The future relies on the next generation of maize short-season products with increase diversity (Carena 2013).

REFERENCES

Barata C and Carena MJ (2006) Classification of North Dakota maize inbred lines into heterotic groups based on molecular and testcross data. **Euphytica 151**: 339-349.

Bari MAA and Carena MJ (2014) Can expired proprietary maize (*Zea may* L.) industry lines be useful for short-season breeding programs? I. Grain quality and nutritional traits. **Euphytica 202**: 157-171.

Carena MJ (2011) Germplasm enhancement for adaptation to climate changes. **Crop Breeding and Applied Biotechnology 11**: 56-65.

Carena MJ (2013) Developing cold and drought tolerant short-season maize products for fuel and feed utilization. **Crop Breeding and Applied Biotechnology 13**: 1-8.

CarenaMJ and Hallauer AR (2001) Response to inbred progeny recurrent selection in Leaming and Midland Yellow Dent populations. **Maydica 46**: 1-10.

Carena MJ and Wanner DW (2003) Registration of ND2000 inbred line of maize. **Crop Science 43**: 1568-1569.

Carena MJ and Wanner DW (2009) Development of genetically broad-based inbred lines of maize for early-maturing (70-80 RM) hybrids. **Journal of Plant Registrations 3**: 107-111.

Carena MJ and Wicks III ZW (2006) Maize early maturing hybrids: An exploitation of U.S. temperate public genetic diversity in reserve. **Maydica 51**: 201-208.

Carena MJ, Wanner DW and Cross HZ (2003) Registration of ND291 inbred line of maize. **Crop Science 43**: 1568.

Carena MJ, Wanner DW and Yang J (2010) Linking pre-breeding for local germplasm improvement with cultivar development in maize breeding for short- season (85-95 RM) hybrids. **Journal of Plant Registrations 4**: 86-92.

Carmer SG and Swanson MR (1971) Detection of differences between means: A Monte Carlo study of five pair wise multiple comparison procedures. **Agronomy Journal 63**: 940-945.

Comstock RE and Robinson HF (1948) The components of genetic variance in the populations of bi-parental progenies and their use in estimating the average degree of dominance. **Biometrics 4**: 254-266.

Dubreuil P, Dufour P, Krejci E, Causse M, Vienne D, Gallais A and Charcosset A (1996) Organization of RFLP diversity among inbred lines of maize representing the most significant heterotic groups. **Crop Science 36**: 796-799.

Eyherabide GH and Hallauer AR (1991) Reciprocal full-sib recurrent selection in maize: II. Contribution of additive, dominance, and genetic drift effects. **Crop Science 31**: 1442-1448.

Fan XM, Chen HM, Tan J, Xu CX, Zhang YM, Huang YX and Kang MS (2008) A new maize heterotic pattern between temperate and tropical germplasms. **Agronomy Journal 100**: 917-923.

Fan XM, Zhang YM, Yao WH, Chen HM, Tan J, Xu CX, Han XL, Luo LM and Kang MS (2009) Classifying maize inbred lines into heterotic groups using a factorial mating design. **Agronomy Journal 101**: 106-112.

Gethi JG, Labate JA, Lamkey KR, Smith ME and Kresovich S (2002) SSR variation in important US maize inbred lines. **Crop Science 42**: 951-957.

Hallauer AR and Carena MJ (2009) Maize Breeding. In Carena MJ (ed) **Handbook of plant breeding: Cereals**. Springer, New York, p. 3-98.

Hallauer AR, Carena MJ and Miranda FJB (2010) **Quantitative genetics in maize breeding.** 3rd edn, Springer, New York, 663p.

Lu H and Bernardo R (2001) Molecular marker diversity among current and historical maize inbreds. **Theoretical Applied Genetics 103**: 613-617.

Melani MD and Carena MJ (2005) Alternative maize heterotic patterns for northern Corn Belt. **Crop Science 45**: 2186-2194.

Melchinger AE (1999) Genetic diversity and heterosis. In Coors JG, Pandey S and Gender JT (eds) **The genetics and exploitation of heterosis in crops**. CSSA-SP, Madison, p. 99-118.

Mendes UC, Miranda-Filho JB, Oliveira AS and Reis EF (2015) Heterosis and combining ability in crosses between two groups of open-pollinated maize populations. **Crop Breeding and applied Biotechnology 15**: 235-243.

Menkir A, Melake-Berhan A, The C, Ingelbrecht I and Adepoju A (2004) Grouping of tropical mid-altitude maize inbred lines on the basis of yield data and molecular markers. **Theoretical Applied Genetics 108**: 1582-1590.

Mikel MA (2006) Availability and analysis of proprietary dent corn inbred lines with expired U.S. plant variety protection. **Crop Science 46**: 2555-2560.

Mikel MA and Dubley JW (2006) Evolution of North American dent corn from public to proprietary germplasm. **Crop Science 46**: 1193-1205.

Nelson PT, Coles ND, Holland JB, Bubeck DM, Smith and Goodman MM (2008) Molecular characterization of maize inbreds with expired U.S. plant variety protection. **Crop Science 48**: 1673-1685.

SAS (2010) **SAS user's guide guide: Statistics**. 4th edn, SAS Institute, Cary, 2861p.

Scott MP, Peterson JM and Hallauer AR (2009) Evaluation of combining ability of quality protein maize derived from U.S. public inbred lines. **Maydica 49**: 303-311.

Sharma S and Carena MJ (2012) NDSU EarlyGEM: Increasing the genetic diversity of northern U.S. hybrids through the development of unique exotic elite lines. **Maydica 57**: 34-42.

Troyer AF (1999) Background of US hybrid corn. **Crop Science 39**: 601-626.

Troyer AF (2006)Adaptedness and heterosis in corn and mule hybrid. **Crop Science 46**: 528-543.

10

Software Selegen-REML/BLUP: a useful tool for plant breeding

Marcos Deon Vilela de Resende[1*]

***Corresponding author:**
E-mail: marcos.deon@gmail.com

[1] Embrapa Florestas, Universidade Federal de Viçosa, Departamento de Engenharia Florestal, 36.570-000, Viçosa, MG, Brazil

Abstract: *The software Selegen-REML/BLUP uses mixed models, and was developed to optimize the routine of plant breeding programs. It addresses the following plants categories: allogamous, automagous, of mixed mating system, and of clonal propagation. It considers several experimental designs, mating designs, genotype x environment interaction, experiments repeated over sites, repeated measures, progenies belonging to several populations, among other factors. The software adjusts effects, estimates variance components, genetic additive, dominance and genotypic values of individuals, genetic gain with selection, effective population size, and other parameters of interest to plant breeding. It allows testing the significance of the effects by means of likelihood ratio test (LRT) and analysis of deviance. It addresses continuous variables (linear models) and categorical variables (generalized linear models). Selegen-REML/BLUP is friendly, easy to use and interpret, and allows dealing efficiently with most of the situations in plant breeding. It is free and available at http://www.det.ufv.br/ppestbio/corpo_docente.php under the author's name.*

Key words: *Linear mixed models, prediction, variance components, BLUP, REML, selection.*

INTRODUCTION

The estimation of variance components and the prediction of genetic values are essential procedures in breeding programs. In the 1990s, there was a qualitative progress in the analytical methodologies of genetic parameters estimation and selection applied to plant breeding. Currently, REML/BLUP (Residual or Restricted Maximum Likelihood/Best Linear Unbiased Prediction) is the standard procedure for the estimation of genetic parameters and optimal selection in several species.

The field trial, as a rule, is associated with imbalance of data due to several reasons, such as loss of plants and plots, unequal quantities of seeds and seedlings available by treatment, experimental net with different numbers of replications per experiment, and different experimental designs, non-evaluation of all combinations of genotypes and environments, among others. As a result, REML/BLUP method is an optimal procedure of genotypic evaluation, and it is also known as mixed model methodology. This procedure naturally deals with the imbalance, leading to more accurate estimates and predictions of genetic parameters and genetic values, respectively.

BLUP is the optimal selection procedure for additive genetic effects (a), of dominance effects (d) and genotypic effects (g), depending on the situation.

BLUP maximizes selective accuracy, and allows the simultaneous use of several information sources, such as those from various experiments installed in one or several locations and evaluated in one or more crops. The individual BLUP uses all the effects of the statistical model, addresses imbalance, considers the genetic relatedness between the evaluated plants, and the coincidence between selection and recombination units.

The main practical advantages of using REML/BLUP is that they: allow comparing individuals or varieties over time (generations, years) and space (location, blocks); allow simultaneous correction for environmental effects, estimation of variance components, and prediction of genetic values; allows dealing with complex data structures (repeated measurements, different years, location and designs); may be applied to unbalanced data and to not orthogonal designs.

At the end of the second millennium, pioneering works in Brazil applied to plant breeding were carried out in the field of mixed linear models fitted under the frequentist approach via REML/BLUP (Resende et al. 1993, 1996, Bueno Filho and Vencovsky 2000, Duarte and Vencovsky 2001), and the Bayesian approach via Gibbs sampler (Resende 1997, Resende and Rosa-Perez 1999, Resende 2000). These models included random regressions for longitudinal and multivariate data (Resende 1997, Resende and Rosa-Perez 1999), non-linear or generalized linear models for categorical variables (Resende 2000), spatial analysis (Resende and Sturion 2001), factor analytical mixed models for multivariate analysis and genotype x environment interaction (Resende and Thompson 2003, 2004), and competition or associative models of social interaction (Resende and Thompson 2003, Resende et al. 2005). Pioneering works with Genomic Selection were also carried out (Resende 2007, Resende et al. 2008). These works are summarized in Resende (2002, 2007, 2015) and Resende et al. (2014), which also include reaction norms, structural equations and survival analysis models for censored data.

The generic theory of BLUP as optimal procedure was widespread from the 1970s by the scientists Charles Henderson, in the United States (Henderson 1973, 1975, 1976) and Robin Thompson, in England (Thompson 1976, 1977, 1979), among others. For the application of BLUP, it is necessary reliable estimates of variance components. REML is the optimal method of estimation of variance components, with unbalanced or not data, which was developed by Patterson and Thompson (1971) and Thompson (1973, 1977, 1980).

The REML/BLUP procedure became popular abroad in animal breeding from the 1980s. In Brazil, the method began to be used in dairy cattle from 1994 (Verneque and Valente 2001). This was due to the development of specific softwares that allow the proper handling of additive genetic relationship matrix among the evaluated individuals. Algorithms to directly write this additive genetic relationship matrix were presented by Henderson (1976), in the USA and by Thompson (1977), in England. The software Selegen-REML/BLUP (Statistical system and computerized genetic selection via linear mixed models) was created in association with the improvement of the genetic selection methodologies from the mathematical and statistical analysis of field experiments data. Prior to the emergence of the softwares ASREML and SAS, for mixed models, which were created 1996, Selegen was released in 1993, together with the use of the mixed linear models (via best unbiased linear prediction) in plant breeding in Brazil, which is currently widely used (Bueno Filho and Vencovsky 2009).

The software Selegen-REML/BLUP and its applications

The software Selegen-REML/BLUP was developed to meet the demands of plant breeding programs, and includes the following categories of plants: allogamous, autogamous, of mixed mating system, and of clonal propagation. It considers various experimental designs, several mating designs, genotype x environment interaction, experiments repeated over sites, repeated measures, progenies belonging to several populations, among other factors. The software not only fits the effects and presents the variance components, but also shows the additive genetic values, the dominance genetic values, and the genotypic genetic values of the individuals, the genetic gain with selection, the effective population size, among other parameters of interest in plant breeding. From a statistical point of view, it is also interesting, since it allows testing the significance of the effects by means of the likelihood ratio test (LRT) and analysis of deviance. It also addresses continuous variables (linear models) and categorical variables (generalized linear models).

Selegen-REML / BLUP is easy to use and interpret, and allows dealing efficiently with the most common situations in plant breeding. It is of free access in universities and public research institutes in Brazil and abroad. In the private sector, it has been used in the breeding of coffee, forage crops, fruit trees, forest trees, eucalyptus, pine, black wattle, teak, corn, soybeans, rice, rubber tree, sugarcane, among other species. In public institutions, it has been used in these same species and also in beans, cashew, acerola, cupuaçu, cocoa, coffee, guarana, palm, peach palm, royal palm, orange, brachiaria, panicum, stylosanthes, leucaena, mate tea, pequi, potato, cassava, açaí, mango, passion fruit, camu-camu, buriti, among others.

Mathematical and computational algorithm

The computational implementation of the mixed model methodology is based on heavily numerical methods, especially in numerical linear algebra in order to obtain iterative solutions of mixed model equations (obtainment of BLUP) and numerical calculus for the maximization/minimization of the functions of several variables, in order to obtain the REML estimates.

Several computational algorithms for the obtainment of variance components by REML have been developed, such as MS (Fisher's Method of Scoring), EM (Expectation-Maximization), and AI-REML (Average Information-REML). The EM algorithm is numerically very stable, showing convergence even if initial values are not entirely adequate. However, its disadvantage is the slowness for estimates close to the limit of the parametric space (for instance, when a variance tends to zero). If positive initial values are used, the convergence to non-negative values is guaranteed (Harville 1977).

The EM algorithm works by means of the obtainment of the expectation (by integration) and maximization (derivation) of the likelihood function of the data, successively. In models of individual plants, in which the order of the mixed model equations usually exceeds the number of observations, the obtainment of estimates by means of first derivative by the EM method requires the inversion of the coefficient matrix of the mixed model equations, increasing computational effort. The methods of Newton-Raphson and of Fisher have quadratic convergence, whereas the EM algorithm presents linear convergence, and is therefore slower.

The AI-REML algorithm is three times faster than the EM, and is an improved derivative procedure that uses the first and second derivative of the likelihood function. This algorithm is based on the use of the information of the mean of the observed and expected second derivatives of the likelihood function, so that the term containing traces of the inverse matrix product is canceled, leaving a simpler expression for computation. Sparse matrices techniques are used in the calculation of the elements of the inverse matrix of the coefficients, which are necessary for the first derivatives of the likelihood function.

However, sometimes the AI algorithm failures in the convergence, and therefore, the EM algorithm would be an alternative, which ensures an increase in the log L in each iteration. The use of the proposed PX-EM algorithm (Parameter expanded EM) is an excellent choice, and it presents stability and good performance. The software Selegen-REML/BLUP combines the methods of Takahashi and the method of Zollenkopf sparse bifactorization in the EM algorithm with PX-EM. Therefore, the algorithm is PX-EM-SB-type (Expanded Parameter Expectation-Maximization/Sparse Bifactorization), and the software is the PX-EM-SB-REML-type.

The methods of Takahashi and Zollenkopf were developed in Electrical Engineering, associated with matrices of impedance in electrical circuits. Both methods are similar, and the Takahashi algorithm is naturally generated by multiplication factors to the left and to the right of the Zollenkopf bifactorization in reverse order. Estimators and predictors used in Selegen-REML/BLUP are described in a series of 10 papers published by Resende in the Journal of Mathematics and Statistics, from 1999 to 2006, and in Resende et al. (1996, 2014) and Resende (1999, 2000, 2002).

The restricted likelihood function to be maximized is given below. A generalized linear mixed model is given by $y = X\beta + Z\tau + \varepsilon$, with the following distributions and structures of means and variances:

$$\tau \sim N(0,\ G) \qquad E(y) = X\beta$$

$$\varepsilon \sim N(0,\ R) \qquad Var(y) = V = ZGZ' + R$$

In which:

y: known vector of observations

β: parametric vector of fixed effects, with incidence matrix X.

τ: parametric vector of random effects, with incidence matrix Z.

ε: unknown vector of errors.

G: variance-covariance matrix of random effects.

R: variance-covariance matrix of errors.

0: null vector.

Assuming *G* and *R* as known, the simultaneous estimation of fixed effects and the prediction of random effects can be obtained by means of mixed model equations (BLUP) given by

$$\begin{bmatrix} X'R^{-1}X & X'R^{-1}Z \\ Z'R^{-1}X & Z'R^{-1}Z+G^{-1} \end{bmatrix} \begin{bmatrix} \hat{\beta} \\ \tilde{\tau} \end{bmatrix} = \begin{bmatrix} X'R^{-1}y \\ Z'R^{-1}y \end{bmatrix}$$

When G and R are not known, the variance components associated with random effects can be efficiently estimated by the REML method (Patterson and Thompson 1971). Except for a constant, the residual likelihood function (in terms of its log) to be maximized is given by:

$$L = -\frac{1}{2}\left(\log|X'V^{-1}X| + \log|V| + v\log\sigma_\varepsilon^2 + y'Py/\sigma_\varepsilon^2\right)$$

$$= -\frac{1}{2}\left(\log|C^*| + \log|R| + \log|G| + v\log\sigma_\varepsilon^2 + y'Py/\sigma_\varepsilon^2\right)$$

in which:

$V = R + ZGZ'$; $P = V^{-1} - V^{-1}X(X'V^{-1}X)^{-1}X'V^{-1}$.

$v = N\text{-}r(x)$: degrees of freedom for the random effects, in which N is the total number of data, and $r(x)$ is the rank of the matrix X.

C^* = matrix of the coefficients of the mixed model equations.

Overall, the generalized model described includes several unique models to each situation.

Statistical, mathematical and genetic procedures of Selegen-REML/BLUP

Selegen-REML/BLUP was developed in Fortran 90 language, and has Windows and DOS operating systems interface. It is suitable for analysis of either balanced or unbalanced experiments, leading to maximum efficiency. There is no need to inform weather the experiment is balanced or not, since it uses optimal and generic mathematical and statistical procedure for any situation.

The files to be analyzed must have .txt extension (MS-DOS text), with a header line. This line is only for user guidance, it is ignored by the program. Thus, the line may contain any names. The program uses classificatory and dependent variables with up to 15 digits. However, for reasons of space, in the program's outputs, the classificatory variables are presented with up to seven digits only. It is recommended the use up to seven decimal digits for these variables. Such classificatory variables may be alpha-numeric.

In statistical terms, the following methods of analysis are performed by Selegen-REML/BLUP:

- General statistics: mean, variance, standard deviation, coefficient of variation, maximum, minimum, kurtosis, asymmetry, covariance, correlation and commonality
- Analysis of variance
- Analysis of covariance and correlation
- Multivariate analysis: principal components and cluster analysis
- Mixed linear models via REML/BLUP

- Mixed linear models via REML/BLUE
- Generalized linear mixed models via REML/BLUP
- Linear mixed models for repeated measures
- Linear mixed models for multiple experiments analysis and G x E interaction
- Mixed linear models with covariates
- Maximization of residual likelihood function for likelihood ratio test (LRT) and analysis of deviance
- Mixed linear models with residual variance heterogeneity within treatments
- Point and interval prediction of random effects (breeding values)
- Computation of residuals for analysis of homogeneity and normality
- Spatial autocorrelation for analysis of residuals
- Hierarchical analysis for population genetics
- Genetic sampling (effective population size)

The current program includes about 200 models of analysis, and in terms of experimentation and genetic improvement, it provides the following results of interest:

- BLUP and REML/BLUP for additive, dominance and genotypic effects
- Heritabilities, genetic and phenotypic correlations, genetic gain
- BLUP under heterogeneity of variance within treatments
- Analysis of deviance
- Experimental designs: randomized, complete blocks, augmented blocks, lattice, row and column
- Groups of experiments, several locations, and genotype x environment interaction
- Mating designs: open and controlled pollination (half-sib, full-sib, factorial, diallel, hierarchical, unbalanced designs, hybrids and selfing generations)
- Clonal tests
- One or more populations
- One or more plants per plot
- One or more repeated measures
- Principal genetic components
- Cluster analysis by genotypic values
- Genetic divergence via genotypic values
- Multi-trait selection index
- Effective population size
- Optimization of selection and restriction in inbreeding
- Stability and adaptability of genotypic values
- Population genetics
- Allogamous and autogamous species, and of mixed mating system, animals

- Associated selection methods: selection of parents (a_g), selection of potentials parents (a), selection of potential clones (g = a + d), selection of crosses (vgc = 0.5 (af + am) + cec), selection of clones. The Selegen-Reml/Blup was designed to maximize the overall efficiency of breeding, and addresses in an intricate way the topics mentioned above, overlapping recurrent selection scheme, crossing design, experimental design, statistical control via covariate, and propagation system of improved material.

Genetic evaluation models in function of the species and of the experimental structure

In plant breeding, there are certain types of cultivars (Table 1). The three types of cultivars considered in this study aim at capturing the genotypes with the two most important characteristics for the production system (sustainable yield and products homogeneity), given the biological conditions (reproduction and propagation systems) of each species. Thus, the breeding strategies (which generate the data to be statistically analyzed) are similar for the species within each of these three types. Clones and hybrids between inbred lines capture both additive genetic effects and dominance effects (which, in presence of genetic diversity, provide heterosis). Inbred lines capture only the additive genetic effects. Breeding programs aiming at inbred lines are similar (hybridization between lines, followed by inbred generations conduction until the selection of new lines) to those aiming at hybrids. However, when the hybrids are the target, the process continues with the cross between lines, in order to identify the superior hybrids. With the advent of double-haploid, the stage of inbred generations conduction has been suppressed, and the obtained lines (so far not tested in the field), have their crosses predicted with the aid of the phenomics and genomics.

Table 1. Types of cultivars considered in plant breeding

Types of cultivars	Mating system of species	Examples
Clones	Allogamous, Autogamous, Asexuals	Perennial and allogamous (sugarcane, eucalyptus, guarana, rubber tree, cocoa, canephora coffee, fruit trees, forage, etc.); Perennial and autogamous (peach); Perennial with annual cycle (cassava, potato)
Lines	Autogamous	Annual and Autogamous (rice, wheat, oat, beans, soybeans, lettuce, pepper); Perennial and autogamous (Coffee arabica)
Hybrids of Lines	Allogamous, Autogamous	Annual and allogamous annual (corn, sunflower, onion, carrot, brassicas, pumpkin); Annual and autogamous (tomato, pepper, tobacco)

General models of genetic evaluation, which can be applied to plant breeding are presented below. The procedures are similar within types, according to the following classification.

a) **Perennial and allogamous plants** (eucalyptus, pine, canephora coffee, sugarcane, fruit trees, forage): Estimation is simultaneously based on heritability, repeatability, covariance structure, and growth curves; genealogy is variable.

b**) Perennial and autogamous plants** (Coffee arabica, peach, apricot, dwarf coconut, lemon, leucaena): Estimation is simultaneously based on heritability, repeatability, covariance structure, and growth curves; genealogy is fixed from each F2 generation to F6 generation.

c) Perennial plants with asexual reproduction (sugarcane, rubber tree, orange, brachiaria, panicum): Estimation is simultaneously based on broad-sense heritability, repeatability, covariance structure, and growth curves; genealogy is constant through clonal stages

d) **Annual and Allogamous** (corn, popcorn, sunflower, broccoli, carrots): Estimation is based on heritability; genealogy is variable.

e) **Annual and autogamous plants** (beans, soybeans, rice, wheat, oat): Estimation is based on the heritability; genealogy is fixed from each F2 generation to F6 generation.

f) **Annual plants with asexual reproduction** (cassava, potato): Estimates based on broad-sense heritability; genealogy is constant through clonal stages

A rough classification of groups of plants regarding the models of genetic evaluation in function of the experimental structure in the field and models of genetic evaluation is presented below:

a) **Annual and vegetable crops** (rice, corn, soybeans, wheat, oat, barley, sorghum, cotton, potato, cassava): data analyzed with one observation per plot and inference at the level of genetic treatment effects (lines, hybrids, clones, families, cultivars, accessions)

b) **Forage and sugarcane:** data analyzed with one observation per plot and repeated measures, inference at level of genetic treatments effects (lines, hybrids, clones, families, cultivars, accessions)

c) **Forest species:** data taken at the level of individual plants, without repeated measures, inference to the level of individuals, parents and clones

d) **Fruit trees, palm trees** (açai, coconut, palm, peach palm, date palm), stimulants (coffee, cocoa, guarana, mate tea), rubber tree: data collected from individual plants, with repeated measures, inference at the level of individuals, parents and clones

The complexity of the models, the difficulty in the genetic evaluation, and the imbalance degree increase from (a) to (d), i.e., complexity increases in the following order: annual and vegetable crops; forage and sugarcane; forest species; fruit trees.

Significance of the effects of the model and complete statistical analysis

Genotypic evaluations addresses the estimation of variance components (genetic parameters) and the prediction of genotypic values. Estimates of genetic parameters, such as heritability and genetic correlations, are fundamental to the design of efficient breeding strategies. REML is an efficient method in the study of the several variation sources associated with the evaluation of field experiments, and allows decomposing the phenotypic variation in their several genetic and environmental components and in genotype x environment interaction components.

In the mixed models analysis with unbalanced data, the effects of the model are not tested via F tests, as it is done in the method of analysis of variance. In this case, for the random effects, likelihood ratio test is recommended (LRT). For fixed effects, an approximate F test can be used. A similar Table to that of analysis of variance can be elaborated. Such table can be called Analysis of deviance (ANADEV), and is established according to the following steps:

a) obtainment of the maximum point of the logarithm of the residual likelihood function (L) for models with and without the effect to be tested

b) obtainment of the deviance $D = -2LogL$ for models with and without the effect to be tested

c) differentiation between deviances for models with and without the effect to be tested, with the obtainment of the likelihood ratio (LR)

d) testing, via LRT, of the significance of this difference using the chi-square test with 1 or 0.5 degrees of freedom.

As an example, the following experiment was carried out in randomized blocks with several plants per plot, and the following model was specified: $y = \mu + g + b + gb + e$, in which g is the vector of random effects of genotypes, b is the vector of fixed effects of blocks, gb is the vector of random effects of plots, and e is the vector of random residuals within plots. The following analysis of deviance (ANADEV) can be carried out (Table 2).

The software Selegen provides (.dev extension files) the deviances when it is processed models with or without (just by reseting to zero the corresponding coefficient of determination c2 on Selegen's display) the effects to be tested. With

Table 2. Analysis of deviance (ANADEV)

Effect	Deviance	LRT(chi-square)	Variance components	Coefficient of determination
Genotypes	647.18	6.55**	0.0329*	h2genotype = 0.045*
Plots	654.13	13.50**	0.0685**	c2plot =0.094**
Residuals	-	-	0.6206	c2residual=0.859
Complet model	640.62	-	-	c2total=1.000
Block	-	F = 7.012**	-	-

Chi-square of 3.84 and 6.63 for 5% and 1% significance, respectively

these deviances, it becomes easy to build the table of analysis of deviance. In the present example, it is verified that the effects of genotype and plots are significant. Consequently, the respective variance components are significantly different from zero, as well as their coefficients of determination (heritability of genotypic effects - h2genotype and coefficient of determination of the effects of plots - c2plot, as obtained by the models 1 and 2 of Selegen). The factor block, considered as fixed effect, was tested by the F test of Snedecor.

In general, a complete statistical analysis involves the following six activities: the estimation of mean components; the estimation of variance components; hypothesis tests; the inference regarding the accuracy (reliability); the bias; and precision of the estimation/prediction. Considering the mixed models, these activities involve BLUP prediction, REML estimation, analysis of deviance, calculation of the prediction accuracy and of the variance of the prediction error, respectively. In the REML/BLUP procedure, bias is assumed as zero, since these estimators/predictors belong to the class of the best linear unbiased estimators/predictors (BLUE/BLUP).

Results generated by the six activities of a complete statistical analysis should be discussed in the papers. In the genetic field, the following script should be followed:

Hypothesis test: inferences on the significance of the genetic variability (Vg), using the analysis of deviance or the F test of the analysis of variance

Variance components and their proportions: inferences on the genetic control (high, moderate, and low, according to Resende 2002), or heritabilities and correlations between traits, coefficients of variation, repeatabilities

Mean components: genetic values and genetic gain

Precision: PEV (variance of the prediction error); ratio PEV/Vg (with parametric space between 0 and 1)

Accuracy: with parametric space between 0 and 1, and classification according to Resende and Duarte (2007)

Bias: given by the regression of y in $\hat{y}$ ($\beta(y/\hat{y})$), in which $\beta = 1$ is the ideal and indicates unbiasedness.

Moreover, studies on diversity (population effective size or Ne; genetic distances, and multivariate groupings) complement the inferences.

Interaction between organisms

There are also estimates based on genotypes of two organisms in a single individual, such as those of rubber tree, orange, peach, mango, which are cultivated in rootstock + graft. This also occurs in the experiments involving plant x pathogen interaction. This type of statistical analysis is addressed in the software Selegen-REML/BLUP. The study on plant x pathogen interactions involves the evaluation of different genotypes (accessions) of the plant species subjected to inoculation with different races, strains or inoculum sources of the microorganisms species (bacteria, fungi, nematodes). Each plant must be inoculated with only one type of microorganism species. The experiment must contain replications of each plant-inoculum combination. Thus, the analysis model will contain the effects of replications (b, for blocks designs), genotypic effects (g) of the plant accessions, genotypic effects (m) of the microorganism strains, plant x pathogen (gm) genotypic interaction, and residual (e). This model, for a variable in a vector y, is given by $y = Xb + Zg + Wm + Tgm + \varepsilon$, in which X, Z, W and T are incidence matrices for the respective effects vectors. Thus, the genetic cause of the disease can be attributed to three factors:

a) Effect on the phenotype, explained by the pathogen genotype

b) Effect on the phenotype, explained by the plant genotype

c) Effect on the phenotype, explained simultaneously by the pathogen and the plant, i.e., the combination of pathogen genotype – plant genotype (is the actual plant – pathogen interaction, or specific combining ability – SCA). The breeding values predicted by the model, for each combination of pathogen genotype – plant genotype, are given by VGC = $\hat{u} + \hat{g}_i + \hat{m}_j + \hat{g}m_{ij}$.

The software Selegen-REML/BLUP has been used internationally (Colombari et al. 2013, Oliveira et al. 2012, Pedrozo et al. 2011, Rosado et al. 2010, Dunlop et al. 2005), and is a very useful tool for plant breeding.

REFERENCES

Bueno Filho JSS and Vencovsky R (2000) Alternativas de análise de ensaios em látice no melhoramento vegetal. **Pesquisa Agropecuária Brasileira 35**: 259-296.

Bueno Filho JSS and Vencovsky R (2009) Selection in several environments by BLP as an alternative to pooled anova in crop breeding. **Ciência e Agrotecnologia 33**: 1342-1350.

Colombari Filho JM, Resende MDV, Morais OP, Castro AP, Guimarães EP, Pereira JA, Utumi MM and Breseghello F (2013) Upland rice breeding in Brazil: a simultaneous genotypic evaluation of stability, adaptability and grain yield. **Euphytica 192**: 117-129.

Duarte JB and Vencovsky R (2001) Estimação e predição por modelo linear misto com ênfase na ordenação de médias de tratamentos genéticos. **Scientia Agricola 58**: 109-117.

Dunlop RW, Resende MDV and Beck SL (2005) Early assessment of first year height data from five *Acacia mearnsii* (black wattle) sub-populations in South Africa using REML/BLUP. **Silvae Genetica 54**: 166-174.

Harville DA (1977) Maximum likelihood approaches to variance component estimation and to related problems. **Journal of the American Statistical Association 72**: 320-328.

Henderson CR (1973) Sire evaluation and genetic trends. In **Animal breeding and genetics symposium in honour of J. Lush.** American Society of Animal Science, Champaign, p. 10-41.

Henderson CR (1975) Use of all relatives in intraherd prediction of breeding values and producing abilities. **Journal of Dairy Science 58**: 1910-1916.

Henderson CR (1975) Best linear unbiased estimation and prediction under a selection model. **Biometrics 31**: 423-449.

Henderson CR (1976) A simple method for computing the inverse of a numerator relationship matrix used in prediction of breeding values. **Biometrics 32**: 69-83.

Oliveira EJ, Resende MDV, Santos VS, Ferreira CF, Oliveira GAF, Silva MS, Oliveira, LA and Aguilar-Vildoso CI (2012) Genome-wide selection in cassava. **Euphytica 187**: 263-276.

Patterson HD and Thompson R (1971) Recovery of inter-block information when block sizes are unequal. **Biometrika 58**: 545-554.

Pedrozo CA, Barbosa MHP, Silva FL, Resende MDV and Peternelli LA (2011) Repeatability of full-sib sugarcane families across harvests and the efficiency of early selection. **Euphytica 182**: 423-430.

Resende MDV (1997) Avanços da genética biométrica florestal. In Bandel G, Vello NA and Miranda Filho JB (eds) **Encontro sobre temas de genética e melhoramento: genética biométrica vegetal.** Esalq/Usp, Piracicaba, p. 20-46.

Resende MDV (2000) **Inferência bayesiana e simulação estocástica (amostragem de Gibbs) na estimação de componentes de variância e valores genéticos em plantas perenes.** Embrapa Florestas, Colombo, 68p.

Resende MDV (2000) **Análise estatística de modelos mistos via REML/BLUP no melhoramento de plantas perenes**. Embrapa Florestas, Colombo, 101p.

Resende MDV (2002) **Genética biométrica e estatística no melhoramento de plantas perenes.** Embrapa Informação Tecnológica, Brasília, 975p.

Resende MDV (2007) Seleção genômica ampla (GWS) e modelos lineares mistos. In Resende MDV (ed) **Matemática e estatística na análise de experimentos e no melhoramento genético**. Embrapa Florestas, Colombo, p. 517-534.

Resende MDV (2015) **Genética quantitativa e de populações.** Suprema, Visconde do Rio Branco, 452p.

Resende MDV and Duarte JB (2007) Precisão e controle de qualidade em experimentos de avaliação de cultivares. **Pesquisa Agropecuária Tropical 37**: 182-194.

Resende MDV, Higa AR and Lavoranti OJ (1993) Predição de valores genéticos no melhoramento de Eucalyptus – melhor predição linear (BLP). **Silvicultura 43**: 144-147.

Resende MDV, Lopes PS, Silva RL and Pires IE (2008) Seleção genômica ampla (GWS) e maximização da eficiência do melhoramento genético. **Pesquisa Florestal Brasileira 56**: 63-78.

Resende MDV, Prates DF, Jesus A and Yamada CK (1996) Estimação de componentes de variância e predição de valores genéticos pelo método da máxima verossimilhança restrita (REML) e melhor predição linear não viciada (BLUP) em Pinus. **Pesquisa Florestal Brasileira 32/33**: 18-45.

Resende MDV and Rosa-Perez JRH (1999) **Genética quantitativa e estatística no melhoramento animal**. Editora UFPR, Curitiba, 494p.

Resende MDV, Silva FFE and Azevedo CF (2014) **Estatística matemática, biométrica e computacional.** Suprema, Visconde do Rio Branco, 881p.

Resende MDV and Sturion JA (2001) **Análise genética de dados com dependência espacial e temporal no melhoramento de plantas perenes via modelos geoestatísticos e de series temporais empregando REML/BLUP ao nível individual**. Embrapa Florestas, Colombo, 80p.

Resende MDV, Stringer JK, Cullis BC and Thompson R (2005) Joint modelling of competition and spatial variability in forest field trials. **Brazilian Journal of Mathematics and Statistics 23**: 7-22.

Resende MDV and Thompson R (2003) **Multivariate spatial statistical analysis of multiple experiments and longitudinal data**. Embrapa Florestas, Colombo, 126p. (Documento 90).

Resende MDV and Thompson R (2004) Factor analytic multiplicative mixed models in the analysis of multiple experiments. **Revista de Matemática e Estatística 22**: 1-22.

Rosado CCG, Guimarães LMS, Titon M, Lau D, Rosse LN, Resende MDV and Alfenas AC (2010) Resistance to ceratocystis wilt (*Ceratocystis fimbriata*) in parents and progenies of *Eucalyptus grandis* x *E. urophylla*. **Silvae Genetica 59**: 99-106.

Thompson R (1973) The estimation of variance and covariance components when records are subject to culling. **Biometrics 29**: 527-550.

Thompson R (1976) Relationship between the cumulative difference and best linear unbiased predictor methods of evaluating bulls. **Animal Production 23**: 15-24.

Thompson R (1977) The estimation of heritability with unbalanced data. **Biometrics 33**: 485-504.

Thompson R (1979) Sire evaluation. **Biometrics 35**: 339-353.

Thompson R (1980) Maximum likelihood estimation of variance components. **Mathematische Operationsforschung und Statistik. Series Statistics 11**: 545-561.

Verneque RS and Valente J (2001) Avaliação genética de vacas e touros. In Valente J, Durães MC, Martinez ML and Teixeira NM (eds) **Melhoramento genético de bovinos de leite**. Vol. 1, Embrapa Gado de Corte, Juiz de Fora, p. 127-154.

11

Relationship between fruit traits and contents of ascorbic acid and carotenoids in peach

Rosana Gonçalves Pires Matias[1], Danielle Fabíola Pereira da Silva[1], Priscila Maria Dias Miranda[1], João Alison Alves Oliveira [1*], Leonardo Duarte Pimentel[1] and Cláudio Horst Bruckner[1]

***Corresponding author:**
E-mail: joao.alison@yahoo.com.br

[1] Universidade Federal de Viçosa (UFV), Departamento de Fitotecnia, Avenida P.H. Rolfs, 36.570-900, Viçosa, MG, Brazil

Abstract: *This study aimed to evaluate the relationship between fruit traits and their direct and indirect effects on the content of ascorbic acid and carotenoids in peaches and nectarines. The traits fruit mass (FM); equatorial diameter (ED); suture diameter (SD); polar diameter (PD); pulp firmness (FIR); soluble solids (SS); titratable acidity (TA); SS/TA ratio; contents of ascorbic acid (AA) and carotenoids (CT); and skin and pulp color were evaluated in 28 peach cultivars, and two nectarine cultivars. The phenotypic correlation coefficients were estimated (rf), and after multicollinearity diagnosis, unfolding was carried out in direct and indirect effects of the explanatory variables in the response variable by using path analysis. The strongest correlations were found between FM, SD, ED, and PD, and between carotenoid content and °h pulp. The traits considered in the path diagrams are not the main determinants of the ascorbic acid content. The yellow color of the pulp has the potential for indirect selection for carotenoid content.*

Key words: Prunus persica, *indirect selection, fruit quality.*

INTRODUCTION

Correlated responses are common in breeding programs for selection of variables which are difficult to be measured, or when the measurements are expensive. Therefore, understanding the relationship between variables is crucial, since obtaining genetic gains and choosing the best genotypes often rely on a set of agronomic and commercial variables. The knowledge of these relationships allows obtaining a main variable of low heritability, and/or of difficult measurement to be selected based on another (s) variable (s), providing the breeder a more rapid progress than that used for direct selection.

Although it is important, the simple correlation coefficient may create misconceptions regarding the relationship between two variables, and may not be a true cause and effect measurement. Thus, a high or low coefficient of correlation between two variables may result from the effect of a third variable or group of variables, without giving the exact relative importance of the direct and indirect effects of these factors (Cruz et al. 2012).

The path analysis (Wright 1921) allows the study of the direct and indirect effects on a response variable, whose estimates are obtained by regression equations using previously standardized variables. The success of the path analysis is based on the most consistent formulation of the cause-effect relationship between variables. Moreover, the split correlation is dependent on the set of variables studied, which is usually determined from prior knowledge of their importance for research, and possible inter-relationships expressed in path diagrams (Cruz et al. 2012, Oliveira et al. 2010).

The consumption of fruits and vegetables has always been valued due to the health benefits of the large amount of vitamins, minerals and fibers they have, contributing to the prevention or delay of the onset of cardiovascular diseases and cancer (Bowen-Forbes et al. 2010, Tsantili et al. 2010). This protective effect has been attributed to the presence of antioxidant phytochemicals. Vitamins C and E, carotenoids, and flavonoids are among the non-enzymatic antioxidants that have received more attention for their possible body beneficial effect (Silva et al. 2010, Wolfe et al. 2008).

Vitamin C stability in foods is affected by heat, light, oxygen, and pH. Furthermore, high-risk handling reagents, such as sulfuric acid, are used in the quantification and, or in determination of ascorbic acid content (Spinola et al. 2013, Tarrago-Trani et al. 2012).

The need for reliable information on food carotenoids is widely recognized in several fields of study. The factors that make this analysis difficult include the large number of naturally occurring carotenoids, the quantitative and qualitative variation of carotenoids in foods, the small amount of pro-vitamin A carotenoids, their varied biopotency, and the fact that carotenoids are highly unsaturated molecules, which can cause isomerization, oxidation and degradation during analysis (Rivera and Canela-Garayoa 2012, Rodriguez-Amaya 2010).

The objective of this study was to evaluate the relationship between traits of peach and nectarine, and their direct and indirect effects on the content of ascorbic acid and carotenoids, using path analysis, aiming to assist selection in breeding programs.

MATERIAL AND METHODS

The plant material evaluated in this work is part of the collection of 56 peach cultivars and 3 nectarine cultivars located in the orchard of the Department of Plant Science, Federal University of Viçosa, Viçosa (lat 20° 45' S, long 42° 51' W, alt 649 m asl), Minas Gerais State, Brazil. However, only 30 cultivars were evaluated due to fruit availability (Table 1).

The orchard was established in the container growing system, in October 2008, spaced 5.0 m between rows, and 3.5 m between plants, with three plants of each cultivar arranged side by side, in an area of about half a hectare. The plant cultivars canopy was obtained by grafting, using the cultivar 'Okinawa' as rootstock. It was carried out cultural practices usually recommended for the culture.

Evaluations were carried out in the years of 2011, 2012 and 2013. Thirty fruits were randomly collected from three trees of each cultivar. Fruits were harvested when the green background color changed to light yellow or white cream, according to the fruit flesh color, and 14 physical and chemical traits were evaluated: fruit mass (FM), in grams (g), was measured to the nearest 0.1 kg with a digital scale; suture diameter (SD) (maximum transversal distance from the suture to the opposite face), equatorial diameter (ED) (maximum transversal distance perpendicular to the suture), and polar diameter (PD) (distance from the apex to the stalk cavity) were measured (mm) using a digital caliper; pulp firmness (FIR) was measured on the equatorial region of one of the faces of each fruit after skin removal, using a 8 mm diameter plunger tip digital penetrometer (TF-011), and was expressed in Newtons (N); soluble solids (SS) was analyzed in the hand squeezed juice from one equatorial face of each fruit, using an ATAGO digital refractometer (Palette PR-101), and was expressed in ºBrix; titratable acidity (TA) was obtained by titrating 5 g ground pulp plus 95 ml distilled water with NaOH 0.1 N solution, and was expressed as percentage of malic acid; soluble solids and titratable acidity (SS/TA) ratio was calculated by dividing values of the soluble solids by the values of titratable acidity; ascorbic acid content in the pulp (AA) was determined by titration, using the Tillman's reagent [2,6dichlorophenolindophenol (sodium salt) 0.1%], according to AOAC (1997), and was expressed in mg of ascorbic acid per 100 g pulp; carotenoid content (CT) was extracted with 80% acetone of about 2g pulp, according to the methodology proposed by Lichtenthaler (1987), and was expressed in mg 100 g^{-1} pulp; skin color (measured in the equatorial region of opposite fruit sides), and pulp color (measured in the central region of one of the pulp faces of the fruit), given by the coordinate b* and the hue angle h°, were determined by reflectometry, using a Minolta reflectometer (Color Reader CR-10), which provides readings of L*, a*, b*, C and °h. The coordinate b* ranges from blue (-60) to yellow (+60), and the hue angle °h [$h = arctg(\frac{b}{a})$] assumes zero value for

Table 1. List of the cultivars, pedigree, origin, pulp color, and botanical variety of cultivars evaluated for fruit quality

Code	Cultivar	Pedigree	Origin[1]	Pulp	Bot. v[2]
1	Aldrighi	Rio Grande do Sul Selection	ECT	A	P
2	Argel	-	-	A	P
3	Aurora 2	Open pollinationOuromel-4	IAC	A	P
4	Baronesa	(Hawaiia x Southland)F3	ECT	A	P
5	Biuti	Halford-2 x Rubi	IAC	A	P
6	Campinas-1	Autof. De Lake City	IAC	A	P
7	Capdebosq	Lake City x S56-87	ECT	A	P
8	Cerrito	(Lake City x S56-32) F2	ECT	A	P
9	Colibri	Self pollination of Cristal	IAC	B	P
10	Coral	Delicioso x Interlúdio	ECT	B	P
11	Cristal	Suber x Pérola de Itaquera	IAC	B	P
12	Delicioso Precoce	Supermel x Rubrosol	IAC	B	P
13	Diamante	Convênio x Pelota 77	ECT	A	P
14	Elberta	-	ECT	A	P
15	Flordaprince	Fla 2-7 x Maravilha	Flórida	A	P
16	Jóia4	Catita x Rubrosol	IAC	B	P
17	Josefina	Open pollination (Ouromel x Rubrosol)	IAC	B	N
18	Lake City	-	ECT	A	P
19	Maciel	Conserva 171 x Conserva 334	ECT	A	P
20	Marli	(Delicioso x Interlúdio) F2	ECT	B	P
21	Minasul	-	-	A	P
22	Olímpia	Bolinha x 7-28	ECT	A	P
23	Pérola de Itaquera	-	-	B	P
24	Real	Lake City x Rei da Conserva	IAC	A	P
25	Rei da Conserva	Discovered in Itaquera (SP)	-	A	P
26	Rubimel	Chimarrita x Flordaprince	ECT	A	P
27	Rubrosol	Open pollination (Southland x Hawaiia)F2	Flórida	A	N
28	Talismã	Rei da Conserva x Jewel	IAC	B	P
29	Tropical	Open pollination of IAC 371-2	IAC	B	P
30	Tropic Beauty	-	Flórida	A	P

[1] IAC (Instituto Agronômico de Campinas); ECT (Embrapa Clima Temperado); Flórida (Peach Improvement Program - United States)
[2] Bot. v: *Prunus persica* (L.) Batsch *vulgaris*, N: *Prunu spersica* (L.) Batsch *nucipersica*

the color red, 90° for yellow, 180° for green, and 270° for blue (Mcguirre 1992).

Estimates of phenotypic correlation coefficients were calculated by the Pearson's method (Steel and Torrie 1960), and tested at 1 and 5% probability by the *t* test, with n-2 degrees of freedom. Phenotypic correlation coefficients were determined for all combinations of traits to provide information on the nature and intensity of the relationship between them. Afterwards, the multicollinearity diagnostics was carried out, and the phenotypic correlation was decomposed into direct and indirect effects using path analysis, according to Cruz et al. (2012).

The multicolinearity test was carried out according to the criteria proposed by Montgomery and Peck (1981), which are based on the determinants of the correlation matrix and on the condition number (CN = ratio between the largest and smallest eigenvalue) of these matrices. According to these authors, as the determinant of the correlation matrix between traits approaches zero, multicollinearity becomes more severe. Besides, if NC <100, multicollinearity is not a serious problem (weak multicolinearity). If 100<CN<1000, multicollinearity is moderate to strong, and when CN>1000, there is evidence of severe multicollinearity. Analysis of the elements of the eigenvectors associated with eigenvalues, as described by Belsley et al. (1980), was performed to detect the traits that contributed to the presence of multicollinearity.

To estimate the path coefficients, it was first used a flow diagram to show the cause-effect relationships using the association between the response variable ascorbic acid content (AA) and the explanatory variables, and then it was used another causal diagram to show the interrelationship between the response variable carotenoid content (CT) and the explanatory variables. All analyses were carried out using the GENES software (Cruz 2013), with means of three years of evaluation.

RESULTS AND DISCUSSION

Table 2 shows the phenotypic correlations between the variables. The strongest positive correlations were observed between FM, SD, ED and PD (above 0.87). Albuquerque et al. (2004) also observed that the phenotypic correlation between fruit mass, equatorial diameter and polar diameter was strong and positive (above 0.83) in two years of evaluation. In the same way, Saran (2007) studied the association between peach traits, and found strong positive correlations between fruit mass, length (PD) and diameter (ED) (above 0.96). Albuquerque et al. (2004) discussed that the improvement of fruit physical traits can be based on selection for fruit diameter and low length (PD)/diameter (ED) ratio. Negreiros et al. (2007) suggested that these correlations indicate that the selection of plants with heavy fruit can be based on the equatorial fruit diameter, in the field, without weighing them, which can greatly facilitate selection in passion fruit.

All low correlations were observed between ascorbic acid content (AA) and the other variables (Table 2), such as the correlation between TA and AA (0.20). It can be inferred that the environment influences the relationship between the studied variables. Similarly, Silva et al. (2013) found weak correlation between TA and AA (0.07) in peaches. However, in acerola, Nunes et al. (2004) reported that the association between TA and AA was strong (0.77).

Using path analysis significant direct association with the carotenoid content of the pulp (°h) (0.92) (Table 2), indicating that fruits with more intense yellow pulp, with lower values of (°h) have higher carotenoid contents. This also evidences the direct influence of the variable carotenoid content on the color of the pulp, regardless of other traits under study. Costa et al. (2010) concluded that total carotenoid content characterizes the yellow color of red mombin pulp (*Spondias purpurea*), by using colorimetric analysis. Meléndez-Martínez et al. (2010) and Meléndez-Martínez et al. (2007) proposed that the variation in the pulp color observed among orange varieties is due to variations in the amount of different carotenoids.

The path analysis constitutes an expansion of multiple regression, when they are involved in complex interrelationships and, or, several causal diagrams, the reliability of the path coefficients may be affected by the effects of existing multicollinearity between the traits that make up the causal diagram due to high variance associated with their estimators (Souza et al. 2014). When multicollinearity increases, the ability to define any effects of variables decreases. It should be noted that some estimators reach very high values, indicating an unreliable estimate (Hair 1998). A recommended solution is to remove one or more independent variables which are highly correlated. A way to remove these variables

Table 2. Phenotypic correlations between 14 traits in 28 peach cultivars and two nectarine cultivars, in 2011, 2012 and 2013

Variables[1]	FM	SD	ED	PD	FIR	SS	TA	SS/TA	AA	CT	b* skin	°h skin	b* pulp	°h pulp
FM	1	0.97**	0.99**	0.94**	-0.40*	-0.27	0.08	-0.21	-0.12	0.20	0.46*	0.38*	0.30	-0.07
SD		1	0.95**	0.89**	-0.30	-0.43*	0.06	-0.22	-0.21	0.11	0.32	0.22	0.31	0.01
ED			1	0.95**	-0.47**	-0.27	0.04	-0.19	-0.14	0.24	0.50	0.42	0.31	-0.09
PD				1	-0.48**	-0.24	0.01	-0.15	-0.17	0.31	0.55**	0.47**	0.28	-0.13
FIR					1	-0.18	0.29	-0.17	0.14	-0.25	-0.61**	-0.57**	-0.02	0.18
SS						1	0.04	0.17	0.34	0.17	0.20	0.38*	-0.29	-0.30
TA							1	-0.92**	0.20	0.52**	0.16	-0.10	0.53**	-0.54**
SS/TA								1	-0.07	-0.63**	-0.19	0.12	-0.70**	0,60**
AA									1	0.00	-0.08	-0.07	0.10	-0.01
CT										1	0.54**	0,27	0.63**	-0.92**
b* skin											1	0,83**	0.30	-0.39*
°h skin												1	-0.03	-0.09
b* pulp													1	-0.49**
°h pulp														1

[1] FM: fruit mass (g); SD: suture diameter (mm); ED: equatorial diameter (mm); PD: polar diameter (mm); FIR: pulp firmness (N); SS: soluble solids (° Brix); TA: titratable acidity (% malic acid 100 g^{-1} pulp); SS/TA: soluble solids and titratable acidity ratio; AA: ascorbic acid; CT: carotenoid content (mg 100 g^{-1} pulp); skin b*, skin °h, pulp b*, and pulp h: coordinates referring to skin and pulp colors

*, ** Significant at 5 and 1% probability by the *t* test, respectively.

is by using the principal component regression through the main components corresponding to the eigenvalues. Thus, it was noted that the diagnosis of multicollinearity between the explanatory variables of the ascorbic acid content (AA) indicated high collinearity, and variables MF, SS/TA, CT, and DS were redundant. The diagnosis of multicollinearity found for the explanatory variables of the carotenoid content (CT) presented high collinearity, and the variables MF, DE and SS/TA were removed from the path analysis (Tables 3 and 4).

The first causal diagram (Table 3) showed high direct effect of SS on the ascorbic acid content (AA); however, the correlation between SS and AA was weak. Moreover, the path analysis showed that the explanatory variables (ED, PD, FIR, SS, TA, skin b*, skin °h , pulp b*, and pulp °h) considered in this model are not the main determinants of the ascorbic acid content (AA), since the coefficient of determination of the model was of low magnitude (R^2 = 0.35) and of high residual effect (0.81). The high instability of vitamins and pro-vitamins increasing local and climate change may result in significant changes in qualitative and quantitative composition of these nutrients. According to Silva et al. (2016), peaches can lose nutrients due to different crop harvesting, and ascorbic acid content has been reported as the most spoilage reaction that occurs between harvests. Ascorbic acid is a synergistic antioxidant and oxygen scavenger. It acts directly with oxygen, forming ascorbic dehydroacetic acid and eliminating the supply of oxygen available for auto-oxidation reactions (Daiuto et al. 2011). Hojo et al. (2011) stated that, during senescence, ascorbic acid of the fruit is

Table 3. Path analysis of the main dependent variable (ascorbic acid content) and independent variables, in 28 peach cultivars and two nectarine cultivars, with decomposition of phenotypic correlations into components of direct (main diagonal in bold) and indirect (off-diagonal) effects

Variables[1]	ED	PD	FIR	SS	TA	b* skin	°hskin	b* pulp	°hpulp	Total
ED	**0.08**	0.05	-0.03	-0.22	0.00	0.11	-0.22	0.13	-0.05	-0.14
PD	0.07	**0.05**	-0.03	-0.20	0.00	0.12	-0.24	0.12	-0.07	-0.17
FIR	-0.04	-0.03	**0.06**	-0.14	0.04	-0.14	0.30	-0.01	0.10	0.14
SS	-0.02	-0.01	-0.01	**0.83**	0.01	0.05	-0.20	-0.13	-0.17	0.34
TA	0.00	0.00	0.02	0.04	**0.13**	0.04	0.05	0.22	-0.30	0.20
b* skin	0.03	0.02	-0.03	0.18	0.02	**0.22**	-0.43	0.13	-0.22	-0.08
°hskin	0.03	0.02	-0.03	0.32	-0.01	0.19	**-0.52**	-0.01	-0.05	-0.07
b* pulp	0.02	0.01	0.00	-0.24	0.07	0.07	0.02	**0.43**	-0.27	0.10
°hpulp	-0.01	-0.01	0.01	-0.25	-0.07	-0.09	0.05	-0.21	**0.56**	-0.01
R^2	0.35									
Residual effect	0.81									

[1] ED: equatorial diameter (mm); PD: polar diameter (mm); FIR: pulp firmness (N); SS: soluble solids (° Brix); TA: titratable acidity (% malic acid 100 g^{-1} pulp); skin b*, skin°h, pulp b*, and pulp h°: coordinates referring to skin and pulp colors
*. ** Significant at 5 and 1% of probability by the *t* test, respectively

Table 4. Path analysis of the main dependent variable (carotenoid content) and independent variables, in 28 peach cultivars and two nectarine cultivars, with decomposition of phenotypic correlations into components of direct (main diagonal in bold) and indirect (off-diagonal) effects

Variables[1]	SD	PD	FIR	SS	TA	AA	b* skin	°h skin	b* pulp	°h pulp	Total
SD	-0.34	0.30	-0.01	0.08	0.00	0.00	-0.06	0.08	0.07	-0.01	0.11
PD	-0.30	**0.33**	-0.01	0.04	0.00	0.00	-0.10	0.17	0.06	0.11	0.31
FIR	0.10	-0.16	**0.02**	0.03	0.01	0.00	0.11	-0.20	0.00	-0.15	-0.25
SS	0.15	-0.09	0.00	**-0.19**	0.00	0.01	-0.03	0.14	-0.07	0.25	0.17
TA	-0.02	0.00	0.01	-0.01	**0.03**	0.00	-0.03	-0.03	0.12	0.45	0.52**
AA	0.07	-0.05	0.00	-0.06	0.01	**0.02**	0.01	-0.03	0.02	0.01	0.00
b* skin	-0.11	0.18	-0.01	-0.04	0.01	0.00	**-0.18**	0.30	0.07	0.33	0.54**
°h skin	-0.07	0.15	-0.01	-0.07	0.00	0.00	-0.15	**0.36**	-0.01	0.07	0.27
b* pulp	-0.10	0.09	0.00	0.05	0.02	0.00	-0.05	-0.01	**0.23**	0.41	0.63**
°h pulp	0.00	-0.04	0.00	0.06	-0.02	0.00	0.07	-0.03	-0.11	**-0.84**	-0.92**
R^2	0.96										
Residual effect	0.21										

[1] SD: suture diameter (mm); PD: polar diameter (mm); FIR: pulp firmness (N); SS: soluble solids (° Brix); TA: titratable acidity (% malic acid 100 g^{-1} pulp); AA: ascorbic acid content (mg 100 g^{-1} pulp); skin b*, skin °h, pulp b*, and pulp h°: coordinates referring to skin and pulp colors
*, ** Significant at 5 and 1% probability by the *t* test, respectively

used in oxidative reactions, which are activated by cellular stresses experienced by the membranes during this period, contributing to the reduction in the ascorbic acid content. Nunes et al. (2004), working with red mombin, found that the variable TA was the main determinant of the ascorbic acid content, which is opposed to reports by Gomes et al. (2000), who recommended the selection based on SS for gains in vitamin C in acerola.

The second causal diagram showed that the coefficient of determination was high (R2= 0.96±0.011), with low residual effect (0.21), meaning that the explanatory variables (SD, PD, FIR, SS, TA, AA, skin b*, skin °h, pulp b*, and pulp h°) used in this model are the main determinants of the response variable carotenoid content (Table 4).

Among the variables in the path diagram, pulp °h presents coefficient of correlation of -0.92, and direct effect of -0.84 (Table 4). These results indicate that the selection based on the variable yellow hue results in gain for carotenoid content. This fact is of great importance for breeding programs, since the evaluation of the yellow hue (pulp °h) with a colorimeter is cheaper and easier. The spectrophotometric methods described in the literature for the determination of carotenoids are very accurate, although they are laborious for the evaluation of large progenies; they require large amount of reagents, and are time consuming (Carvalho et al. 2005). In tomato, studies on the intensity of the correlation between chromaticity and pigment concentrations have shown good correlation between fruit color and lycopene content, a predominant carotenoid in tomatoes (George et al. 2011, Meléndez-Martínez et al. 2007).

The other explanatory variables had, in general, significant correlations with carotenoid content; however, they were associated with direct effects of low magnitude, and the largest indirect effects observed were through °h of pulp, as for TA, skin b*, and pulp b*.

The highest correlation coefficients were found between the traits FM, SD, ED, PD, and between carotenoid content and °h of pulp. The yellow hue of the pulp is associated with the carotenoid content of peaches and nectarines. The physical and chemical traits considered in the path diagrams (ED, PD, FIR, SS, TA, skin b*, skin °h, pulp b*, pulp °h) are not the main determinants of ascorbic acid content. The yellow hue of the pulp (pulp °h) has potential to be used as indirect selection for carotenoid content.

AKNOWLEDGEMENTS

The authors thank CNPq, CAPES and FAPEMIG for the financial support to this research.

REFERENCES

Albuquerque AS, Bruckner CH, Cruz CD, Salomão LCC and Neves JCL (2004) Repeatability and correlations among peach physical traits. **Crop Breeding and Applied Biotechnology 4**: 441-445.

AOAC (1997) **Official methods of analysis of the Association of the Official Analytical Chemists International.** 16th edn, Patricia Cunniff, Washinghton, 1683p.

Belsley DA, Kuh E and Welch RE (1980) **Regression diagnostics: identifying data and sources of collinearity.** John Wiley, New York, 588p.

Bowen-Forbes CS, Zhang Y and Nair MG (2010) Anthocyanin content, antioxidant, anti-inflammatory and anticancer properties of blackberry and raspberry fruits. **Journal of Food Composition and Analysis 23**: 554-560.

Carvalho W, Fonseca MEN, Silva HR, Boiteux LS and Giordano LB (2005) Estimativa indireta de teores de licopeno em frutos de genótipos de tomateiro via análise colorimétrica. **Horticultura Brasileira 23**: 819-825.

Costa MGP, Figueiredo FJ, Silva QJ and Lima VLAG (2010) Carotenoides totais e caracterização cromática de polpas de frutos de genótipos de cirigueleiras cultivadas no banco de germoplasma do IPA. In X **Jornada de Ensino, Pesquisa e Extensão**. UFRPE, Recife, p. 114-119.

Cruz CD (2013) GENES – a software package for analysis in experimental statistics and quantitative genetics. **Acta Scientiarum. Agronomy 35**: 271-276.

Cruz CD, Regazzi AJ and Carneiro PCS (2012) **Modelos biométricos aplicados ao melhoramento genético**. 4th edn, Editora UFV, Viçosa, 514p.

Daiuto ER, Vieites RL and Carvalho LR (2011) Avaliação sensorial do guacamole com adição de á-tocoferol e ácido ascórbico conservado pelo frio. **Revista Ceres 58**: 140-148.

George S, Tourniaire F, Gautier H, Goupy P, Rock E and Caris-Veyrat C (2011) Changes in the contents of carotenoids, phenolic compounds and vitamin C during technical processing and lyophilisation of red and yellow tomatoes. **Food Chemistry 124**: 1603-1611.

Gomes JG, Perecin D, Martins ABG and Almeida EJ (2000) Variabilidade fenotípica em genótipos de acerola. **Pesquisa Agropecuária Brasileira 35**: 2205-2211.

Lichtenthaler HK (1987) Chlorophylls and carotenoids: Pigments of photosynthetic biomembranes. **Methods in Enzymology 148**: 349-382.

Hair JF (1998) **Multivariate data analysis.** 5th edn, Prentice-Hall, New Jersey, 730p.

Hojo ETD, Durigan JF and Hojo RH (2011) Uso de tratamento hidrotérmico e ácido clorídrico na qualidade de lichia 'Bengal'. **Revista Brasileira de Fruticultura 33**: 386-393.

Meléndez-Martínez AJ, Escudero-Gilete ML, Vicario IM and Heredia FJ (2010) Effect of increased acidity on the carotenoid pattern and colour of orange juice. **European Food Research and Technology 230**: 527-532.

Meléndez-Martínez AJ, Britton G, Vicario IM and Heredia FJ (2007) Relationship between the colour and the chemical structure of carotenoid pigments. **Food Chemistry 101**: 1145-1150.

Meléndez-Martínez AJ, Vicario IM and Heredia FJ (2007) Review: Analysis of carotenoids in orange juice. **Journal of Food Composition and Analysis 20**: 638-649.

Mcguire RG (1992) Reporting of objective color measurements. **HortScience 27**: 1254-1260.

Montgomery DC and Peck EA (1981) **Introduction to linear regression analysis.** John Wiley, New York, 1012p.

Negreiros JRS, Álvares VS, Bruckner CH, Morgado MADO and Cruz CD (2007) Relação entre características físicas e o rendimento de polpa de maracujá-amarelo. **Revista Brasileira de Fruticultura 29**: 546-549.

Nunes ES, Carneiro PCS, Couto FAA and Braz VB (2004) Importância das características físicas e químicas na determinação do teor de vitamina C em frutos de aceroleira. **Revista Ceres 51**: 657-662.

Oliveira EJ, Lima DS, Lucena RS, Motta TBN and Dantas JLL (2010) Correlações genéticas e análise de trilha para número de frutos comerciais por planta em mamoeiro. **Pesquisa Agropecuária Brasileira 45**: 855-862.

Rivera SM and Canela-Garayoa R (2012) Analytical tools for the analysis of carotenoids in diverse materials. **Journal of Chromatography A 1224**: 1-10.

Rodriguez-Amaya DB (2010) Quantitative analysis, in vitro assessment of bioavailability and antioxidant activity of food carotenoids - A review. **Journal of Food Composition and Analysis 23**: 726-740.

Saran PL (2007) Association analysis in peach (*Prunus persica* L.) genotypes. **Progressive Horticulture 39**: 49-53.

Silva MLC, Costa RS, Santana AS and Koblitz MGB (2010) Compostos fenólicos, carotenoides e atividade antioxidante em produtos vegetais. **Semina 31**: 669-681.

Silva DFP, Matias RGP, Silva JOC and Bruckner CH (2016) Characterization of white-fleshed peach cultivars grown in the Zona da Mata area of Minas Gerais State, Brazil. **Comunicata Scientiae 7**: 149-153.

Silva DFP, Silva JOC, Matias RGP, Ribeiro MR and Bruckner CH (2013) Correlação entre características quantitativas e qualitativas de frutos de pessegueiros na geração F_2 cultivados em região subtropical. **Revista Ceres 60**: 053-058.

Spinola V, Berta B, Câmara JS and Castilho PC (2013) Effect of time and temperature on vitamin c stability in horticultural extracts. UHPLC-PDA vs. iodometric titration as analytical methods. *LWT* - **Food Science and Technology 50**: 489-495.

Souza TV, Silveira SC and Scalon JD (2014) Análise de trilha na relação entre características morfológicas do milho e sua produtividade de grãos. **Revista da Estatística da Universidade Federal de Ouro Preto 3**: 124-128.

Steel RGD and Torrie JH (1960) **Principles and procedures of statistics.** McGraw-Hill, New York, 640p.

Tarrago-Trani MT, Phillips KM and Cotty M (2012) Matrix specific method validation for quantitative analysis of vitamin c in diverse foods. **Journal of Food Composition and Analysis 26**: 12-25.

Tsantili E, Shin Y, Nock JF and Watkins CB (2010) Antioxidant concentrations during chilling injury development in peaches. **Postharvest Biology and Technology 57**: 27-34.

Wolfe KL, Kang X, He X, Dong M, Zhang Q and Liu RH (2008) Celular antioxidant activity of common fruits. **Journal of Agricultural and Food Chemistry 56**: 8418-8426.

Wright S (1921) Correlation and causation. **Journal of Agricultural Research 20**: 557-585.

12

Tissue culture efficiency of wheat species with different genomic formulas

Olga Alikina[1], Mariya Chernobrovkina[2], Sergey Dolgov[1,2] and Dmitry Miroshnichenko[1,2*]

***Corresponding author:**
E-mail: miroshnichenko@bibch.ru

[1] Russian Academy of Sciences, Branch of Shemyakin and Ovchinnikov Institute of Bioorganic Chemistry, Science Ave 6, Pushchino, Moscow Region, 142290, Russian Federation
[2] All Russian Research Institute of Agricultural Biotechnology, Timiryazevskaja 42, Moscow, 127550, Russian Federation

Abstract: *Ancient wheats are increasingly considered as valuable resources for genes of interest which could be analyzed and introduced into cultivated varieties by genetic engineering technologies. The first stage of biotechnological crop improvement consists of successful in vitro plant regeneration. Twelve wheat germplasms with different genomic formulas (AA, AABB, AAGG, AABBDD, AADDGG genomes) were examined with the use of two explant types (immature vs. mature embryos). All of the tested germplasms were able to regenerate plants, although the morphogenic ability of immature embryos was higher. The highest rate of embryogenic/regenerable structure formation was found in immature embryo cultures of tetraploid species* (T. polonicum, T. turgidum, T. carthlicum, and T. dicoccum) *as well as of hexaploid* T. spelta. *At the same time, diploid einkorn wheat (*T. monococcum*) and polyploid species with G chromosomes (*T. timopheevii and T. kiharae*) were characterized by low* embryogenesis *and by the presence of albino plantlets among shoots.*

Key words: *Callus induction, somatic embryogenesis, plant regeneration, albino plants.*

INTRODUCTION

Wheat (*Triticum* L.) is one of the most important food grains used around the globe for human food and livestock feed. Over the last decade, wild and cultured ancient wheats were increasingly involved in modern wheat breeding programs, as donors of genes conferring resistance to both biotic and abiotic stresses (Nevo 2011, Longin and Reif 2014). However, the exploitation of these resources is time-consuming and limited by cross-incompatibility barriers and linkage drags. In addition, the problem of determining the function of genes responsible for enhanced levels of resistance/tolerance is an obstacle. The transgenic approach, along with the exploration of natural or artificial mutant phenotypes, may be used as a powerful tool to detect, reduce or knock out the expression of candidate genes to clarify their functions (Repellin et al. 2001). Thereafter, this biotechnological tool can accelerate wheat breeding by the creation of artificially improved cis-genic lines, by introducing the discovered genes of interest into the existing cultivar, avoiding interspecific crossing barriers. Unfortunately, transgenic studies are basically focused on the two most common species: bread wheat (*T. aestivum* L., AABBDD genome) and pasta (durum) wheat (*T. turgidum* L., AABB genome), which is the reason for the current lack of genetic transformation protocols for ancient cultivated wheats (Mamrutha et al. 2014, Jones 2015).

In the development of wheat species transformation techniques, the critical factor is the capacity of routine regeneration of whole plants from *in vitro* tissue cultures. Wheat includes more than 20 cultivated species (Goncharov 2011), however substantial research effort has been dedicated mainly to durum and bread wheat, owing to their prevalence in the world production (Fennel et al. 1996, Machii et al. 1998, Barro et al. 1999, Bohorova et al. 2001, He and Lazzeri 2001, Zale et al. 2004, Vendruscolo et al. 2008, Yin et al. 2011). Until recently, many wheat species were not analyzed for plant regeneration ability *in vitro*; but nowadays, the interest in such studies is on the rise (Chang et al. 2012, Yang et al. 2015, Özgen et al. 2015). Over the years, several studies have demonstrated the possibility to regenerate plants from different tissues of tetraploid emmer wheat (AABB genome) (Eapen and Rao 1982, Chauhan et al. 2007, Bi et al. 2007, Chang et al. 2012, Yang et al. 2015), diploid *T. monococcum* and *T. urartu* (AA genome) (Eudes et al. 2003, Yang et al. 2015). Proliferation was induced in somatic cell cultures of tetraploid timopheevii wheat (AAGG genome) and hexaploid spelt wheat (AABBDD genome), although no shoot differentiation was observed (Lazar et al. 1983) or the number of regenerated plants was low (Yang et al 2015, Özgen et al. 2015). Some attempts to establish reliable plant regeneration protocols using various tissues of diploid wheat species (AA genome) failed to produce positive results (Lazar et al. 1983, Zale et al. 2004, Bi et al. 2007, Yin et al. 2011). According to Yang et al. (2015), plantlet regeneration by means of somatic embryogenesis represents a great obstacle for various germplasms, e.g., *T. carthlicum, T. macha* and *T. polonicum*. Thus, even after many years of research, during which various tissues, culture media and environmental conditions were tested, the screening of *in vitro* response of germplasms/genotypes is still important for biotechnological applications.

Immature embryos are currently the most frequently and successfully used explants for the initiation of regeneration of wheat species (Machii et. al 1998, Barro et al. 1999, Bohorova et al. 2001, He and Lazzeri 2001, Pellegrineschi et al. 2002, Eudes et al. 2003, Vendruscolo et al. 2008, Miroshnichenko et al. 2013) and for reliable and efficient genetic engineering (Mamrutha et al. 2014, Jones 2015). However, the cultivation of donor plants to ensure a regular supply of immature embryos is labor-intensive and requires a lot of time and space. Mature dry seeds can be used as alternative explants, providing low-cost and year-round accessibility. Efficient *in vitro* culture of mature wheat embryos was established by two main techniques. Embryos should be isolated from mature in the same way as from immature seeds, followed by placing the entire (Chauhan et al. 2007, Özgen et al. 2015), or longitudinally bisected (Zale et al. 2004, Yu et al. 2008) or multiple divided embryos (Delporte et al. 2014) on callus induction medium. Alternatively, *in vitro* plant regeneration under endosperm-supported culture may be achieved by placing the whole seed with the wounded mature embryo on induction medium (Özgen et al. 1998, Filippov et al. 2006).

The aim of the current investigation was to analyze a set of promising wheat germplasms with different genomic formulas (genomes AA, AABB, AAGG, AABBDD, AADDGG) in order to clarify which kind of germplasms might have direct application in molecular breeding and genetic transformation programs and which ones require further optimization of the *in vitro* regeneration technique. To this end, the immature embryo cultures and endosperm-supported mature embryo cultures were compared.

MATERIAL AND METHODS

Immature and mature embryo tissues of 12 spring wheat germplasms, namely diploid (2n=2x=14), tetraploid (2n=4x=28) and hexaploid (2n=6x=42) were used. Diploid wheat species were *T. monococcum* L. (einkorn, AA) and *T. sinskajae* A.Filat. et Kurk (sinskajae wheat, AA). Tetraploid wheat species were *T. dicoccum* (Schrank) Schubl. (emmer wheat 'Runo', AABB), *T. carthlicum* Nevski (Persian wheat, AABB), *T. turgidum* L. (cone wheat, AABB), *T. polonicum* L. (Polish wheat, AABB) and *T. timopheevii* Zhuk (timopheevii wheat, AAGG). Hexaploid wheat species were *T. spelta* L. (spelt wheat, AABBDD), two germplasms of *T. compactum* Host. (white and red club wheats, AABBDD), *T. sphaerococcum* Perciv. (shot wheat, AABBDD) and *T. kiharae* Dorof. et Migusch (kiharae wheat, AADDGG).

To induce plant regeneration from immature embryo culture we used a conventional two-step protocol that included 4-week cultivation of explants in the dark on media containing 2,4-D (2,4-dichlorophenoxyacetic acid) following plant differentiation in the light on media without growth regulators (Fennel et al. 1996, Machii et al. 1998, Pellegrineschi et al. 2002, Chauhan et al. 2007, Miroshnichenko et al. 2013). Immature embryos were isolated from caryopses of greenhouse-grown wheat plants 11-15 days after anthesis. The caryopses were placed in 70% ethanol solution for 3 min, thereafter soaked for 20-25 min in a 20% solution of commercial bleach (5.25% sodium hypochlorite) and then rinsed in sterile distilled water. Wheat embryos with a size of 1-2 mm were extracted under a binocular microscope, with subsequent placement on callus induction medium, scutellum side up. After 30 days of culturing at 25 °C in the dark, the number of explants that produced embryogenic/nonembryogenic callus was scored. Endosperm-supported mature embryo culture was performed according to a protocol based on the combined application of 3,6-dichloro-o-anisic acid (Dicamba) and indoleacetic acid (IAA), for embryogenic callus induction (Filippov et al. 2006, Miroshnichenko et al. 2011). Prepared seeds were placed furrow down in Petri dishes and incubated at 25 °C in the dark for 25 days. After that the number of explants producing embryogenic/nonembryogenic callus was scored. The callus induction medium for immature embryos contained mineral salts and vitamins according to Murashige and Skoog (1962) and was supplemented with 30 g L^{-1} sucrose, 150 mg L^{-1} asparagine, 7 g L^{-1} agarose, and 2 mg L^{-1} 2,4-D. The medium for callus induction from mature embryo tissues contained the same mineral salts and vitamins, 20 g L^{-1} sucrose, 7 g L^{-1} agarose, 12 mg L^{-1} Dicamba, and 0.1 mg L^{-1} IAA. For plant regeneration, embryogenic calli produced by mature or immature embryos were transferred into culture flasks on hormone-free medium containing MS mineral salts and vitamins, 20 g L^{-1} sucrose, and 7 g L^{-1} agarose. Calli were cultured under a photoperiod of 16h/8h (day/night) for 30 days at 26 °C. At the end of culturing, the number of plantlets regenerated from embryogenic calli was calculated.

Each experiment was considered a completely randomized design. The treatments were repeated at least three times. Each Petri dish was considered one replication. Every dish contained 10-12 mature seeds or 20-23 immature embryos. Statistical analysis was performed with the use of ANOVA and Duncan's Multiple Range Test (determination of the significance of results with the use of LSD_{05}). We analyzed the frequency of callus induction per 100 explants, the frequency of embryogenic callus induction per 100 explants, the frequency of callus regeneration per 100 explants, the number of regenerated shoots per one embryogenic callus and the number of albino shoots per one embryogenic callus. On the basis of these indices, we calculated *in vitro* culture efficiency as the number of produced green shoots per one initial explant. Software Statistica 10.0 was used for statistical calculations.

RESULTS AND DISCUSSION

Callus growth and embryogenesis were observed in the cultured tissues of all tested wheat species (Tables 1 and 2). The appearance of primary calli on the explant surface was observed after two to four days of culturing. Mixtures of the embryogenic callus surrounded by amorphous callus were observed in most of the wheat species after 10-15 days of cultivation (Figure 1). Non-regenerating amorphous callus was friable, soft and translucent. Immature embryo tissues produced yellowish to white, nodular embryogenic callus, characterized by the formation of leaf-like structures, green shoots and rooted shoots after transferring on the regeneration medium. Unlike the immature embryos, in which the scutellar tissue was clearly identified as a source for embryogenesis, the embryogenic callus of mature embryos was originated from divided cells of the coleoptile base, first node and epiblast. Therefore the embryogenic callus of

Table 1. In vitro culture response of immature embryo tissue of different wheat species[1]

Species	Genome	No. of cultured explants	Callus induction (%)[2]	Embryogenic callus induction (%)[3]	Regeneration efficiency (%)[4]	No. of regenerated shoots[5]		Culture efficiency[6]
						Total	Albino plants	
T. monococcum (einkorn wheat)	A	180	92.2 ab	2.8 i	1.7 i	10.0 d	0.33 b	0.2 h
T. sinskajae (sinskajae wheat)	A	115	93.9 ab	83.5 c	82.6 c	6.9 e	0.22 b	5.5 f
T. dicoccum (emmer wheat 'Runo')	AB	150	97.3 ab	94.0 ab	94.0 ab	12.8 c	0.00 b	12.0 bc
T. carthlicum (persian wheat)	AB	183	97.8 ab	86.3 bc	86.3 bc	16.4 ab	0.01 b	14.1 a
T. polonicum (polish wheat)	AB	168	100.0 a	98.2 a	96.4 a	11.3 cd	0.00 b	10.6 c
T. turgidum (cone wheat)	AB	192	85.4 bc	82.8 c	81.8 c	15.2 b	0.00 b	12.3 b
T. timopheevii (timopheevii wheat)	AG	140	77.1 c	14.3 h	12.9 h	7.4 e	0.39 b	0.9 h
T. compactum (white club wheat)	ABD	227	86.3 bc	63.4 e	63.4 e	12.4 c	0.00 b	7.9 e
T. compactum (red club wheat)	ABD	158	94.3 ab	54.4 f	54.4 f	17.5 a	0.10 b	9.4 d
T. spelta (spelt wheat)	ABD	179	97.2 ab	73.2 d	73.2 d	17.2 a	0.00 b	12.6 b
T. sphaerococcum (shot wheat)	ABD	203	94.6 ab	20.2 h	20.2 h	11.1 cd	0.00 b	2.2 g
T. kiharae (kiharae wheat)	ADG	226	79.2 c	36.7 g	36.7 g	16.3 ab	1.41 a	5.5 f

[1] Different letters, in each column, indicate statistical differences by the Duncan's Multiple Range Test ($p < 0.05$); [2] Percentage of initial explants produced callus; [3] Percentage of initial explants produced embryogenic callus; [4] Percentage of initial explants produced plantlets after the transferring to regeneration medium; [5] Number of regenerated plantlets per one embryogenic callus; [6] Average number of green plants regenerated per one initial explant.

Table 2. In vitro culture response of mature embryo tissue of different wheat species

Species	Genome	No. of cultured explants	Callus induction (%)[2]	Embryogenic callus induction (%)[3]	Regeneration efficiency (%)[4]	No. of regenerated shoots[5]		Culture efficiency[6]
						Total	Albino plants	
T. monococcum (einkorn wheat)	A	139	97.8	0.7 e	0.7 f	2.0 f	0.0	0.01 f
T. sinskajae (sinskajae wheat)	A	82	100.0	8.5 cd	6.1 cdef	6.2 e	0.0	0.38 e
T. dicoccum (emmer wheat 'Runo')	AB	161	100.0	28.6 a	28.6 a	11.9 b	0.0	3.41 a
T. carthlicum (persian wheat)	AB	147	100.0	18.4 b	12.9 c	11.4 b	0.0	1.48 c
T. polonicum (polish wheat)	AB	106	100.0	25.5 a	21.7 b	10.7 b	0.0	2.32 b
T. turgidum (cone wheat)	AB	186	100.0	4.3 de	3.8 def	8.3 cd	0.0	0.31 ef
T. timopheevii (timopheevii wheat)	AG	172	98.8	12.2 bc	11.0 cd	6.5 de	0.0	0.72 d
T. compactum (white club wheat)	ABD	181	100.0	6.1 cde	4.4 def	8.5 c	0.0	0.38 e
T. compactum (red club wheat)	ABD	158	100.0	5.1 de	4.4 def	16.4 a	0.0	0.73 d
T. spelta (spelt wheat)	ABD	172	100.0	2.9 de	2.3 ef	5.8 e	0.0	0.13 ef
T. sphaerococcum (shot wheat)	ABD	159	100.0	5.7 cde	5.7 cdef	6.7 cde	0.0	0.38 e
T. kiharae (kiharae wheat)	ADG	147	100.0	8.8 cd	8.8 cde	16.2 a	0.0	1.43 c

[1] Different letters, in each column, indicate statistical differences by the Duncan's Multiple Range Test ($p < 0.05$); [2] Percentage of initial explants produced callus; [3] Percentage of initial explants produced embryogenic callus; [4] Percentage of initial explants produced plantlets after the transferring to regeneration medium; [5] Number of regenerated plantlets per one embryogenic callus; [6] Average number of green plants regenerated per one initial explant.

mature embryos had a smaller size (Figure 2) and delayed differentiation.

The ability to form embryogenic calli depended on the wheat species and varied significantly even between wheats of the same ploidy (Tables 1 and 2). In this study, both immature and mature embryo-derived tissue cultures of tetraploid wheats with AB genome generally had a greater embryogenic potential and efficiency of plant regeneration than the wheats with other genomic formulas. Efficiencies of immature embryo-derived cultures for AB-genome wheats ranged from 82.8% (*T. turgidum*) to 98.2% (*T. polonicum*). By the end of culture period, the entire explant surfaces of *T. carthlicum*, *T. turgidum* and *T. polonicum* were usually covered with numerous embryogenic structures (Figure 1C, 1D, 1E). *Triticum dicoccum* 'Runo' was distinguished by somatic embryos formation predominantly at the scutellum edge (Figure 1H). The immature embryos of AABBDD hexaploid wheats formed a higher portion of nonembryogenic calli (Figure 1F, 1G, 1I, 1K), thus embryogenic callus induction ranged from only 20.2% (*T. sphaerococcum*) to 73.2% (*T. spelta*). Low frequencies were observed for wheats with G genome; where the percentage of embryogenic callus formation in tetraploid *T.timopheevii* was lower than in hexaploid *T. kiharae* (14.3% and 36.7%, respectively).

The diploid species *T. monococcum* and *T. sinskajae* were characterized by a similar pattern of embryogenic structure formation in the apical part of scutellum (Figure 1A, B), but they differed extremely in the ability to form embryogenic callus in immature-embryo-derived culture.

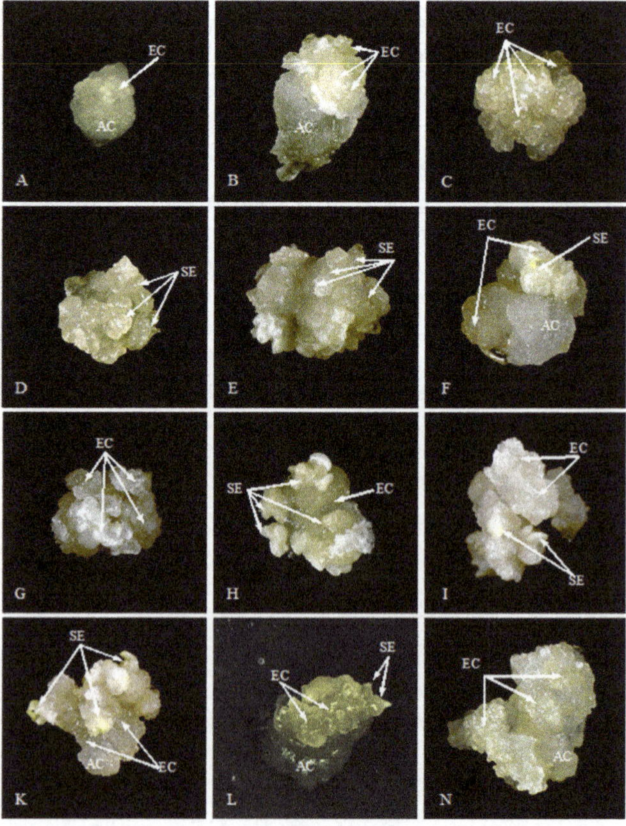

Figure 1. Embryogenic callus formation from immature embryo tissues after 4 weeks of culture in (A) *T. monococcum*, (B) *T. sinskajae*, (C) *T. polonicum*, (D) *T. turgidum*, (E) *T. carthlicum*, (F) *T. compactum* var. white, (G) *T. compactum* var. red, (H) *T. dicoccum 'Runo'*, (I) *T. spelta*, (K) *T. sphaerococcum*, (L) *T. timopheevii*, (N) *T. kiharae*, SE – somatic embryos, EC – embryogenic callus, AC - amorphous (non-regenerating) callus.

Triticum sinskajae, regarded as a natural naked mutation of *T. monococcum* (Goncharov 2011), has demonstrated a high ability to form embryogenic callus in immature-embryo-derived culture (83.5%). In contrast, its progenitor *T. monococcum* showed the weakest *in vitro* response both in immature and mature embryo-derived tissue cultures. Explants of *T. monococcum* formed callus readily (92.2-97.8%), but their embryogenic capacity (0.7-2.8%) and plant regeneration efficiency (0.7-1.7%) were extremely low. This observation is in agreement with previous reports that showed either nonembryogenic callus formation or only occasional plant regeneration from cultured explants of *T. monococcum* (Lazar et al. 1983, Zale et al. 2004, Yin et al. 2011).

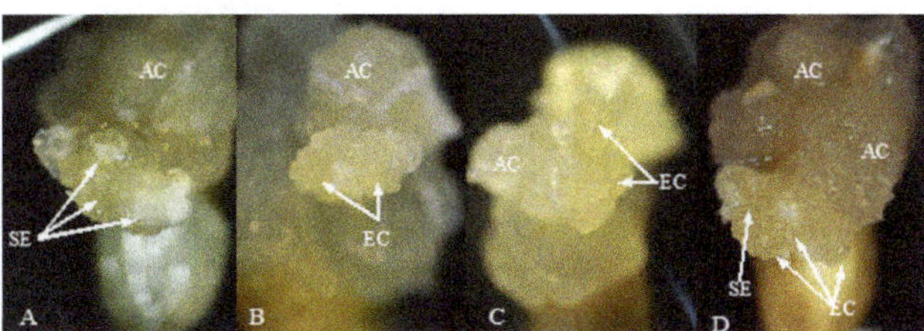

Figure 2. Embryogenic callus formation from mature embryo tissues after 4 weeks of culture in (A) *T. kiharae*, (B) *T. carthlicum*, (C) *T. polonicum* and (D) *T. dicoccum*, SE – somatic embryos, EC – embryogenic callus; AC - amorphous (non-regenerating) callus.

Figure 3. Regeneration of albino plants (indicated by arrows) from embryogenic callus of immature embryo tissues: (A) *T. kiharae*, (B) *T. timopheevii*.

In this study, the *in vitro* performance of immature embryos of *T. carthlicum* (AABB genome) was best (14 green shoots per initial explant). Likewise, the culture efficiency of immature embryo tissues of the other wheat species with AABB genome (*T. dicoccum* 'Runo', *T. turgidum* and *T. polonicum*) as well as of hexaploid *T. spelta* (AABBDD genome) was good (about 10-12 green plant per initial explants). This performance is usually regarded as sufficient for inclusion in genetic engineering programs for polyploid wheat cultivars (Pellegrineschi et al. 2002, He et al. 2010). The results of our study contradict findings of Yang at al. (2015), who reported that the immature embryo culture of *T. carthlicum, T. turgidum* and *T. polonicum* displayed very low regeneration rates, while *T. monococcum* produced a high portion of differentiated callus. However, the results of this report are not directly comparable since other plant growth regulators/medium compositions were used and there are no data concerning the number of regenerated plants. In our study, the lowest culture efficiency (less than one green plant per initial explant) was observed for *T. monococcum* and *T. timopheevii*. This result was caused mainly by the insufficient capacity for embryogenesis and also by the albinism of regenerated plants.

Our studies revealed that six wheat species had produced albino-regenerants from immature tissue cultures, though in mature-tissue cultures this phenomenon was not observed. The higher portion of albino-regenerants was primarily attributed to G genomic formula wheats (Figure 3), such as *T. kiharae* (on average 8.5 albino regenerants per 100 plants) and *T. timopheevii* (on average 5.2 albino-regenerants per 100 plants). The other species, including *T. carthlicum, T. compactum, T. monococcum,* and *T. sinskajae,* produced a much lower number of albino plants (0.01-0.33 per embryogenic explant). In cereals, the regeneration of albino plants is a major challenge in routine anther and microspore culture experiments. For immature and mature embryo-derived cultures however, albinism is not a common problem. In androgenic cultures of wheat and barley, more than 60-80% of the variations associated with albino plant regeneration is genotype-dependent (Kumari et al. 2009, Makowska and Oleszczuk 2014). Nevertheless, the recovery of green vs. albino plants in anther cultures of cereals could be improved by optimized regeneration protocols (Kumari et al. 2009, Makowska and Oleszczuk 2014). In accordance with this assumption, our latest research showed that the changes in culture medium composition, namely of the concentrations/types of plant growth regulators and carbohydrate content, provided a more effective control of somatic embryogenesis and a lower production of albino plants in *T. kiharae* (Miroshnichenko et al. 2016).

Evidently, the two-step protocol used in our study to induce immature embryo-derived embryogenic callus of polyploid wheats with different genomes (*T. monococcum, T. compactum, T. sphaerococcum, T. timopheevii,* and *T. kiharae*) is not optimal. Moreover, some wheat germplasms with high efficiency of embryogenic callus formation (*T. sinskajae, T. dicoccum,*

T. polonicum) formed less plantlets than the germplasms with low embryogenic ability (*T. kiharae, T. compactum*). These indicate that adequate changes in medium composition are required to achieve a satisfactory response at the different stages of somatic embryogenesis, including induction, maintenance and shoot development. For example, the application of the alternative 5-step protocol proposed by Eudes et al. (2003) involved a much larger number of media components/growth regulators and resulted in satisfactory plant regeneration induction from scutellum of *T. monococcum, T. durum* and *T. aestivum,* without callus phase. There are numerous reports that describe the substantial increase in regeneration response in cultures of hexaploid bread wheat with AABBDD genome by modifications of the medium composition (Fennel et al. 1996, Machii et al. 1998, Barro et al. 1999, Bohorova et al. 2001, He and Lazzeri, 2001, Chauhan et al. 2007, Vendruscolo et al. 2008, Miroshnichenko et al. 2013). However, in most of the cited publications, the different manipulations were frequently inefficient in overcoming the strong genotypic effect and many varieties had to be screened to detect responsive wheat genotypes. Obviously, this indicates that no universal recipe can be given for an efficient culture of wheat species with different genomic formulas, and some optimization is likely required.

In contrast to the *in vitro* mature-embryo culture of other cereals, as of barley (He and Jia 2008) and rice (Li et al. 2015), the rate of somatic embryogenesis of the analyzed wheat species, with the exception of *T. diccocum*, was low, and they were unable to form more than one green plant per initial explant (Table 2). The callus regeneration ability of all cultures of AB genome tetraploid wheats, except for *T. turgidum,* was higher (11.0-28.6%) than that of hexaploid (2.3-8.8%) and diploid wheats (0.7-6.1%). Due to the relatively high frequency of embryogenic structure formation, *T. diccocum* 'Runo' had the highest culture efficiency of mature embryo tissues (3.41 plants/per initial explant). Generally speaking, this observation agrees with previous reports of Bi et al. (2007), who found that the response of mature embryo-derived callus formation of *T. diccocum* was better than that of the other tetraploid (*T. durum*) and hexaploid (*T. aestivum*) wheats. On the other hand, in several other reports (Chauhan et al. 2007, Chang et al. 2012, Yang et al. 2015) the difference between wheats with different genome composition (AABB vs. AABBDD) including *T. diccocum* was not clearly evident. In addition, conflicting data have been presented for diploid species. In this study as well as in that of Yin et al. (2011), the embryogenesis/regeneration ability of diploid wheats with AA genome was found to be very low. Recently, Özgen et al. (2015) reported that mature-embryo cultures of *T. monococcum* had a rather high callus regeneration capacity. However, in that study nodular calli with only green spots were taken into consideration, instead of the number of regenerated plants, so a correct comparison was not possible.

Mature wheat embryos are thought to be more recalcitrant to tissue culture than immature embryos, due to differences in the physiological and biochemical tissue status (Özgen et al. 1998, Filippov et al. 2006, Delporte et al. 2014, Yang et al. 2015). The acquisition of embryogenic competence in tissues of zygotic wheat embryos was shown to depend significantly on the endogenous level of growth regulators and total contents of phenols and soluble sugars (Jiménez and Bangerth 2001, Yang et al. 2015). With regard to the dormancy of the mature embryos at the stage of culture initiation, the particular interaction of the medium and other environmental parameters with genotype-specific endogenous factors was frequently required to achieve efficient morphogenesis, especially in endosperm-supported culture (Filippov et al. 2006, Bi et al. 2007, Chauhan et al. 2007, Yu et al. 2008, Miroshnichenko et al. 2011, Yin et al. 2011, Chang et al. 2012, Delporte et al. 2014). With the media tested here however, immature embryos are the preferable explant material. Regardless of the low culture efficiency of the studied germplasms, the ability of mature embryo-derived cultures to generate entire green plants ought to be more deeply investigated, along with immature embryo and anther cultures. Currently, albinism still significantly delays breeding programs using doubled haploid technologies, so some comparative investigations may help discover genetic and developmental mechanisms responsible for the reduction of albino plant formation.

Thus, we conclude that conventional 2-step protocols using 2,4-D as main exogenous inducer can be used to generate a readily sustainable supply of morphogenic material from immature embryos of *T. carthlicum*, *T. dicoccum, T. polonicum, T. spelta,* and *T. turgidum*. On the other hand, the endosperm–supported protocol, which is originally developed for mature embryos of bread wheat, might fail to meet the specific requirements of wheat species with various genetic backgrounds to provide sufficient plant regeneration. Some modifications of protocols should however be performed, e.g., of the type and concentration of micronutrients, growth regulators, sugars, and organic compounds in the medium. This is particularly true for cultures of diploid (*T. monococcum, T. sinskajae*), hexaploid (*T. compactum, T. sphaerococcum*) and polyploid germplasms with G chromosomes (*T. timopheevii, T. kiharae*), to provide a more effective control of somatic embryogenesis and to improve the quality of regenerated plants for developing efficient recombinant DNA techniques.

ACKNOWLEDGEMENTS

The authors thank Prof. Ludmila A. Bespalova, Krasnodar Lukyanenko Research Institute of Agriculture and Prof. Gennady I. Karlov, Center for Molecular Biotechnology, Russian State Agrarian University - Moscow Timiryazev Agricultural Academy, for kindly providing the seeds of various wheat germplasms.

REFERENCES

Barro F, Martin A, Lazzeri PA and Barceló P (1999) Medium optimization for efficient embryogenesis and plant regeneration from immature inflorescences and immature scutella of elite cultivars of wheat, barley and tritordeum. **Euphytica 108:** 161-167.

Bi RM, Kou M, Chen LG, Mao SR and Wang HG (2007) Plant regeneration through callus initiation from mature embryo of *Triticum*. **Plant Breeding 126:** *9-12.*

Bohorova NE, Pfeiffer WH, Mergoum M, Crossa J, Pacheco M and Estanol P (2001) Regeneration potential of CIMMYT durum wheat and triticale varieties from immature embryos. **Plant Breeding 120:** 291-295.

Chang CM, Penna S and Bhagwat SG (2012) Callus induction and plant regeneration from different *Triticum* species. **The Asian and Australasian Journal of Plant Science and Biotechnology 6:** 56-62.

Chauhan H, Desai SA and Khurana P (2007) Comparative analysis of the differential regeneration response of various genotypes of *Triticum aestivum*, *Triticum durum* and *Triticum dicoccum*. **Plant Cell Tissue and Organ Culture 91:** 191-199.

Delporte F, Pretova A, du Jardin P and Watillon B (2014) Morpho-histology and genotype dependence of *in vitro* morphogenesis in mature embryo cultures of wheat. **Protoplasma 251:** 1455-1470.

Eapen S and Rao PS (1982) Plant regeneration from callus cultures of durum and emmer wheat. **Plant Cell Reports 1:** 215-218.

Eudes F, Acharya S, Laroche A, Selinger LB and Cheng KJ (2003) A novel method to induce direct somatic embryogenesis, secondary embryogenesis and regeneration of fertile green cereal plants. **Plant Cell Tissue and Organ Culture 73:** 147-157.

Fennell S, Bohorova N, Ginkel MV, Crossa J and Hoisington D (1996) Plant regeneration from immature embryos of 48 elite CIMMYT bread wheats. **Theoretical and Applied Genetics 92:** 163-169.

Filippov M, Miroshnichenko D, Vernikovskaya D and Dolgov S (2006) The effect of auxins, time exposure to auxin and genotypes on somatic embryogenesis from mature embryos of wheat. **Plant Cell Tissue and Organ Culture 84:** 213-222.

Goncharov NP (2011) Genus *Triticum* L. taxonomy: the present and the future. **Plant Systematics and Evolution 295:** 1-11.

He GY and Lazzeri PA (2001) Improvement of somatic embryogenesis and plant regeneration from durum wheat *(Triticum turgidum* var. *durum* Desf.) scutellum and inflorescence cultures. **Euphytica 119:** 369-376.

He TF and Jia JF (2008) High frequency plant regeneration from mature embryo explants of highland barley (*Hordeum vulgare* L. var. nudum Hk. f.) under endosperm-supported culture. **Plant Cell Tissue and Organ Culture 95**: 251-254.

He Y, Jones HD, Chen S, Chen XM, Wang DW, Li KX, Wang DS and Xia LQ (2010) Agrobacterium-mediated transformation of durum wheat (*Triticum turgidum* L. var. *durum* cv Stewart) with improved efficiency. **Journal of Experimental Botany 61:** 1567-1581.

Jiménez VM and Bangerth F (2001) Endogenous hormone concentrations and embryogenic callus development in wheat. **Plant Cell Tissue and Organ Culture 67**: 37-46.

Jones HD (2015) Wheat biotechnology: current status and future prospects. In Azhakanandam K, Silverstone A, Daniell H and Davey RM (eds) **Recent advancements in gene expression and enabling technologies in crop plants**. Springer New York, New York, p. 263-290.

Kumari M, Clarke, Small I and Siddique KHM (2009) Albinism in plants: a major bottleneck. **Critical Reviews in Plant Science 28:** 393-409.

Lazar MD, Collins GB and Vian WE (1983) Genetic and environmental effects on the growth and differentiation of wheat somatic cell cultures. **Journal of Heredity 74:** 353-357.

Li D, Xu H, Sun X, Cui Z, Zhang Y, Bai Y, Wang X and Chen W (2015) Differential transformation efficiency of Japonica rice varieties developed in northern China. **Crop Breeding and Applied Biotechnology 15:** 162-168.

Longin CF and Reif JC (2014) Redesigning the exploitation of wheat genetic resources. **Trends in Plant Science 19:** 631-636.

Machii H, Mizuno H, Hirabayashi T and Hagio T (1998) Screening wheat genotypes for high callus induction and regeneration capability from anther and immature embryo cultures. **Plant Cell Tissue and Organ Culture 53:** 67-74.

Makowska K, Oleszczuk S (2014) Albinism in barley androgenesis. **Plant Cell Reports 33:** 385-392.

Mamrutha HM, Kumar R, Venkatesh K, Sharma P, Kumar R, Tiwari V and Sharma I (2014) Genetic transformation of wheat – Present status and future potential. **Journal of Wheat Research 6:** 107-119.

Miroshnichenko DN, Filippov MV and Dolgov SV (2013) Medium optimization for efficient somatic embryogenesis and *in vitro* plant regeneration of spring common wheat varieties. **Russian Agricultural Sciences 39**: 24-28.

Miroshnichenko DN, Poroshin G N and Dolgov SV (2011) Genetic transformation of wheat using mature seed tissues. **Applied**

Biochemistry and Microbiology 47: 767-775.

Miroshnichenko D, Chernobrovkina M and Dolgov S (2016) Somatic embryogenesis and plant regeneration from immature embryos of *Triticum timopheevii* Zhuk. and *Triticum kiharae* Dorof. et Migusch, wheat species with G genome. **Plant Cell Tissue and Organ Culture 125:** 495-508.

Murashige T and Skoog F (1962) A revised medium for rapid growth and bioassays with tobacco tissue cultures. **Physiology Plantarum 15:** 473-497.

Nevo E (2011) Triticum. In Kole C (ed) **Wild crop relatives: genomic and breeding resources, cereals**. Springer Verlag, Berlin, p. 407-456.

Özgen M, Türet M, Altmok S and Sancak C (1998) Efficient callus culture induction and plant regeneration from mature embryo culture of winter wheat (*Triticum aestivum* L) genotypes. **Plant Cell Reports 18**: 331-335.

Özgen M, Birsin MA and Benlioglu B (2015) Biotechnological characterization of a diverse set of wheat progenitors (*Aegilops* sp. and *Triticum* sp.) using callus culture parameters. **Plant Genetic Resources, First View**: 1-6, Published online: 14 August 2015, DOI: 10.1017/S1479262115000350.

Pellegrineschi A, Noguera LM, Skovmand B, Brito RM, Velazquez L, Salgado MM, Hernandez R, Warburton M and Hoisington D (2002) Identification of highly transformable wheat genotypes for mass production of fertile transgenic plants. **Genome 45:** 421-430.

Repellin A, Baga M, Jauhar PP and Chibbar RN (2001) Genetic enrichment of cereal crops via alien gene transfer: new challenges. **Plant Cell Tissue and Organ Culture 64:** 159-183.

Vendruscolo ECG, Schuster I, Negra ES and Scapim CA (2008) Callus induction and plant regeneration by Brazilian new elite wheat genotypes. **Crop Breeding and Applied Biotechnology 8:** 195-201.

Yang S, Xu K, Wang Y, Bu B, Huang W, Sun F, Liu S and Xi Y (2015) Analysis of biochemical and physiological changes in wheat tissue culture using different germplasms and explant types. **Acta Physiology Plantarum 37**: 1-10.

Yin GX, Wang YL, She MY, Du LP, Xu HJ, MA JX and Ye XG (2011) Establishment of a highly efficient regeneration system for the mature embryo culture of wheat. **Agricultural Sciences in China 10:** 9-17.

Yu Y, Wang J, Zhu ML and Wei ZM (2008) Optimization of mature embryo-based high frequency callus induction and plant regeneration from elite wheat cultivars grown in China. **Plant Breeding 127:** 249-255.

Zale JM, Borchardt-Wier H, Kidwell KK and Steber CM (2004) Callus induction and plant regeneration from mature embryos of a diverse set of wheat genotypes. **Plant Cell Tissue and Organ Culture 76:** 277-281.

13

QTL mapping for yield components and agronomic traits in a Brazilian soybean population

Josiane Isabela da Silva Rodrigues[1*], Fábio Demolinari de Miranda[2], Newton Deniz Piovesan[1], Adésio Ferreira[3], Marcia Flores da Silva Ferreira[2], Cosme Damião Cruz[4], Everaldo Gonçalves de Barros[5] and Maurilio Alves Moreira[6]

***Corresponding author:**
E-mail: josianeisabela@gmail.com

[1] Universidade Federal de Viçosa (UFV), Departamento de Bioquímica e Biologia Molecular, Instituto de Biotecnologia Aplicada à Agropecuária (BIOAGRO), 36.571-000, Viçosa, MG, Brazil.
[2] Universidade Federal do Espírito Santo (UFES), Departamento de Biologia, 29.500-000, Alegre, ES, Brazil
[3] UFES, Departamento de Produção Vegetal, 29.500-000, Alegre, ES, Brazil
[4] UFV, Departamento de Biologia Geral, BIOAGRO
[5] Universidade Católica de Brasília, Quadra SGAN 916, módulo B, bloco C, sala 213, 70.790-160, Brasília, DF, Brazil
[6] In memoriam

Abstract: *The objective of this work was to map QTL for agronomic traits in a Brazilian soybean population. For this, 207 $F_{2:3}$ progenies from the cross CS3035P-TA276-1-5-2 x UFVS2012 were genotyped and cultivated in Viçosa-MG, using randomized block design with three replications. QTL detection was carried out by linear regression and composite interval mapping. Thirty molecular markers linked to QTL were detected by linear regression for the total of nine agronomic traits. QTL for SWP (seed weight per plant), W100S (weight of 100 seeds), NPP (number of pods per plant), and NSP (number of seeds per plant) were detected by composite interval mapping. Four QTL with additive effect are promising for marker-assisted selection (MAS). Particularly, the markers Satt155 and Satt300 could be useful in simultaneous selection for greater SWP, NPP, and NSP.*

Key words: *Glycine max, yield, quantitative trait locus, microsatellite markers.*

INTRODUCTION

Soybean is by far the main export product in Brazil, and it is grown almost everywhere in the country. In the 2014/2015 season, its production accounted for 96.243 million tons, which corresponds to 46% of total grain yield in the country (CONAB 2015). Due to its economic importance in the Brazilian agricultural scenario, soybean breeding programs seek to develop more productive cultivars for the various Brazilian conditions.

Traditionally, productive parental lines are used to obtain new cultivars (Carter et al. 2004), which contribute to the narrowing of the genetic base of the improved germplasm, and raises difficulties to obtain further gains in yield. In addition to the low variability in the improved germplasm (Hyten et al. 2006), other factors also hinder the selection of productive cultivars, such as the environmental influence (Ainsworth et al. 2012), which reduces the efficiency of selection of superior genotypes, and the existing negative correlation between grain yield and protein content (Popovic et al. 2012), since another purpose of breeding programs is the increase in the protein content of the grain. Facing these difficulties, the knowledge of the genetic control involved with grain yield may direct more effective strategies for selection, such as marker-assisted selection (MAS).

In soybeans, several QTL have recently been mapped for traits related to grain yield, although many of them may be repetitive. Altogether, at least 99

QTL are reported for traits related to weight and size of the seeds (Liu et al. 2011, Xu et al. 2011, Han et al. 2012, Sun et al. 2012, Hu et al. 2013, Liu et al. 2013, Pathan et al. 2013, Yesudas et al. 2013, Kato et al. 2014). Moreover, at least six QTL have been reported for grain yield, specifically (Palomeque et al. 2009, Sebastian et al. 2010, Kim et al. 2012, Fox et al. 2015), besides the several QTL mapped for other grain yield components from different populations (Chen et al. 2007, Guzman et al. 2007, Zhang et al. 2010, Liu et al. 2011, Liu et al. 2013).

According to the studies mentioned above, the genetic control of grain yield components involves several loci, and has strong environmental influence. For this reason, molecular markers have been used in the identification and location of QTL for strategies involving MAS, which can be useful in breeding for allowing indirect selection of agronomic traits from early generations and early stages of plant development. By using MAS, unfavorable alleles can be eliminated or greatly reduced at early generations, which allows the evaluation and selection of a small number of plants in the field. In another application, MAS can facilitate the introgression of favorable alleles in commercial materials from non-adapted germplasm sources.

In spite of the number of QTL available in the literature for grain yield components in soybean, none of the QTL has been mapped for Brazilian cultivars and/or in Brazilian soil and climate conditions. Thus, this study aims to map QTL for yield components and agronomic traits from an $F_{2:3}$ soybean populations, from the cross between a line and a Brazilian cultivar, aiming at implementing MAS in breeding programs in the country.

MATERIAL AND METHODS

From the cross between the line CS3035PTA276-1-5-2 and the variety UFVS2012, it was obtained a population of 207 F_2 plants. The F_2 progenies were cultivated in a greenhouse, and young leaves were collected, frozen in liquid nitrogen, and stored at -80 °C for subsequent DNA extraction. Phenotypic evaluation was carried out in F_3 families cultivated in the field, using a randomized block design with three replications. In each block, it was collected phenotypic values of five individuals per family. Cultivars BARC-8 and Monarca, and the parents CS3035PTA276-1-5-2 and UFVS2012 were also cultivated as controls. This trial was carried out in the Experimental Field Diogo Alves de Mello (lat 20° 45′ S, long 42° 52′ W, alt 650m asl), Viçosa, in the state of Minas Gerais, Brazil. The soil of the region is classified as clayey dystrophic red-yellow latosol.

In the F_3 generation, the following agronomic traits were evaluated: number of days to flowering (NDF), from the emergence until 50% flowered plants in the row; number of days to maturity (NDM), from the emergence until 95% pods with maturation color; plant height at maturity (PHM), expressed in centimeters, from the ground level up to the last node on the main stem; height of the first pod (HFP), expressed in centimeters; number of nodes at maturity (NNM), counted on the main stem from the cotyledon node, at the R8 stage; number of pods per plant (NPP); seed weight per plant (SWP); weight of one hundred seeds (W100S); and number of seeds per plant (NSP). It was carried out analysis of variance of each variable using the GENES software (Cruz 2013). The statistical model used (family trial with intercalated controls) is illustrated below.

$$Y_{ij} = \mu + g_i + b_j + \varepsilon_{ij}$$

in which: Y_{ij} is the value of the trait for the i-th treatment in the j-th block; m, is the general mean of the treatments; g_i, the effect of the i-th treatment (i = 1,2, ..., t); b_j, is the random block effect (j = 1,2, .., r), and ε_{ij} is the random error of the control, being ε_{ij}~NID (0, s^2).

Broad-sense heritability (h^2) and the coefficient of experimental variation (CVe) were obtained by the following estimators:

$$h^2 = (MSF - MSR)/MSF$$

$$CV_e = 100(MSF)^{0.5}/\mu$$

in which: MSF and MSR refer to the mean square of the families and mean square of the residue, respectively, and μ is the general mean. The components of residual variance (σ^2), of genotype (σ_g^2) and of controls (Φ_{te}) were estimated as follows:

$\sigma^2 = MSR/r$

$\sigma_g^2 = (MSF - QMR)/r$

$\Phi_{te} = (MSC - QMR)/r$

in which: MSR is the mean square of the residue; MSF is the mean square of the families; MSC is the mean square of the controls; and r is the number of replications. The hypothesis of normal distribution of the nine traits evaluated was tested by the Lilliefors test.

DNA extraction was carried out by the CTAB method, according to Doyle and Doyle (1990). Microsatellite markers used in the experiment were developed by Cregan et al. (1999), and the respective sequences of the pair of primers are available in the database Soybase (http://soybase.org/). PCR reactions were carried out in a total volume of 15 µL containing: Tris-HCl 10 mM, pH 8.3; KCl 50 mM; $MgCl_2$ 2 mM; Triton X100 0.1%; 100 µM of each deoxynucleotide; 0.3 µM of each microsatellite primer; a unit of *Taq* DNA polymerase (Phoneutria); and 30 ng DNA. PCR reactions had initial step at 94 °C for 4 min; 30 cycles of 94 °C for 1 min; 55 °C for 1 min, and 72 °C for 2 min; and a final step at 72 °C for 7 min. Amplification products were separated by electrophoresis in 10% polyacrylamide vertical gels, using 1X TAE buffer (40 mM Tris-acetate and 1 mM EDTA), and running time of three hours at 140 volts. Gels were stained with ethidium bromide (1µg mL^{-1}) and recorded with the aid of a photodocumentator device *Eagle Eye* II (Stratagene).

The individual segregation of molecular markers was tested by the chi-square test using the Bonferroni criteria ($p < 0.05$). Molecular markers with an expected segregation (1:2:1) were grouped using a minimum LOD score of 3.0, and maximum recombination frequency of 30%. The Kosambi mapping function was used as a distance measure, and the Rapid Chain Delineation algorithm (RCD) (Doerge 1996) was used to define the order of the markers in the linkage groups. For the construction of the genetic map, it was used the GQMOL software (http://www.ufv.br/dbg/gqmol/gqmol.htm).

The association between the trait and molecular marker was determined by the single marker analysis, using linear regression (i.e., by the regression of the values of each quantitative trait in function of the scores related to the genotypes of the markers). The presence of the QTL in the intervals of the linkage map was evaluated by composite interval mapping for the nine traits (Jansen 1993, Zeng 1994). Selection of cofactors was based on the stepwise regression analysis, at 5% and 10% significance level for the input and output of independent variables, respectively. For some of the intervals, it was used cofactors of the linkage group (LG) and a significant association with the variable by the single marker analysis. The estimates of the additive and dominance value, the coefficient of determination corresponding to the peak of higher statistical significance of the QTL, and the position of each QTL were declared when the maximum likelihood ratio values (LR) exceeded the cut-off critical values ($\alpha = 0.05$) in each LG. The critical LR values were determined by performing 1000 permutations. QTL analyses were carried out with the aid of the GQMOL (http://www.ufv.br/dbg/gqmol/gqmol/htm).

RESULTS AND DISCUSSION

The analysis of variance of the agronomic traits indicated genetic variability for the nine traits at 1% probability, evidencing the potential of the population for QTL analysis (Table 1). The suitability of the population is also evidenced by the contrast observed between the means of the parents for most traits (Table 2). Although there was no difference between the means of the parents for the variables NDF, NDM and W100S, these traits presented transgressive segregation between the F_3 families. The ratio CVg/CVe was high for all traits, especially for NDM, NNM and W100S, indicating that most part of the variation was due to genetic variance. The coefficients of variation presented acceptable values, which were close to those observed by other authors in Brazilian conditions (Pires et al. 2012, Rocha et al. 2012, Barbosa et al. 2013, Torres et al. 2014), indicating good precision in controlling the causes of experimental variation. Broad-sense heritability values between traits ranged from 44.92 to 86.59%, including medium and high values. Estimates close to those found for the same yield traits are reported in other studies on soybeans (Pooprompan et al. 2006, Vieira et al. 2006, Kim et al. 2012). Higher heritability coefficient was observed for the traits NDM, NNM and W100S, indicating that the phenotypic selection efficiency can be high, since the heritability expresses the proportion of total variance, which is attributed to the genetic variation (Falconer and Mackay 1996). High heritability values for the same traits are also reported by Eskandari et al. (2013) and Palomeque et al. (2009).

Forty-eight pairs of microsatellite primers were evaluated in the segregating F_2 population. Segregation of each

Table 1. Analysis of variance, means, coefficients of variation, and genetic parameter for agronomic traits in F_3 generation of the cross between CS3032PTA276-1-5-2 and UFVS2012

Sources of variation	df	NDF	NDM	PHM	HFP	NNM	NPP	SWP	W100S	NSP
						QM				
Block	2	66.66	83.03	2413.27	33.90	8.62	2219.09	4450.41	105.18	23105.47
Treatment	210	6.47**	15.46**	358**	14.18**	5.50**	1025.57**	4135.42**	6.52**	4632.02**
F_3 Family	206	4.87**	8.83**	324.77**	14.04**	5.00**	967.78**	3919.46**	6.00**	4526.59**
Controls	3	85.86**	335.44**	2608.97**	24.22*	41*	3225.41**	14285.80**	23.65**	8891.22*
Fam vs Control	1	97.56**	421.57**	451.35*	12.04	2.47	6329.60**	18173.40*	61.54**	13572.77*
Residue	420	2.43	2.03	156.77	6.391	0.806	410.34	1702.93	0.80	2493.14
General mean		64.07	138.03	86.99	16.32	14.95	111.66	195.70	16.98	213.81
Family mean		64.12	138.15	87.1	16.34	14.95	112.10	196.43	17.02	214.45
Control mean		61.25	132.16	80.91	15.33	14.5	88.91	157.14	14.74	180.5
$\sigma^2 g$		0.81**	2.26**	55.99**	2.55	1.40**	185.81**	738.34**	1.73**	677.81**
CVg		1.40	1.08	8.59	9.77	7.90	12.15	13.83	7.73	12.14
Heritability (%)		49.96	77	51.72	54.5	83.88	57.6	56.55	86.59	44.92
CV (%)		2.43	1.03	14.39	15.48	6.00	18.14	21.08	5.28	23.35
CVg/CVe		0.57	1.05	0.59	0.63	1.31	0.67	0.65	1.46	0.52

*, ** Significant at 5 and 1% probability by the F test; NDF = number of days to flowering; NDM = number of days to maturity; PHM = plant height at maturity; HFP = height of the first pod; NNM = number of nodes at maturity; NPP = number of pods per plant; SWP = seed weight per plant (g); W100S = weight of one hundred seeds (g); and NSP = number of seeds per plant.

Table 2. Descriptive analysis of the agronomic traits in the F_3 generation, and means of the parents CS3032PTA276-1-5-2 (P1) and UFVS2012 (P2)

Traits	Unit	Population			Parents mean	
		Mean	SD	Min-Max	P1	P2
NDF	days	64.12	1.28	60.5-67.6	64.3	64.6
NDM	days	138.15	7.52	88.5-141.8	137.5	138.3
PHM	cm	87.1	11.53	53.4-147.04	102.3	76.3
HFP	cm	16.34	2.51	7.0-23.2	15	19.3
NNM	-	14.95	1.40	6.5-17.5	18	13
NPP	-	112.1	18.40	57.4-158.0	100.3	112.3
SWP	g	196.43	36.11	109.9-291.9	195.4	179.3
W100S	g	17.02	1.54	8.2-20.8	16.3	15
NSP	-	214.45	38.83	129.1-394.5	226	203.6

SD = standard deviation; Min/Max = minimum and maximum value of the trait; NDF = number of days to flowering; NDM = number of days to maturity; PHM = plant height at maturity; HFP = height of the first pod; NNM = number of nodes at maturity; NPP = number of pods per plant; SWP = seed weight per plant (g); W100S = weight of one hundred seeds (g); and NSP = number of seeds per plant.

marker was tested using the chi-square test, and three molecular markers presented distorted segregation (Satt352, Satt429, and Satt454). However, these molecular markers did not occur in any of the intervals with significant association with the studied traits. In the linkage analysis, it was obtained nine LG by the grouping of 25 microsatellite markers, representing part of the LG A1, B1, D1a, G, I, M, and O (Figure 1). The position of the molecular markers in each LG coincided with the consensus map of the species, with the exception of the marker Satt370, which was located in LG I, while in the reference map (Song et al. 2004), the same marker is located was the LG D1a. As expected, 23 molecular markers were not grouped in any of the LG, since the distances between the markers were greater than 30 cM, according to the consensus map published by Song et al. (2004). The grouping of the molecular markers used LOD=3 and r=0.30 as criteria, but for eight of the nine LG, grouping also occurred with LOD=5, which shows the consistency of the groups.

The association between trait and molecular markers was evaluated by means of linear regression, in which the significant effect indicates the existence of an association between a marker and a trait, due to genetic linkage between a marker and a QTL (Schuster and Cruz 2004). The associations detected in this method can be attributed to QTL of major effect and relatively distant from the marker, or to a QTL of minor effect and close to the marker. In total, 59 significant associations between markers and traits were observed in the single marker analysis using linear regression (Table 3).

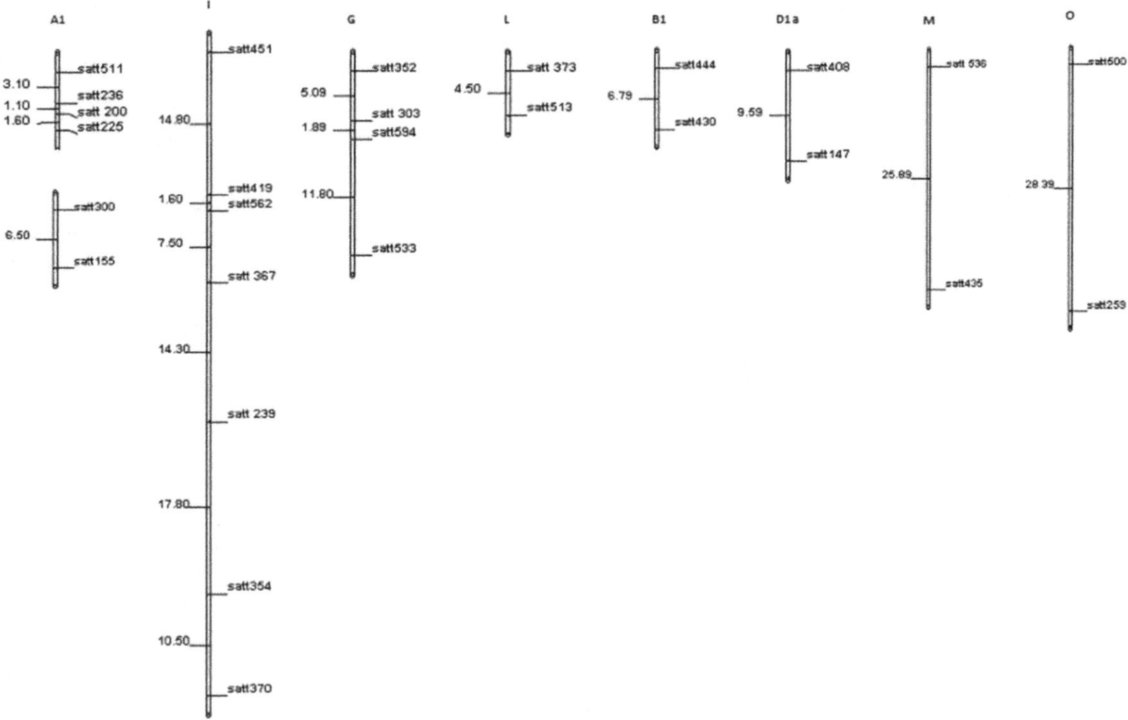

Figure 1. Linking Groups from the analysis of the F2 population with 48 microsatellite markers based on the criteria of LODmin = 3.0 and rmax = 0.30.

Thirty molecular markers presented association with at least one variable, and the number of significant markers for each variable ranged from 2 to 13. Fifteen molecular markers also presented association with more than one trait. QTL for the respective trait in the same linkage groups have been reported (Reinprecht et al. 2006, Chen et al. 2007, Gai et al. 2007, Zhang et al. 2010, Liu et al. 2011, Kim et al. 2012). Other QTL related to these traits can also be found in the *Glicyne max* (http://soybase.org/) genomic database.

QTL analysis in the intervals of the linkage map was carried out by composite interval mapping. In this method, additional markers are included as cofactors in the analysis, increasing the accuracy of the estimates of position and effect of the QTL. The inclusion of these markers reduces the effects caused by other QTL out of interval, which increases the power of the test, according to Schuster and Cruz (2004).

In the QTL analysis by composite interval mapping, it was detected five QTL for four of the nine studied traits (Table 4). Two QTL were identified for SWP, in the linkage groups A1 and D1a, which explained 12.32% and 9.03% of the total phenotypic variation, respectively (Table 4). Another QTL associated with W100S was mapped in LG I. This QTL located in the region of the locus Satt239 explained 13.47% of the variation. Two other QTL associated with NPP and NSP were mapped between the markers Satt300 and Satt155 in LG A1, in the same position of the QTL mapped for SWP. The variation in NPP and NSP explained by the QTL was of 9.43 and 7.19%, respectively. Probably, these QTL may be very close in the region of the marker Satt300, or alternatively, they can be a single QTL associated simultaneously to three variables, according to the hypothesis of pleiotropic effect.

QTL related to yield in LG A1 and I have been reported by other authors. In the same region of the molecular marker Satt300, in LG A1, Chen et al. (2007) mapped one QTL, which explained 12.56% of the variation in seeds weight per plant, with estimated additive effect of 7.1 g. Another QTL associated with yield in the region of the same marker was mapped by Guzman et al. (2007), which explained 18% of the variation in yield, with estimated additive effect of 60 kg ha^{-1}. Another QTL for yield in the same region was reported by Palomeque et al. (2009).

Table 3. QTL analysis by single marker for agronomic traits from an $F_{2:3}$ population derivative from a cross between CS3035PTA276-1-5-2 and UFVS2012

Marker	Prob (F)	Mean (2)	Mean (1)	Mean (0)	R^2 (%)	LG
			NDF			
Satt191	0.033*	138.164	137.447	134.628	3.386	G
Satt594	0.038*	138.518	137.505	138.947	12.785	G
R8 NDM						
Satt191	0.033*	138.164	137.447	134.628	3.386	G
Satt594	0.038*	138.518	137.505	138.947	12.785	G
			PHM			
Satt191	0.046*	87.940	86.940	82.831	3.054	G
Satt436	0.003**	89.885	85.819	82.155	5.745	D1a
Satt408	0.0192*	89.139	87.278	82.81	4.823	D1a
Satt454	0.0494*	79.569	86.501	86.813	6.011	A1
Satt354	0.048	85.66	86.066	93.301	6.001	I
			HFP			
Satt263	0.044*	15.643	15.766	16.799	3.191	E
Satt367	0.002**	16.733	15.448	16.847	6.7	I
Satt536	0.004**	17.090	15.483	16.059	6.7	M
Satt429	0.001**	14.260	16.674	16.129	17.109	A2
Satt300	0.023*	15.290	16.245	16.274	3.943	A1
Satt436	0.028*	15.423	15.978	16.768	3.637	D1a
Satt512	0.031*	15.863	15.371	16.983	6.27	-
Satt562	0.003**	16.993	15.506	16.260	5.861	I
Satt419	0.022*	16.767	15.573	16.148	3.903	I
Satt354	0.022*	16.555	15.679	17.349	7.076	I
			NNM			
Satt373	0.0**	15.255	14.999	14.246	7.724	L
Satt367	0.033*	14.411	14.997	15.115	3.611	I
Satt513	0.001**	15.822	15.128	14.255	19.156	L
Satt436	0.026*	15.101	14.968	14.400	3.712	D1a
Satt454	0.016*	14.364	15.136	14.912	4.54	A1
Satt562	0.003**	14.286	15.069	15.077	5.795	I
Satt451	0.0292*	14.347	15.085	15.038	4.954	I
			NPP			
Satt251	0.005**	107.456	110.795	118.983	5.496	B1
Satt536	0.016*	108.958	116.464	108.494	4.296	M
Satt529	0.049*	110.975	109.600	117.381	2.987	J
Satt300	0.005**	114.957	113.039	104.574	5.467	A1
Satt301	0.009**	111.988	105.140	96.329	11.023	D2
Satt454	0.001**	102.027	119.254	119.401	13.217	A1
Satt155	0.004**	113.941	112.883	103.422	5.515	A1
			SWP			
Satt263	0.033*	191.433	203.992	189.669	3.468	E
Satt529	0.038*	197.958	190.766	206.762	3.232	J
Satt500	0.009**	177.315	200.872	209.133	10.942	O
Satt429	0.043*	205.774	178.528	187.373	8.503	A2
Satt300	0.0**	205.464	200.736	178.380	9.759	A1
Satt301	0.039*	197.070	183.369	170.175	7.699	D2
Satt454	0.0002**	175.282	211.663	214.926	15.87	A1
Satt562	0.009**	182.372	201.088	199.987	4.833	I
Satt147	0.045*	199.828	202.437	187.526	3.313	D1a
Satt155	0.0**	201.599	201.621	177.287	8.32	A1
			W100S			
Satt154	0.0**	17.420	17.164	15.858	11.836	D2
Satt263	0.0**	17.553	17.077	16.173	10.782	E
Satt367	0.002**	16.349	17.257	17.190	6.642	I
Satt239	0.0**	16.239	17.169	17.507	9.496	I
Satt536	0.034*	16.674	17.299	16.746	3.557	M
Satt303	0.037*	16.614	17.244	17.213	3.742	G
Satt529	0.041*	16.451	17.106	17.246	3.157	J
Satt444	0.0387*	16.669	16.684	17.278	3.511	B1
Satt191	0.011*	16.641	16.851	17.483	4.473	G
Satt573	0.002**	16.646	16.802	17.642	6.571	E
Satt562	0.009**	16.401	17.203	16.973	4.805	I
Satt419	0.004**	16.385	17.265	16.874	5.648	I
Satt370	0.001**	16.329	16.946	17.576	9.409	D1a
			NSP			
Satt300	0.013*	220.925	217.502	201.346	4.502	A1
Satt454	0.0009**	196.754	230.963	230.601	13.379	A1
Satt155	0.015*	218.578	217.510	199.305	4.276	A1

*,** Significant at 5 and 1% probability by the F test, respectively. (2) homozygous for parent P1 CS3035PTA276-1-5-2) (AA), (0) homozygous for the parent P2 (UFVS2012) (aa) and (1) respective heterozygous (Aa). NDF = number of days to flowering; NDM = days to maturity; PHM = plant height at maturity; HFP = height of the first pod; NNM = number of nodes at maturity; NPP = number of pods per plant; SWP = seed weight per plant (g); W100S = weight of one hundred seeds (g); and NSP = number of seeds per plant.

Table 4. QTL analysis by composite interval mapping for agronomic traits from an $F_{2:3}$ soybean population

Trait	LG	Interval/Region	LR	R^2 (%)	a	d
SWP	A1	Satt300-Satt155	20.65**	12.32	15.511	20.403
	D1a	Satt408-Satt147	13.40**	9.03	10.051	28.281
W100S	I	Satt367-Satt239	21.63**	13.47	-0.602	1.242
NPP	A1	Satt300-Satt155	13.01**	9.43	7.167	7.292
NSP	A1	Satt300-Satt155	11.08*	7.19	13.38	19.387

LG = linking group according to Song et al. (2004); R^2 = proportion of phenotypic variation explained by the QTL; a = estimated additive effect for the QTL; d = estimated dominance effect for the QTL; *,** LR values that exceed the cut-off critical values at 5% and 1% probability, respectively. Critical LR values were determined by 1000 permutations. SWP = seed weight per plant (g); W100S = weight of one hundred seeds (g); NPP, number of pods per plant; and NSP = number of seeds per plant.

In LG I, Sebolt et al. (2000) mapped one QTL related to grain yield in two environments in the region of the marker Satt127, at approximately 2 cM from the interval of the QTL mapped in LG I in the present study, according to the reference map (Song et al. 2004). Moreover, from the combined data of five environments, Csanádi et al. (2001) mapped one QTL for the weight of one thousand seeds in the region of the marker Satt562, at approximately 5 cM from the same interval. Reinprechet et al. (2006) also observed an association of the molecular marker Satt367 with the weight of one hundred seeds in three environments. In addition, another marker in the LG I located at approximately 18 cM from Satt367, Satt354, was identified in two of the environments, according to the author. Another QTL in the LG I related to grain yield (g plant^{-1}) is reported by Du et al. (2009) between the molecular markers Satt102 and Sat419, at approximately 5 cM from the QTL described for W100S in the present study. Palomeque et al. (2009) and Palomeque et al. (2010) also found another marker in the LG I associated with grain yield (kg ha^{-1}) in multiple environments (Satt162). However, this marker is located in another region of LG I. In addition, another molecular marker related to yield was found in LG D1a, (Sat036), as reported by Orf et al. (1999), explaining 6% of the variation in seed weight (mg seed^{-1}). This marker is found at approximately 30 cM from the interval of the QTL mapped in this study.

In the present study, it was identified several molecular markers associated with nine agronomic traits, evidencing a large number of QTL involved in the control of the traits, and complexity in the genetic base related to grain yield in soybeans. Although several QLT related to grain yield have been described in the scientific literature, maybe only a few of them have already been confirmed (Kassem et al. 2006, Bernardo 2008, Sebastian et al. 2010). QTL, which have already been selected in breeding programs and/or transferred to new cultivars are rare (Ainsworth et al. 2012). Thus, further studies are still necessary, so that additive QTL was manipulated by MAS.

REFERENCES

Ainsworth EA, Yendrek CR, Skoneczka JA and Long SP (2012) Accelerating yield potential in soybean: potential targets for biotechnological improvement. **Plant, cell & environment 35**: 38-52.

Barbosa MC, Braccini AL, Scapim CA, Albrecht LP, Piccinin GG and Zucareli C (2013) Desempenho agronômico e componentes da produção de cultivares de soja em duas épocas de semeadura no arenito Caiuá. **Semina: Ciências Agrárias 34**: 945-960.

Bernardo R (2008) Molecular markers and selection for complex traits in plants: Learning from the last 20 years. **Crop Science 48**: 1649-1664.

Carter Jr TE, Nelson RL, Sneller CH and Zhanglin C (2004) Genetic diversity in soybean. In Boerma HR and Specht JE (eds) **Soybeans: improvement, production, and uses.** 3rd edn, American Society of Agronomy, Madison, p. 303-416.

Chen Q, Zhang Z, Liu C, Xin D, Qiu H, Shan D, Shan C and Hu G (2007) QTL Analysis of major agronomic traits in soybean. **Agricultural Sciences in China 6**: 399-405.

CONAB – Companhia Nacional de Abastecimento (2015) Acompanhamento da safra brasileira/grãos. V.2 - Safra 2014/15. N.12 - Décimo segundo levantamento, Setembro/2015. 136p.

Cregan PB, Jarvik T, Bush AL, Shoemaker RC, Lark KG, Kahler AL, Kaya N, Van Toai TT, Lohnes DG, Chung J and Specht JE (1999) An integrated genetic linkage map of the soybean genome. **Crop Science 39**: 1464-1490.

Cruz CD (2013) GENES - a software package for analysis in experimental statistics and quantitative genetics. **Acta Scientiarum Agronomy 35**: 271-276.

Csanádi G, Vollmann J, Stift G and Lelley T (2001) Seed quality QTLs identified in a molecular map of early maturing soybean. **Theoretical and Applied Genetics 103**: 912-919.

Doerge RW (1996) Constructing genetic maps by rapid chain delineation. **Journal of Quantitative Trait Loci 2**: 121-132.

Doyle JJ and Doyle JL (1990) A rapid DNA isolation procedure from small quantities of fresh leaf tissues. **Phytochemical Bulletin 19**: 11-15.

Du W, Wang M, Fu S and Yu D (2009) Mapping QTLs for seed yield and drought susceptibility index in soybean (*Glycine max* Merril L.) across different environments. **Journal of Genetics and Genomics**

36: 721-731.

Eskandari M, Cober ER and Rajcan I (2013) Genetic control of soybean seed oil: II. QTL and genes that increase oil concentration without decreasing protein or with increased seed yield. **Theoretical and Applied Genetics 126**: 1677-1687.

Falconer DS and Mackay TFC (1996) **Introduction to quantitative genetics**. Longman, Edinburgh, 464p.

Fox CM, Cary TR, Nelson RL and Diers BW (2015) Confirmation of a seed yield QTL in soybean. **Crop Science 55**: 992-998.

Gai J, Wang Y, Wu X and Chen S (2007) A comparative study on segregation analysis and QTL mapping of quantitative traits in plants with a case in soybean. **Frontiers of Agriculture in China 1**: 1-7.

Guzman PS, Diers BW, Neece DJ, St Martin SK, Leroy AR, Grau CR and Nelson RL (2007) QTL associated with yield in three backcross-derived populations of soybean. **Crop Science 47**: 111-122.

Han Y, Li D, Zhu D, Li H, Li X, Teng W and Li W (2012) QTL analysis of soybean seed weight across multi-genetic backgrounds and environments. **Theoretical and Applied Genetics 125**: 671-683.

Hu Z, Zhang H, Kan G, Ma D, Zhang D, Shi G and Yu D (2013) Determination of the genetic architecture of seed size and shape via linkage and association analysis in soybean (*Glycine max* L. Merr.). **Genetica 141**: 247-254.

Hyten DL, Song Q, Zhu Y, Choi IY, Nelson RL, Costa JM, Specht JE, Shoemaker RC and Cregan PB (2006) Impacts of genetic bottlenecks on soybean genome diversity. **Proceedings of the National Academy of Science - USA 103**: 16666-16671.

Jansen RC (1993) Interval mapping of multiple quantitative trait loci. **Genetics 135**: 205-211.

Kassem MA, Shultz J, Meksem K, Cho Y, Wood AJ, Iqbal MJ and Lightfoot DA (2006) An updated 'Essex'by 'Forrest'linkage map and first composite interval map of QTL underlying six soybean traits. **Theoretical and Applied Genetics 113**: 1015-1026.

Kato S, Sayama T, Fujii K, Yumoto S, Kono Y, Hwang TY and Ishimoto M (2014) A major and stable QTL associated with seed weight in soybean across multiple environments and genetic backgrounds. **Theoretical and Applied Genetics 127**: 1365-1374.

Kim KS, Diers BW, Hyten DL, Mian MAR, Shannon JG and Nelson RL (2012) Identification of positive yield QTL alleles from exotic soybean germplasm in two backcross populations. **Theoretical and Applied Genetics 125**: 1353-1369.

Liu W, Kim MY, Van K, Lee YH, Li H, Liu X and Lee SH (2011) QTL identification of yield-related traits and their association with flowering and maturity in soybean. **Journal of Crop Science and Biotechnology 14**: 65-70.

Liu YL, Li YH, Reif JC, Mette MF, Liu ZX, Liu B and Qiu LJ (2013) Identification of quantitative trait loci underlying plant height and seed weight in soybean. **The Plant Genome 6**: 1-11.

Orf JH, Chase K, Jarvik T, Mansur LM, Cregan PB, Adler FR and Lark KG (1999) Genetics of soybean agronomic traits: comparison of three related recombinant inbred populations. **Crop Science 39**: 1642-1651.

Palomeque L, Li-Jun L, Li W, Hedges B, Cober ER and Rajcan I (2009) QTL in mega-environments: I. Universal and specific seed yield QTL detected in a population derived from a cross of high-yielding adapted x high-yielding exotic soybean lines. **Theoretical and Applied Genetics 119**: 417-427.

Palomeque L, Liu LJ, Li W, Hedges BR, Cober ER, Smid MP and Rajcan I (2010) Validation of mega-environment universal and specific QTL associated with seed yield and agronomic traits in soybeans. **Theoretical and Applied Genetics 120**: 997-1003.

Pathan SM, Vuong T, Clark K, Lee JD, Shannon JG, Roberts CA and Sleper DA (2013) Genetic mapping and confirmation of quantitative trait loci for seed protein and oil contents and seed weight in soybean. **Crop Science 53**: 765-774.

Pires LPM, Peluzio JM, Cancellier LL, Ribeiro GR, Colombo GA and Afférri FS (2012) Desempenho de genótipos de soja, cultivados na região centro-sul do estado do Tocantins, safra 2009/2010. **Bioscience Journal 28**: 214-223.

Pooprompan P, Wasee S, Toojinda T, Abe J, Chanprame S and Srinives P (2006) Molecular marker analysis of days to flowering in vegetable soybean (*Glycine max* (L.) Merrill). **Kasetsart Journal 40**: 573-581.

Popovic V, Jaksic S, Glamoclija D, Grahovac N, Djekic V and Stefanovic VM (2012) Variability and correlations between soybean yield and quality components. **Romanian Agricultural Research 29**: 131-138.

Reinprecht Y, Poysa VW, Yu K, Rajcan I, Ablett GR and Pauls KP (2006) Seed and agronomic QTL in low linolenic acid, lipoxygenase-free soybean (*Glycine max* (L.) Merrill) germplasm. **Genome 49**: 1510-1527.

Rocha RS, Silva JAL, Neves JA, Sediyama T and Teixeira RC (2012) Desempenho agronômico de variedades e linhagens de soja em condições de baixa latitude em Teresina-PI1. **Revista Ciência Agronômica 43**: 154-162.

Schuster I and Cruz CD (2004) **Estatística genômica aplicada a populações derivadas de cruzamentos controlados**. Editora UFV, Viçosa, 568p.

Sebastian SA, Streit LG, Stephens PA, Thompson JA, Hedges BR, Fabrizius MA, Soper JF, Schmidt DH, Kallem RL, Hinds MA, Feng L and Hoeck JA (2010) Context-specific marker-assisted selection for improved grain yield in elite soybean populations. **Crop Science 50**: 1196-1206.

Sebolt AM, Shoemaker RC and Diers BW (2000) Analysis of a quantitative trait locus allele from wild soybean that increases seed protein. **Crop Science 40**: 1438-1444.

Song QJ, Marek LF, Shoemaker RC, Lark KG, Concibido VC, Delannay X, Specht JE and Cregan PB (2004) A new integrated genetic linkage map of soybean. **Theoretical and Applied Genetics 109**: 122-128.

Sun YN, Pan JB, Shi XL, Du XY, Wu Q, Qi ZM and Chen QS (2012) Multi-environment mapping and meta-analysis of 100-seed weight in soybean. **Molecular Biology Reports 39**: 9435-9443.

Torres FE, Silva EC, Teodoro PE (2014) Desempenho de genótipos de soja nas condições edafoclimáticas do ecótono Cerrado-Pantanal. **Interações 15**: 71-78.

Vieira AJD, Oliveira DA, Soares TCB, Schuster I, Piovesan ND, Martínez CA, Barros EG and Moreira MA (2006) Use of the QTL approach to the study of soybean trait relationships in two populations of recombinant inbred lines at the F_7 and F_8 generations. **Brazilian Journal of Plant Physiology 18:** 281-290.

Xu Y, Li HN, Li GJ, Wang X, Cheng LG and Zhang YM (2011) Mapping quantitative trait loci for seed size traits in soybean (*Glycine max* L. Merr.). **Theoretical and Applied Genetics 122**: 581-594.

Yesudas CR, Bashir R, Geisler MB and Lightfoot DA (2013) Identification of germplasm with stacked QTL underlying seed traits in an inbred soybean population from cultivars Essex and Forrest. **Molecular Breeding 31**: 693-703.

Zeng ZB (1994) Precision mapping of quantitative trait loci. **Genetics 136**: 1457-1468.

Zhang D, Cheng H, Wang H, Zhang H, Liu C and Yu D (2010) Identification of genomic regions determining flower and pod numbers development in soybean (*Glycine max* L.). **Journal of Genetics and Genomics 37**: 545-556.

14

Cytoembryological evaluation, meiotic behavior and pollen viability of *Paspalum notatum* tetraploidized plants

Karine Cristina Krycki[1*], Carine Simioni[1] and Miguel Dall'Agnol[1]

Abstract: *This study evaluated the mode of reproduction, the meiotic behavior and the pollen viability of three tetraploid plants (2n=4x=40) originated from somatic chromosome duplication of Paspalum notatum plants. The plant WKS 3 changed the mode of reproduction after duplication and became apomictic. The plants WKS 63 and WKS 92 confirmed sexual mode of reproduction identical to that of the original genotype. The analyzed plants presented meiotic abnormalities related to tetraploidy, and the chromosome pairing were variable, but it did not hinder the meiotic products, which were characterized by regular tetrads and satisfactory pollen fertility, ranging from 88.7 to 95.7%. Results show that all plants are meiotically stable and that they can be used in intraspecific crosses in the breeding program of Paspalum notatum.*

Key words: *Chromosome duplication, cytogenetic analysis, genetic breeding, intraspecific crosses.*

***Corresponding author:**
E-mail: carine.simioni@ufrgs.br

[1] Universidade Federal do Rio Grande do Sul (UFRGS), Departamento de Plantas Forrageiras e Agrometeorologia, Avenida Bento Gonçalves, 7712, 91.501-970, Porto Alegre, RS, Brazil

INTRODUCTION

The accelerated degradation of natural pastures of the state of Rio Grande do Sul has led to the loss of genetic diversity of forage species and of quality forage supply to cattle in the state (Macedo 2009). Among the several forage species that form native pastures, *Paspalum notatum* Flugge stands out for having excellent forage value and for being present in all natural pastures of the state (Nabinger and Dall'Agnol 2008). The breeding of this species is an alternative to the use of old and/or exotic varieties, with the search of selected materials; also, they are eligible for registration at the Ministry of Agriculture, Livestock and Supply (MAPA) and subsequent seed commercialization.

In the *Paspalum* genus, there is close correlation between ploidy level and mode of reproduction; diploidy is correlated with sexual reproduction, and allogamy and tetraploidy are correlated with apomixis (Quarin 1992). Apomictic genotypes preserve the genetic diversity, which is made available, and enables crosses with sexual genotypes. The new gene combinations allow selecting individuals that solve problems related to these species, since they are more adapted to different environments, consequently mitigating risk where biotic agents, especially pests and diseases threaten the development and the production. The possibility of artificial chromosome duplication of sexual diploid plants from natural populations of variety Pensacola of *P. notatum* and their use in intraspecific crosses schemes in breeding programs (Burton and Forbes 1961) makes it possible to generate superior genotypes of apomictic reproduction. Therefore, it is possible to protect the developed cultivar.

Weiler et al. (2015) artificially duplicated chromosomes of three individuals of *P. notatum* (Pensacola bahiagrass) by immersing flower buds and seeds for different exposure time in various concentrations of colchicine, which is an antimitotic agent responsible for chromosome duplication. The three duplicated plants, WKS 3, WKS 63, and WKS 92, confirmed the tetraploid level of ploidy and the complete euploidy by the root tip analysis and meiotic cells diakinesis. The objective of this study was to evaluate these duplicated plants regarding the mode of reproduction, the meiotic behavior, and the pollen viability in order to use them as parents in intraspecific hybridization schemes in the breeding program.

MATERIAL AND METHODS

This work was carried out in the Cytogenetics Laboratory of the Department of Forage Plants and Agrometeorology of the Department of Agronomy of the Federal University of Rio Grande do Sul. The three plants tetraploidized by Weiler et al. (2015) were evaluated, nominated WKS 3, WKS 63, and WKS 92, regarding the cytoembryological, cytogenetic and pollen viability analysis.

Cytoembryological analysis

Analyses of the mode of reproduction of the plants were carried out with inflorescences in anthesis. Flowers were dissected and fixed in FAA (95% ethanol: 40 mL; distilled water: 14 mL; 40% formalin: 3 mL; and glacial acetic acid: 3 mL) for 24 hours. After that, they were stored in 70% alcohol under refrigeration until ovaries extraction, which went through clearing process by means of a series of alcohol dehydration with methyl salicylate, following the protocol of Young et al. (1979), modified by Acuña et al. (2007), and were stored in methyl salicylate solution (100%) until the analysis in interference contrast optical microscope. At least 30 ovaries per plant were analyzed to determine the mode of reproduction.

Cytogenetic analysis

For the analysis of meiotic behavior, inflorescences of plants were collected at several development stages, and were fixed in an absolute ethanol solution: glacial acetic acid (3: 1) for 24 hours, transferred to 70% ethanol, and stored under refrigeration (Araújo et al. 2005, Dahmer et al. 2008). For the preparation of the slides, inflorescences were dissected, stained with 1% propionic carmine, and analyzed in optical microscope. It was sought to observe cells at different stages of meiotic division, as well as the arrangement of chromosomes. All meiotic abnormalities were considered. For the verification of chromosome pairing, analyses were carried out in at least 20 cells per plant at diakinesis and metaphase I stages (Dahmer et al. 2008, Simioni and Valle 2011).

Pollen viability analysis

Pollen grains viability was estimated in the anthers collected from inflorescences at mature stage, fixed in 3: 1 solution (absolute ethanol: glacial acetic acid), at room temperature for 24 hours, and stored in 70% alcohol until analysis. In the preparation of the slides, pollen grains were extracted from flowers, stained with 1% propionic carmine, and observed in optical microscope. Pollen grains were considered fertile when full and well stained, while those unstained or weakly stained, were considered sterile (not viable unviable) (Singh 1993). One thousand mature pollen grains were counted in four flowers per plant, following the protocol already established and widely used (Dahmer et al. 2008, Guerra et al. 2013).

RESULTS AND DISCUSSION

Mode of reproduction analysis

The duplicated plants WKS 63 and WKS 92 confirmed having sexual mode of reproduction (Table 1), with Polygonum type embryo sac: a single meiotic embryo sac, two polar nuclei, and a cluster of antipodal cells toward the chalazal (Figure 1a). WKS 3 presented modifications in its mode of reproduction after chromosome duplication, and became apomictic. This plant had ovaries with multiple aposporic embryo sacs, which are characterized by the egg cell, one or two synergids, a binucleated central cell, and absence of antipodes (Figure 1b). Quarin et al. (2001) recorded this phenomenon by analyzing the mode of reproduction of three *P. notatum* plants artificially duplicated; two of them had

Table 1. Number and percentage (%) of meiotic (S), apomictic (A), unidentifiable (U), shriveled (Sh), embryo sacs total analyzed ovaries (T), and mode of reproduction (MR) of the three *P. notatum* tetraploidized plants

Plant	S (%)	A (%)	U	Sh	T	MR
WKS 63	30 (75)	0	8	2	40	sexual
WKS 92	24 (80)	0	5	1	30	sexual
WKS 3	0	26 (87)	4	0	30	apomictic

facultative apomictic reproduction. Based on these results, the authors state that the apomixis gene is present at the diploid level; however, it is not expressed in the plant. The ploidy-dependence may occur at a locus that controls the apomixis by means of a secondary factor that requires higher dosage of alleles to affect the expression of major locus. It is likely that the expression of apomixis in this duplicated plant is a gene dosage effect. Simioni and Valle (2011) observed the confirmation of the sexual mode of reproduction in three plants obtained by somatic chromosome duplication of sexually reproducing diploid genotype of *Brachiaria decumbens*.

Analyses of the mode of reproduction of parents and progenies are fundamental in breeding programs aimed at enabling intraspecific hybridizations. For the choice of the parents, this evaluation allows identifying the plants that will be used as female parents (plants of sexual reproduction) and as male parents (apomictic plants).

Analysis of meiotic behavior

Chromosome associations were observed at the diakinesis and metaphase I stages, which are the best stages for visualization of chromosome configurations (Dahmer et al. 2008, Simioni and Valle 2011). The three plants confirmed tetraploidy (2n=4x=40), as described by Weiler et al. (2015). Chromosome pairing was typical of tetraploidy, with univalent, bivalent, trivalent and quadrivalent chromosome associations (Table 2). WKS 63 (Figure 2) presented mostly quadrivalent associations. According to Ramsey and Schemske (2002), genotypes

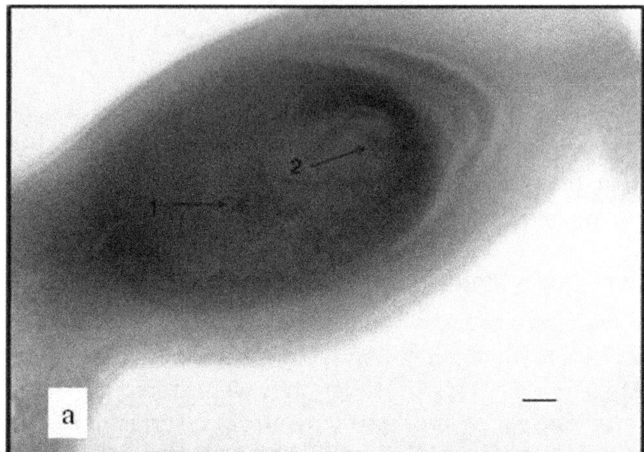

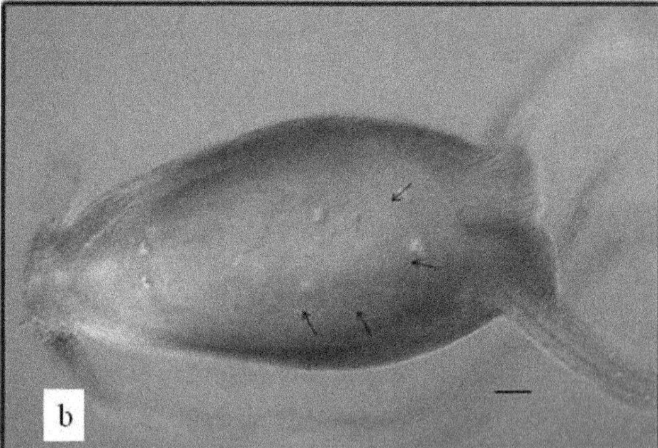

Figure 1. a) Cytoembryological aspect of a sexual ovary (duplicated plant WKS 63): antipodes (1) and polar nuclei (2); b) Cytoembryological aspect of apomictic ovary (duplicated plant WKS 3), multiple sacs (arrows). Scale: 10 μm.

Table 2. Meiotic chromosome configurations at diakinesis phase (prophase I) of three *P. notatum* tetraploidized plants

Plant	Chromosomic number	N. of analyzed cells	Mean n. of observed associations (per cell) (reach*)			
			I	II	III	IV
WKS 63	40	58	0	0.59	0	9.66
			0	(0-6)	0	(8-10)
WKS 92	40	31	2.23	15.26	0.16	1.71
			(0-34)	(3-20)	(0-2)	(0-4)
WKS 3	40	21	0	9.86	0.29	4.86
			0	(2-17)	(0-2)	(0-9)

* Minimum and maximum limits of associations observed in the total cells analyzed.

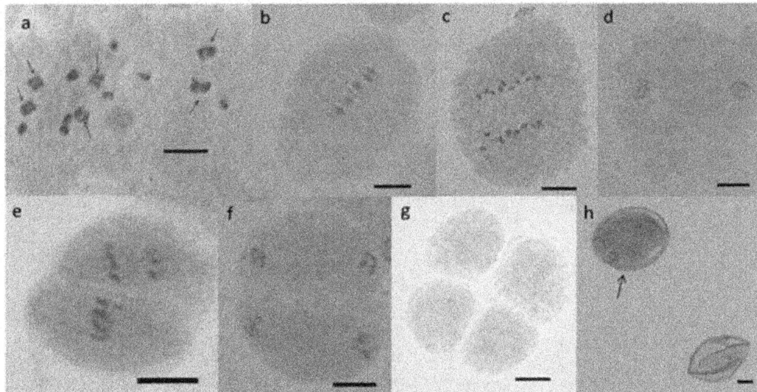

Figure 2. Meiotic aspects of the tetraplidized plant WKS 63. a) Diakinesis. It is observed the presence of quadrivalent associations (arrows). b) Normal metaphase I. c) Normal anaphase I. d) Normal telophase I. e) Anaphase II with asynchrony. f) Normal telophase II. g) Microsporocyte (Tetrad). h) Viable pollen grain (stained; arrow), and unviable pollen grain (non-stained). Scale: 10 µm.

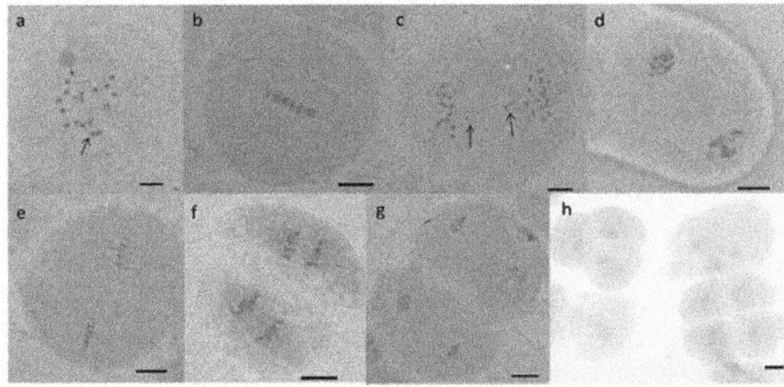

Figure 3. Meiotic aspects of the tetraploidized plant WKS 92. a) Diakinesis. It was observed the presence of one quadrivalent association (arrow). b) Normal metaphase. c) Anaphase I with the presence of laggard chromosomes (arrows). d) Normal telophase I. e) Normal metaphase II. f) Normal anaphase II. g) Normal telophase II. h) Microsporocyte (triad and tetrad). Scale: 10 µm.

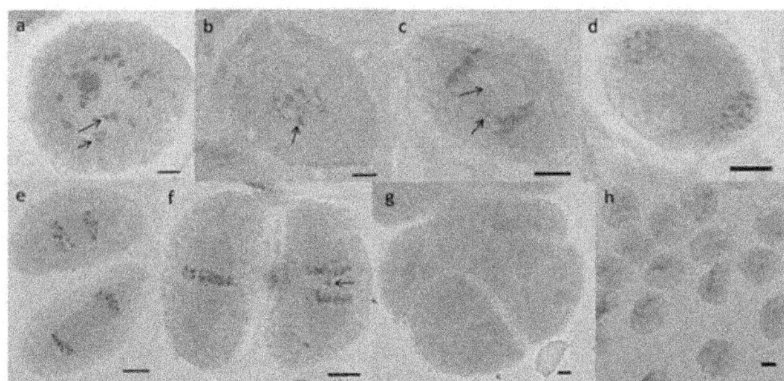

Figure 4. Meiotic aspects of the tetraploidized plant WKS 3. a) Diakinesis. It was observed the presence of two quadrivalents associations (arrows). b) Metaphase I. Presence of chromosomes in early ascension (arrow). c) Anaphase I with the presence of laggard chromosomes (arrows). d) Normal telophase I. e) Normal anaphase II. f) Anaphase II with asynchrony. It was observed the presence of laggard chromosome (arrow). g) Microsporocyte (polyads with micronuclei). h) Viable pollen grains (stained). Scale: 10 µm.

with polysomic inheritance have tendency for multivalent formation.

The plants WKS 92 (Figure 3) and WKS 3 (Figure 4) presented less meiotic abnormalities when compared with WKS 63, with most of the chromosomes in bivalent and sporadic quadrivalent associations, showing tendency for regularization of the chromosome pairing and genetic control of the pairing in these newly formed tetraploids. According to Dahmer et al. (2008), in the case of apomictic ecotypes, there is the need of meiotic regularity, since they are pseudogamic, so that to it is possible to ensure sufficient pollen fertility to form the endosperm. This is the case of WKS 3 in the present experiment.

Several studies on tetraploid species and accessions have registered regular meiosis and wide variability of the emergence of several uni, bi, tri and tetravalent chromosome associations. The predominance of bivalent associations were observed in 36 (Dahmer et al. 2008) and five (Moraes-Fernandes et al. 1973) *P. notatum* accessions, 24 accessions of different *Paspalum* species (Pagliarini et al. 2001), 53 *Paspalum nicorae* accessions (Reis et al. 2008), three polyploidized plants of *Brachiaria decumbens* (Simioni and Valle 2011), six accessions of different *Brachiaria* species (Araújo et al. 2005), one *Brachiaria ruziziensis* accession (Risso-Pascotto et al. 2005) and a *Paspalum durifolium* accession (Quarin 1994). In contrast, other authors reported accessions in which most of the associations were uni- or multivalent: *B. brizantha*, *B. decumbens* and *B. ruziziensis* (Valle and Savidan 1996), *B. decumbens* cv. Basilisk (Junqueira-Filho et al. 2003), and *Panicum maximum* (Caetano et al. 2006, Pessim et al. 2010).

Table 3 shows a total of 4082 microsporocytes analyzed for meiotic behavior in the three duplicated plants in this experiment. In 442 cells, there were no irregularities from prophase I to the meiotic products, representing a mean percentage of meiotic abnormalities of 10.83% in the three plants. Tetraploid plants generally have meiotic abnormalities related to irregular chromosomes segregation, which generates genetically unbalanced microspores, and thus hinders the fertility of pollen grains (Pagliarini and Pozzobon 2004). Multiple associations in diakinesis and abnormalities related to irregular chromosome segregation (early ascension) were the most frequent abnormalities in both divisions in the three plants. It was also noted laggard chromosomes and asynchrony in the three plants. Few bridges and micronucleus were observed. The meiotic products were mostly normal, with few dyads, triads and polyads, which resulted in excellent pollen viability of these plants.

Table 3. Meiotic abnormalities recorded in the three *P. notatum* tetraploidy plants

Individual	Stage	N. of analyzed cells	N. of abnormal cells (%)	Main abnormalities (N. of cell and %)
WKS 63	Diakinesis	58	58(100)	Multiple associations: 58 (100)
	Metaphase I	502	69 (13.75)	Early ascension: 38 (7.57)
				Laggard: 31 (6.18)
	Anaphase I	89	19 (21.35)	Laggard: 13 (14.61)
				Bridges: 3 (3.37)
				Asynchrony: 3 (3.37)
	Telophase I	170	1 (0.59)	Laggard: 1 (0.59)
	Metaphase II	162	18 (11.11)	Early ascension: 4 (2.47)
				Laggard: 2 (1.23)
				Asynchrony: 12 (7.41)
	Anaphase II	34	14 (41.18)	Laggard: 1 (2.94)
				Bridges: 1 (2.94)
				Asynchrony: 12 (35.29)
	Telophase II	156	6 (3.85)	Asynchrony: 6 (3.85)
	Meiotic Products	279	10 (3.58)	Polyads: 1 (0.36)
				Triads: 9 (3.22)
	Total number of cells	1450	195 (13.45)	
WKS 92	Diakinesis	31	30 (96.78)	Multiple associations: 30 (96. 78)
	Metaphase I	585	38 (6.5)	Early ascension: 34 (5.81)
				Laggard: 4 (0.69)
	Anaphase I	136	11 (8.09)	Laggard: 10 (7.35)
				Bridges: 1 (0.74)
	Telophase I	70	4 (5.71)	Laggard: 2 (2.86)
				Micronucleus: 1 (1.43)
				Asynchrony: 1 (1.43)
	Metaphase II	115	9 (7.83)	Early ascension: 2 (1.74)
				Asynchrony: 7 (6.09)
	Anaphase II	33	7 (21.21)	Laggard: 3 (9.09)
				Asynchrony: 4 (12.12)
	Telaphase II	48	1 (2.08)	Asynchrony: 1 (2.08)
	Meiotic Products	812	9 (1.11)	Dyads: 2 (0.25)
				Triads: 7 (0.86)
	Total number of cells	1830	109 (5.96)	
WKS 3	Diakinesis	21	21 (100)	Multiple associations: 21 (100)
	Metaphase I	226	118 (52.21)	Laggard: 1 (0.44)
				Micronucleus: 1 (0.44)
	Anaphase I	72	7 (9.72)	Laggard: 5 (6.94)
				Bridges: 2 (2.78)
	Telophase I	91	2 (2.2)	Bridges: 2 (2.2)
	Metaphase II	28	1 (3.57)	Asynchrony: 1 (3.57)
	Anaphase II	88		
	Telophase II	56	4 (7.14)	Asynchrony: 4 (7.14)
	Meiotic Products	251	15 (5.98)	Polyads: 11 (4.38)
				Triads: 4 (1.59)
	Total number of cells	833	168 (20.17)	

Podio et al. (2012) analyzed five natural apomictic tetraploid accessions and three artificially induced sexual tetraploid of *P. notatum,* and found that, in the apomictics, 55.6% of the cells at anaphase I are normal, and in the sexual accessions, 70.3% of the cells at anaphase I are normal. Abnormalities were mostly laggard chromosomes, chromatin bridges, and the presence of micronuclei, which appeared in 44.3% of the apomictic and in 29.66% of the sexual plants. In telophase I, both apomictic and sexual accessions presented chromosomes clustered in the poles of the cells, as well as small size micronuclei, suggesting they were composed of chromosome fragments. It can be inferred that meiotic behavior is characteristic of each genotype, as previously reported by Stein et al. (2004) for the species.

Pessim et al. (2010) observed high meiotic stability in hybrid genotypes and parents of *P. maximum*, with abnormalities ranging from 6.7 to 14.2%, such as irregular chromosome segregation, chromosome stickiness, and absence of cytokinesis. However, they did not affect pollen viability.

Analyses of pollen viability

The three duplicated plants showed high pollen viability: WKS 3, WKS 63 and WKS 92 recorded 92.3%, 88.7% and 95.7% of stained pollen grains, respectively. Studies reported pollen viability of ecotypes and of native apomictic accessions of *P. notatum* in the state of Rio Grande do Sul: Dahmer et al (2008) found pollen viability ranging from 81.0 to 91.47% in the ecotype "Bagual", and of 86.0 to 98.0% in the ecotype "André da Rocha". Moraes-Fernandes et al. (1973) reported pollen fertility ranging from 0 to 84.3%; and Reis et al. (2008) found pollen viability ranging from 88.99 to 95.06% in 53 accessions, despite the numerous meiotic irregularities found. The high pollen viability of apomictic plants is expected, since seed formation occurs only if there is fertilization of the polar nuclei of the embryo sac by one of the gametic nuclei of pollen grain, due to pseudogamy, typical in species of *Paspalum* and *Brachiaria,* which present this mode of reproduction (Pagliarini and Pozzobon 2004).

Cytological analysis are important tools in breeding programs for the selection of compatible and fertile parents which do not present meiotic abnormalities that may hinder gametes viability (Simioni and Valle 2011). This work allowed observing satisfactory meiotic regularity of the three duplicated plants, and made them viable as parents in the breeding program.

Guerra et al. (2016) also recorded cytological stability of 35 cherry trees accessions (*Eugenia involucrata* DC) collected in the state of Rio Grande do Sul; the average of meiotic cells considered normal was 82.12% and the average pollen viability was 92.44%. The authors concluded that such native access can be used directly in commercial orchards, and also as male parents in directed crosses in breeding programs such as those presented in this work.

With the study of the mode of reproduction, it was possible to direct the plants for intraspecific crosses: WKS 3, of apomictic reproduction, was used as pollen donor, and WKS 63 and WKS and 92 were used as female parents in the hybridizations. In further stages of the program, the hybrid progeny will be evaluated in agronomic trials under field conditions, in order to select genotypes that meet the demands of increased forage yield and that can contribute to the preservation of southern fields, preventing their degradation. This work represents a progress for the development of the Brazilian southern farming, with the use of well-adapted materials, diversifying forage production.

ACKNOWLEDGMENTS

The authors thank SULPASTO, CNPq and CAPES for the financial support, and the Scientific Initiation students Miguel Godinho Verran and Marília Wieser Paz.

REFERENCES

Acuña CA, Blount AR, Quesenberry KH, Hanna WW and Kenworthy KE (2007) Reproductive characterization of bahiagrass germplasm. **Crop Science 47**: 1711-1717.

Araújo ACG, Nóbrega JM, Pozzobon MT and Carneiro VTC (2005) Evidence of sexuality in induced tetraploids of *Brachiaria brizantha* (Poaceae). **Euphytica 144**: 39-40.

Burton GW and Forbes I (1961) Cytology of diploids, natural and induced tetraploids, and intraspecies hybrids of bahiagrass,*Paspalum notatum* Flugge. **Crop Science 1**: 402-406.

Caetano CM, Bonfá BRCN and Canto MW (2006) Autotetraploidia e número cromossômico em uma cultivar de *Panicum maximum* Jacq (Gramineae/Poaceae). **Acta Agronomica 55**: 62-66.

Dahmer N, Schifino-Wittmann MT, Dall'Agnol M and Castro B (2008) Cytogenetic data for *Paspalum notatum* Flugge accessions. **Scientia**

Agricola 65: 381-388.

Guerra D, Schifino-Wittmann MT, Schwarz SF, Souza, PVD and Campos SS (2013) Reproductive characteristics of citrus rootstocks grown under greenhouse and field environments. **Crop Breeding and Applied Biotechnology 13:** 186-193.

Guerra D, De Souza PVD, Schwarz SF, Schifino-Wittmann MT, Werlang CA and Veit PA (2016) Genetic and cytological diversity in cherry tree accessions (*Eugenia involucrata* DC) in Rio Grande do Sul. **Crop Breeding and Applied Biotechnology 16:** 219-225.

Junqueira-Filho RG, Mendes-Bonato MB, Pagliarini MS, Bione NCP, Valle CB and Penteado MIO (2003) Absence of microspore polarity, symmetric divisions and pollen cell fate in *Brachiaria decumbens* (Gramineae). **Genome 46**: 83-86.

Macedo MCM (2009) Integração lavoura e pecuária: o estado da arte e inovações tecnológicas. **Revista Brasileira de Zootecnia 38**: 133-146.

Moraes-Fernandes MIB, Barreto IL and Salzano FM (1973) Cytogenetic, ecologic and morphologic studies in Brazilian forms of *Paspalum notatum*. **Canadian Journal of Genetics and Cytology 15**: 523-531.

Nabinger C and Dall'Agnol M (2008) Principais gramíneas nativas do RS: características gerais, distribuição e potencial forrageiro. In Dall'Agnol M, Nabinger C and Santos RJ (eds) **Anais do 3° simpósio de forrageiras e produção animal**. UFRGS, Porto Alegre, p.7-54.

Pagliarini MS and Pozzobon MT (2004) Meiose vegetal: um enfoque para a caracterização de germoplasma. In Peñaloza APS (ed) **Anais do II curso de citogenética aplicada a recursos genéticos vegetais**. EMBRAPA, Brasília, p. 24-41.

Pagliarini MS, Carraro LR, Freitas PM, Adamowski EV, Batista LA and Valls JFM (2001) Cytogenetic characterization of Brazilian *Paspalum* accessions. **Hereditas 135**: 27-34.

Pessim C, Pagliarini MS, Jank L, Kaneshima MAS and Mendes-Bonato MB (2010) Meiotic Behavior in *Panicum maximum* Jacq. (Poaceae: Panicoideae: Paniceae): hybrids and their genitors. **Acta Scientiarum 32**: 417-422.

Podio M, Siena LA, Hojsgaard D, Stein J, Quarin CL and Ortiz JPA (2012) Evaluation of meiotic abnormalities and pollen viability in aposporous and sexual tetraploid *Paspalum notatum* (Poaceae). **Plant Systematics and Evolution 298**: 1625-1633.

Quarin CL (1992) The nature of apomixis and its origin in Panicoid grasses. **Apomixis Newslett 5**: 7-15.

Quarin CL (1994) A Tetraploid cytotype of *Paspalum durifolium*: cytology, reproductive behavior and this relationship to diploid *P. intermedium*. **Hereditas 121**: 115-118.

Quarin CL, Espinoza F, Martinez EJ, Pessino SC and Bovo OA (2001) A rise of ploidy level induces the expression of apomixis in *Paspalum notatum*. **Sexual Plant Reprodution 13**: 243-249.

Ramsey J and Schemske DW (2002) Neopolyploid in flowering plants. **Annual Review of Ecology and Systematics 33**: 589-631.

Reis CAO, Schifino-Wittmann MT and Dall'agnol M (2008) Chromosome numbers, meiotic behavior and pollen fertility in a collection of *Paspalum nicorae* Parodi accessions. **Crop Breeding and Applied Biotechnology 8**: 212-218.

Risso-Pascotto C, Pagliarini MS and Valle CB (2005) Multiple spindle sand cellularization during microsporogenesis in an artificial induced tetraploid accession of *Brachiaria ruziziensis* (Gramineae). **Plant Cell Reports 23**: 522-527.

Simioni C and Valle CB (2011) Meiotic analysis in induced tetraploids of *Brachiaria decumbens* Stapf. **Crop Breeding and Applied Biotechnology 11**: 43-49.

Singh RJ (1993) **Plant cytogenetics**. CRC Press, Boca Ratton, 391p.

Stein J, Quarin CL, Martinez EJ, Pessino SC and Ortiz JPA (2004) Tetraploid races of *Paspalum notatum* show polysomic inheritance and preferential chromosome pairing around the apospory-controlling locus. **Theoretical and Applied Genetics 109**: 186-191.

Valle CB and Savidan YH (1996) Genetics, cytogenetics and reproductive biology of *Brachiaria*. In Miles JW, Maass BL and Valle CB (eds) ***Brachiaria*: biology, agronomy and improvement**. CIAT/EMBRAPA, Cali, p. 147-163.

Weiler RL, Krycki KC, Guerra D, Simioni C and Dall'Agnol M (2015) Chromosome doubling in *Paspalum notatum* var. *saure* (cultivar Pensacola). **Crop Breeding and Applied Biotechnology 15**: 106-111.

Young BA, Sherwood RT and Bashaw EC (1979) Cleared-pistil and thick-sectioning techniques for detecting aposporus apomixis in grasses. **Canadian Journal of Botany 57**: 1668-1672.

15

Tomato second cycle hybrids as a source of genetic variability for fruit quality traits

Pereira da Costa JH[1,2*], Rodríguez GR[1,2], Liberatti DR[2], Mahuad SL[2], Marchionni Basté E[2], Picardi LA[2,3], Zorzoli R[2,3] and Pratta GR[1,2]

***Corresponding author:**
E-mail: jpereira@unr.edu.ar

[1] IICAR-CONICET (Instituto de Investigaciones en Ciencias Agrarias de Rosario - Consejo Nacional de Investigaciones Científicas y Técnicas), Universidad Nacional de Rosario, Facultad de Ciencias Agrarias, Campo Experimental JF Villarino CC 14 (S2125ZAA), Zavalla, Santa Fe, Argentina2 Universidad Nacional de Rosario
3 IICAR-CONICET, CIUNR (Consejo de Investigaciones de la Universidad Nacional de Rosario), Universidad Nacional de Rosario

Abstract: *The objective of this study was to investigate the phenotypic and molecular variability in a F_2 generation derived from a SCH (Second Cycle Hybrid) in order to detect QTLs for some fruit traits of tomato. Genome coverage at different levels was achieved by three types of molecular markers (polypeptides, sequence-related amplified polymorphism-SRAP and amplified restriction fragment polymorphism - AFLP). Different degrees of polymorphism were detected by SRAP and AFLP at the DNA structure level and also by polypeptides at the DNA expression level. The first two markers, associated with phenotypic variation, detected QTLs involved in important agronomic traits such as fruit shelf life, soluble solids content, pH, and titratable acidity. New gene blocks originated by recombination during the first cycle of crossing were detected. This study confirmed that the observed phenotypic differences represent a new gene rearrangement and that these new gene blocks are responsible for the presence of the genetic variability detected for these traits.*

Keywords: *Recombinant inbred lines, Solanum lycopersicum, genetic recombination, gene blocks.*

INTRODUCTION

In the last 30 years, significant progress has been made in tomato breeding by replacing open-pollinated varieties by hybrids (Figueiredo et al. 2015). A physiological *plateau* was reached when both yield and fruit quality had been improved (Grandillo et al. 1999). However, the advent of genomics has transformed breeding strategies. As a result, the improvement of tomato cultivars is expected to continue in the future. Recombinant Inbred Lines (RILs) allow increase the genetic variability by recombination and chromosome rearrangements, introducing new gene blocks. One way to profit from these new rearrangements is to develop Second Cycle Hybrids (SCH). These genotypes are generated by hybridization among RILs (Hills et al. 2003, Ipsilandis et al. 2006). According to Pratta et al. (2003) and Pereira da Costa et al. (2009), *Solanum pimpinellifolium L.* could be an appropriate parental genotype to improve fruit quality traits. These authors found that its hybrids with cultivated tomato had longer fruit shelf life compared to commercial cultivars and also better fruit quality in terms of color, texture and flavor. Though *S. lycopersicum L.* and S. *pimpinellifolium* have significant phenotypic differences, only minor variation was found among them at the genomic level (about 0.6 %) (Tomato Genome Consortium 2012). It can be postulated that these differences are due to protein functions (Michael and Alba 2012) and consequently to transcriptome regulation. As Giovannoni (2004) pointed out, the fruit quality could be conditioned by the developmental stage of the fruits at harvest time.

Different techniques have generated large amounts of data of gene expression during the development of climacteric and non-climacteric tomato fruits (Aharoni and O'Connell 2002, Henniget al. 2004, Grimpletet al. 2005, Moyle et al. 2005, Terrier et al. 2005). Carotenoid pigment accumulation and fruit softening distinguish at least two ripening stages: mature green and red ripe (Rick 1978, Giovannoni 2004). Although the polymorphism of polypeptide profiles is relatively low, they have been successfully used as molecular markers in various species to characterize genotypes or biological processes (Castro et al. 2006, Rodríguez et al. 2011).

Sequence-related amplified polymorphism markers (SRAPs) are DNA markers that preferentially amplify the expressed genomic regions (Li and Quiros 2001). These markers have been previously used in different species (Ruiz et al. 2005, Cravero et al. 2007, Mahuad et al. 2013). Other commonly used DNA markers are AFLP (Amplified Fragment Length Polymorphism), uniformly distributed in the tomato genome (Saliba-Colombaniet al. 2001), with a high chromosome coverage. Therefore, they were used to generate various inter- and intraspecific maps as well as to assess the genetic diversity of tomato cultivars, for genotyping (fingerprinting) and to detect QTLs of different fruit traits in interspecific crosses (Zhang and Stommel 2001, Lecomte et al. 2004, Pratta et al. 2011). Due to the different nature of genome coverage (expressed regions vs. random regions) of these three types of molecular markers (polypeptides and SRAP vs. AFLP), they could be interesting molecular tools to detect QTLs for fruit quality traits. As pointed out, the genetic recombination that occurs during the generation of the SCHs would produce genetic variability for these traits. Consequently, the aim of this study was to assess the amount of genetic variability at the phenotypic and molecular levels in the F_2 generation derived from a SCH and also to detect QTLs for some fruit quality traits. In addition, we demonstrated that new gene blocks were originated by recombination during the first cycle of crossing and that they were preserved even in the F_2 derived from the selfed SCH.

MATERIAL AND METHODS

Plant material

By antagonistic-divergent selection for weight and fruit shelf life, Rodríguez et al. (2006) obtained 17 RILs. These were derived from the F_2 generation of an interspecific cross between LA722 of *S. pimpinellifolium* and the Argentinean cultivar Caimanta of *S. lycopersicum* (first-cycle parental genotypes). The RILs ToUNR1, ToUNR8, ToUNR9, ToUNR15, and ToUNR18 were selected based on their combining ability. Then the five RILs were crossed according to a diallel cross (Model II without reciprocal crosses) to select the best SCH among the 10 possible F_1s to initiate a new breeding process. Due to the high values of specific combining ability for fruit weight and shelf life and the significant values of general combining ability of its parents, the hybrid ToUNR9 x ToUNR15 was selected to obtain the segregating F_2 population by selfing (Marchionni Basté et al. 2010). Fifteen plants of each parental genotype and the F_1 and 180 F_2 plants were assessed in a greenhouse in a completely randomized design.

Phenotypic analysis

The following fruit traits were evaluated: weight (W; g), diameter (D; cm), height (H; cm), shape (Sh; height/diameter), and shelf life (SL; in days from harvest to fruit softening or excessive wrinkling), following the methodology proposed by Garg et al. (2008). In addition, the soluble solids content (SS, °Brix), measured with a hand refractometer, pH, titratable acidity of the homogenized juice (TA, g), pericarp thickness (PT, mm), locule number (LN), and fruit firmness (F), on the equatorial plane in two opposite areas of the fruit measured with a durometer type Shore A (Durofel DFT100, 0.10 cm^2) were assessed.

Molecular analysis

Pericarp polypeptide profiles

Total pericarp proteins were extracted from fruits harvested at two ripening stages: Mature Green (MG, at least 10% of the fruit surface red) and Red Ripe (RR, 90% of the fruit surface red). Three independent pericarp samples (fruits from three different plants) per parental genotype and F_1, and one sample per F_2 plant were extracted from fruit harvested at each ripening stage. Proteins were extracted and resolved on SDS-PAGE following the protocol proposed by Rodríguez et al. (2008). Equal amounts of polypeptides (20 ug) were run for 1.5 h at a constant current of 35 mA on denaturing polyacrylamide gels (12 % v/v). Gels were stained with a 0.1% solution of Coomassie Brilliant Blue R-250 and destained with boiling water, scanned and analyzed with software Gelpro Analyzer 3.0.

DNA Markers

Young leaves were collected from the parents, F_1 and each F_2 plant and stored at -80 °C. The DNA was extracted with a commercial kit (Wizard ® Genomic DNA Purification Kit of Promega ®). PCR amplifications were performed in duplicate for parental genotypes and F_1.

Sequence-Related Amplified Polymorphism (SRAP)

Four of the primer combinations previously selected for their high level of polymorphisms (Mahuad et al. 2013) were used to characterize the parental genotypes, and F_1 and F_2 plants. The amplification protocol proposed by Li and Quiros (2001) was used with some modifications, as described by Mahuad et al. (2013). The bands were codified with a number (SI, SII, SIII, or SIV), according to the primer combination, followed by another number indicating the order of the band on the gels (e.g. SIV.26).

Amplified Fragment Length Polymorphism (AFLP)

Six AFLP primer combinations selected for their high level of polymorphism among parental genotypes (Liberatti et al. 2013) were used. The AFLP profiles were obtained following the amplification protocol proposed by Liberatti et al. (2013). The bands were codified with a capital letter indicating the primer combination followed by a number indicating the position of the band on the gels (e.g. X4)

Statistical Analysis

Normal distribution of each trait was tested by the Shapiro-Wilk test (Shapiro and Wilk 1965). The mean values of parental genotypes and F_1 were compared with a *t*-Student test (Snedecor 1964). Broad-sense heritability (H^2, $H^2 = \sigma^2_g / \sigma^2_p$), where σ^2_g is genetic variance and σ^2_p is phenotypic variance) was calculated using variance components from ANOVA for a completely randomized design, according to Falconer and Mackay (1996). All statistical analyses were performed with software InfoStat Version 1.0 (Di Rienzo et al. 2001). With a view to the dominant nature of all molecular markers used in this study (polypeptide profiles, SRAP and AFLP), a X^2 test (Snedecor 1964) was used to verify the expected segregation of 3:1 (presence: absence) in the F_2 generation. Only bands with Mendelian inheritance in parents and F_2 were taken into account for the QTL detection. The *de novo* bands (present in the first-cycle but not second-cycle parental genotypes and *vice versa*) were evaluated according to Liberatti et al. (2013). The association between molecular markers and fruit quality traits was analyzed by the single point method (single point analysis) (Tanksley1993). One-way ANOVA was performed with markers as classification variables. All markers with Mendelian inheritance were used to determine the existence of gene blocks. The LOD threshold for accepting a linkage group (LG) was 3.0, estimated by software JoinMap4.0, with the following settings: Rec = 0.40, LOD = 3.0 and Jump = 5. Recombination values were converted to genetic distances using the Kosambi (1943) mapping functions.

RESULTS AND DICUSSION

All phenotypic traits had normal distribution. The second cycle parental genotypes (ToUNR15 and ToUNR9) were significantly different for all traits except for fruit shelf life and firmness (Table 1). The results for weight and shelf life were consistent with those reported by Rodríguez et al. (2006), who characterized ToUNR9 as a genotype with short shelf life and low fruit weight and ToUNR15 as one with short shelf life and high fruit weight (Table 1). However, the frequency distribution in the F_2 (ToUNR15 x ToUNR9) generation for fruit shelf life (Figure 1e) shows wide phenotypic variation despite the similarity of the parental genotypes values. Even though H^2 was 0.38 ± 0.02 for shelf life, transgressive segregation was observed in the F_2 generation (Figure 1e), indicating that QTLs for shelf life, and other attributes with very high H^2 values (SS, pH and TA), could be detected in this population (see Table 1, shown in gray). When the polypeptide

Table 1. Mean values, standard error for each evaluated traits in second cycle parental genotypes (ToUNR15 and ToUNR9), F_1 and F_2 generation and broad-sense heritability (H^2)

Fruit traits	ToUNR15	ToUNR9	F_1 (15x9)	F_2 (15x9)	H^2
D (cm)	3.42 ± 0.09c	1.95 ± 0.06a	2.85 ± 0.04b	2.66 ± 0.03	0.63 ± 0.02***
H (cm)	2.90 ± 0.07c	1.75 ± 0.04a	2.53 ± 0.03b	2.39 ± 0.02	0.62 ± 0.06***
Sh	0.85 ± 0.02a	0.90 ± 0.01b	0.90 ± 0.01b	0.91 ± 0.01	0.24 ± 0.01***
W (g)	21.59 ± 1.04c	4.49 ± 0.32a	12.95 ± 0.32b	10.84 ± 0.36	0.63 ± 0.02***
SL (days)	15.13 ± 1.49a	15.39 ± 0.43a	16.37 ± 0.84a	16.22 ± 0.28	0.38 ± 0.02***
SS (°Brix)	6.84 ± 0.11b	5.07 ± 0.06a	7.20 ± 0.07b	8.38 ± 0.07	0.85 ± 0.02***
pH	4.84 ± 0.05b	4.40 ± 0.02a	4.74 ± 0.04ab	4.68 ± 0.01	0.73 ± 0.02***
TA (g)	0.31 ± 0.01a	0.43 ± 0.02b	0.31 ± 0.02a	0.41 ± 0.01	0.78 ± 0.06***
F	46.70 ± 0.96a	50.50 ± 0.61ab	51.76 ± 1.50b	51.80 ± 0.28	0.52 ± 0.03***
LN	4.42 ± 0.22c	2.00 ± 0.00a	2.95 ± 0.11b	2.60 ± 0.06	0.48 ± 0.02***
PT	0.39 ± 0.02c	0.19 ± 0.01a	0.29 ± 0.02b	0.25 ± 0.01	0.67 ± 0.02***

D: diameter. **H**: height. **Sh**: shape. **W**: weight. **SL**: shelf life. **SS**: soluble solids content. **TA**: titratable acidity. **F**: firmness. **LN**: locule number. **PT**: pericarp thickness. Different letters indicate significant differences ($p < 0.05$)
***$p < 0.001$

profiles of the second cycle parental genotypes ToUNR15 and ToUNR9 were compared, 27% and 8% of polymorphism were found at the mature-green and red-ripe stages, respectively. A high number of monomorphic bands were found at both ripening stages. Perhaps the selection process to obtain these RILs fixed the same alleles in some loci in the parental genotypes ToUNR15 and ToUNR9. No bands with Mendelian inheritance were found at the mature-green stage, but only one polypeptide of 75.6 kDa with Mendelian inheritance was found at the red-ripe stage, though without any association with a phenotypic trait. These results indicated that the phenotypic differences between parents have no correlation with the polymorphism in polypeptide profiles. These kinds of molecular markers were useless to detect associations among polypeptide profiles and phenotypic traits in this genetic background generated by crossing ToUNR15 with ToUNR9. Other authors (Rodríguez et al. 2011, Pereira da Costa et al. 2014) detected QTLs by polypeptide profiles in other genetic backgrounds. Probably the parental genotypes for those studies were genetically more divergent than the RILs studied here.

A different situation was observed when SRAP markers were analyzed. A total of 214 SRAP bands were detected between the second cycle parental genotypes and the F_2 generation. A mean of 53 bands was detected by primer combinations. The combinations 2 and 4 were the most polymorphic (percentages ranged from 29.7% to 22.4%). The mean number of bands detected by SRAP primer combinations was consistent with those reported by Mahuad et al. (2013), Pereira da Costa et al. (2014) and Ruiz et al. (2005). Nine bands with Mendelian inheritance were used for the association analysis. A total of 9 QTLs ($p < 0.01$) were found (Table 2A). The proportion of phenotypic variance explained by each QTL ranged between 5 and 10%. QTLs for soluble solids content, pH, pericarp thinness and fruit height were detected (Table 2A). These results support the hypothesis of Mahuad et al. (2013), suggesting that it would be possible to find SRAP markers associated with tomato fruit quality traits because a high consensus was observed between molecular and phenotypic diversity when a diallel design of 5 RILs and their hybrids was evaluated. The presence of fragment SIV.7 ($p <0.002$), originated from ToUNR15, produced a decrease in SS and pH. A high soluble solids content is a desirable trait in breeding programs for being associated with a better fruit flavor (Stevens et al. 1979). In all cases, the SRAP bands of ToUNR9 were associated with an increase in sugar content, while the SRAP bands inherited from ToUNR15 had the opposite effect (Table 2A). These results were expected according to the parental values for the traits. For fruit height, the effect of band SI.55 was consistent with the phenotype observed in second-cycle parental genotypes. Accordingly, ToUNR9 had fruits with smaller height and size (Table1) and contributed to one SRAP band that diminished the mean values of these traits. These results demonstrate that the selection by RILs development fixed alleles in ToUNR15 andToUNR9 with opposite effects on fruit size and soluble solids content.

A total of 735 AFLP bands were detected of which 496 (67%) were polymorphic. Mendelian inheritance was confirmed for 19 bands (4%) and 13 QTLs were detected by AFLP markers. One QTL (X4, $p=0.001$) for fruit shelf life decreased the mean value in the plants with this band. This QTL was contributed by To UNR15 (Table 2B).

Most associations detected by AFLP markers were related to the traits with greatest genetic variability in F_2, i.e. SS,

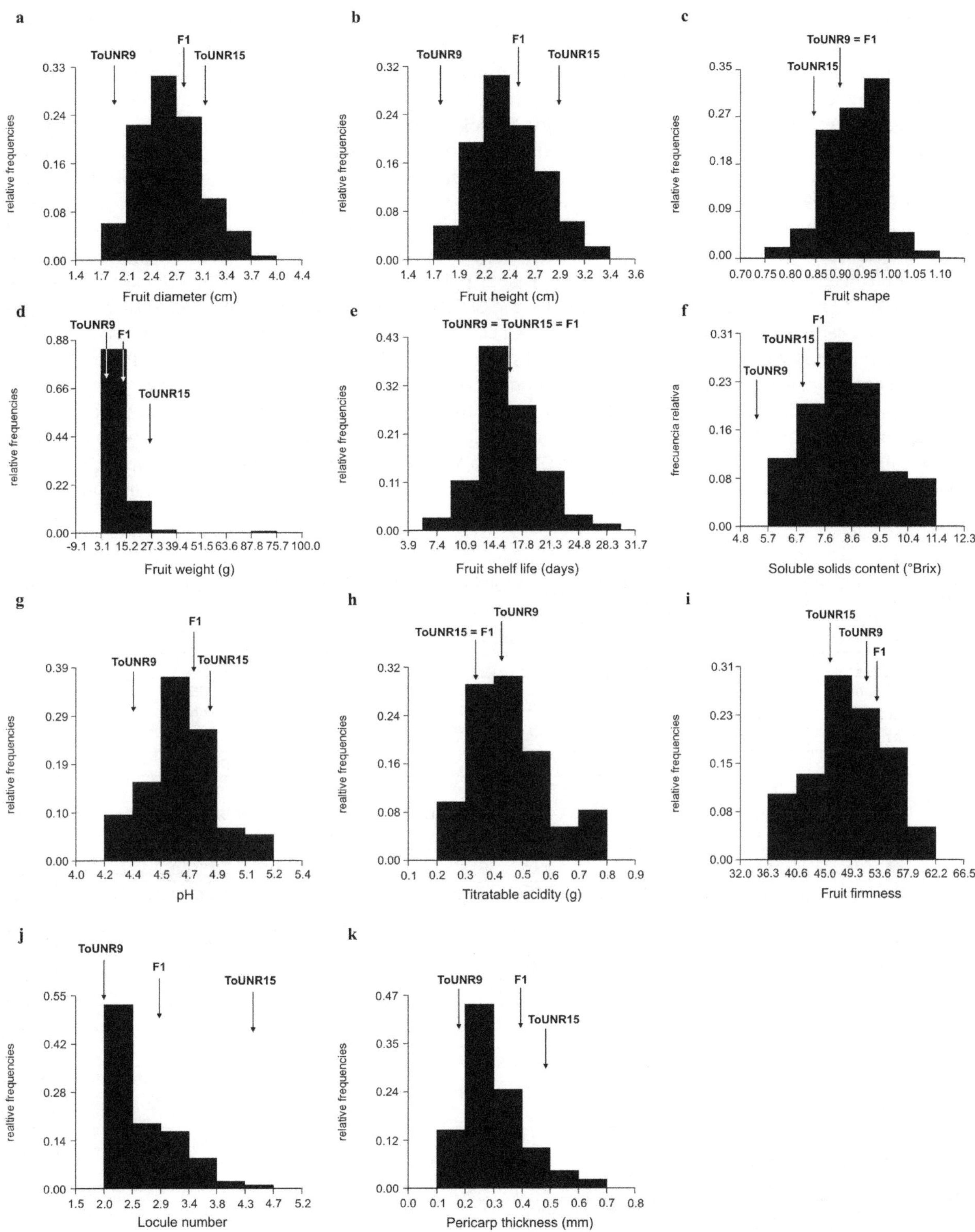

Figure 1. Frequency distributions for each phenotypic trait evaluated in F_2 (ToUNR15 x ToUNR9) generation. Arrows indicates mean values of each trait in parents (ToUNR15 and ToUNR9) and the F_1.

pH, and TA (Tables1 and 2). However, associations for pericarp thickness, firmness and locule number, with intermediate values of genetic variability, were also found. A similar result was found for fruit shelf life in spite of being a trait with low H^2. The occurrence of a band associated with more than one trait could be explained by pleiotropic effects or by a strong linkage between two or more QTLs underlying these different traits (Kearsey and Pooni 1996). Co-localized associations are frequent. In fact, co-localized association for soluble solid content and sugar content, fruit weight and locule number were also found by Xu et al. (2013) in an association mapping study for fruit quality traits of cultivated tomato and related species. Results from Tables 3 and 4 suggest that both types of DNA markers (SRAP and AFLP) allowed QTL detection. Polypeptide profiles were not useful to mark genetic regions that could be involved in the quantitative variation in this segregating generation derived from a SCH. They would probably be more valuable if the parental genotypes were more divergent. DNA markers are more effective because they cover a higher proportion of the genome and can distinguish a larger amount of molecular polymorphism. For same traits QTLs were found using both SRAP and AFLP. From an operational point of view, the SRAP markers were simpler, less expensive and quicker. They can detect the same extent of associations but with a considerably smaller number of total bands compared to AFLP markers (214 against 735, respectively). Gene blocks due to recombination events together with a tight linkage were found by genetic linkage analysis (Figure 2). The genomic regions belonging to the same linkage group (LG) were produced by fragments inherited from first cycle parental genotypes. Nevertheless, some bands assessed in ToUNR9 and ToUNR15 and conserved in generation F_1 (Mendelian inheritance in F_2), were defined as *de novo* bands (Liberatti et al. 2013). Even if they were absent or monomorphic in Caimanta and LA722 (first cycle parental genotypes), inheritance was stabilized in the second cycle parental genotypes as well as in the segregating generation (see examples in LG1, LG2, LG3, and LG4 in Figure 2).

Table 2. Detected associations in F_2 (ToUNR15 x ToUNR9) with SRAP (Sequence-related Amplified Polymorphism) markers (**A**) and AFLP (Amplified Fragment Length Polymorphism) markers (**B**)

A

Bands	Trait	Mean P	nP	Mean A	nA	R^2	p-value	Origin[1]
SI.55	H (cm)	2.36	110	2.55	30	0.05	0.0062	ToUNR9
	SS (°Brix)	8.57	66	7.72	22	0.08	0.0009	
	PT (mm)	0.24	67	0.29	21	0.06	0.0026	
SI.60	LN	2.54	73	2.91	15	0.05	0.0063	ToUNR15
SIV.7	SS (°Brix)	8.13	73	9.19	16	0.10	0.0002	ToUNR15
	pH	4.64	61	4.82	13	0.10	0.0001	
	PT (mm)	0.26	73	0.21	16	0.07	0.0020	
SIV.26	SS (°Brix)	8.48	68	7.80	21	0.05	0.0074	ToUNR9
SIV.41	pH	4.63	58	4.79	16	0.09	0.0002	ToUNR15

B

Band	Traits	Mean P	nP	Mean A	nA	R^2	p-value	Origin[1]
X4	SL (days)	15.36	69	18.64	21	0.12	0.0010	ToUNR15
X14	TA (g)	0.41	54	0.50	13	0.07	0.0020	ToUNR15
X67	SS (°Brix)	8.07	51	8.66	32	0.05	0.0100	ToUNR15
X91	SS (°Brix)	8.10	65	8.99	18	0.08	0.0009	ToUNR15
	pH	4.61	53	4.79	17	0.13	0.0001	
	TA (g)	0.45	51	0.37	16	0.06	0.0042	
	PT (mm)	0.26	66	0.22	20	0.05	0.0080	
X137	F	47.58	60	50.94	31	0.07	0.0015	ToUNR15
	PT (mm)	0.26	63	0.21	23	0.07	0.0016	
R79	SS (°Brix)	8.04	67	9.15	23	0.14	0.0001	ToUNR15
P7	SS (°Brix)	8.68	24	6.89	6	0.24	0.0007	ToUNR9
	LN	2.42	26	3.05	6	0.19	0.0030	
	PT (mm)	0.22	26	0.34	6	0.27	0.0003	

[1] Origin (parental genotype) of DNA amplification fragments are indicated. **Mean P**: mean value for F_2 plants with presence of band, **nP**: number of plant with presence of band, **Mean A**: mean value for F_2 plants with absence of band, **nA**: number of plants with absence of band, **R^2**: fraction of phenotypic variation, **p-value**: probability associated, **H**: fruit height, **SS**: soluble solids content, **PT**: pericarp thickness, **LN**: locule number, **SL**: fruit shelf life, **TA**: titratable acidity.

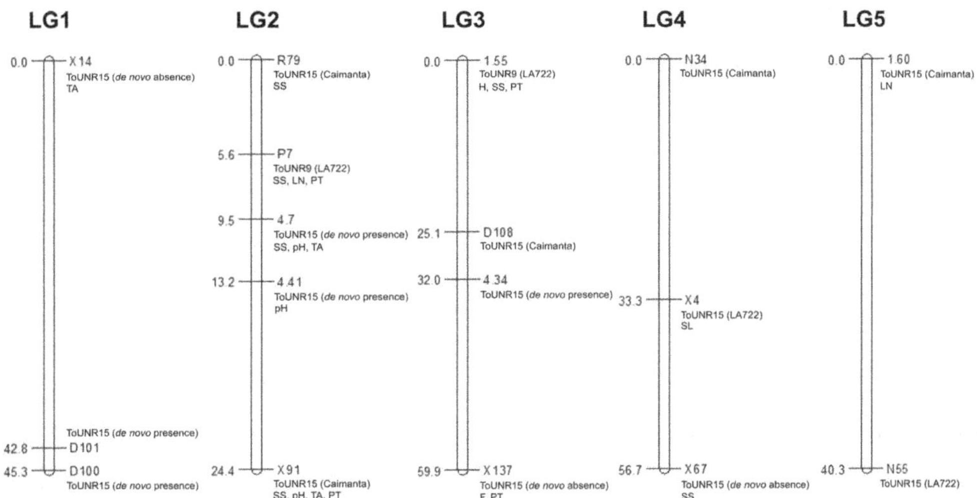

Figure 2. Linkage group (LG) constituted by molecular markers with Mendelian inheritance. The origin of molecular markers in second cycle parental genotypes (ToUNR15 and ToUNR9) and firstcycle parental genotypes (between brackets, Caimanta of *S. lycopersicum* and LA722 of *S. pimpinellifolium*) are indicated. The *de novo* bands were defined according to Liberatti et al. (2013). The phenotypic traits associated with the molecular markers are also indicated. **TA**: titratable acidity, **SS**: soluble solids content, **LN**: locule number, **PT**: pericarp thickness, **H**: height, **F**: firmness, **SL**: shelf life.

The five linkage groups spanned 226.6 cM (about 15% of the whole tomato genome, based on previous maps reported by Khialparast et al. 2013 and Tanksley et al. 1992), with a mean interval length of 13.3 cM between markers. It was found that 17 of 26 (65%) markers with Mendelian segregation belong to some linkage groups. The number of markers per linkage group ranged from 2 (LG5) to 5 (LG2). On average, 3.4 markers per group were detected. The LG3 was the largest (59.9 cM) and LG2 had the smallest genetic linkage distances. Although all polymorphic polypeptides with Mendelian segregation were considered in the analysis they do not belong to any linkage group. The most interesting group was LG2, for being the shortest. During the first cycle of crosses, new gene blocks from chromosome regions of Caimanta and LA722 were created by recombination. They were conservatively inherited and located at LG2. Moreover, the markers belonging to this LG were associated with fruit quality traits (SS, pH, TA, LN and PT) in robust agreement with those observed in the parental RILs for these traits. This fact implies that ToUNR9 and ToUNR15 genotypes fixed alleles with opposite effects during the selection process. A similar result was found for pericarp thickness, a trait for which the QTL detected in LG3 had opposite effects on the linkage markers (X137 increases its mean value and SI.55 decreases it). These results are consistent with the breeding approach proposed by Bai and Lindhout (2007), who stated that early selection for fruit quality traits is rarely performed in crosses derived from wild germplasm, as many generations may be needed to remove the deleterious genes that accompany the introduced genes, due to linkage drag. When the parental lines reach a high level of homozygosity (F_4 to F_6), crosses are made to test hybrids.

Phenotypic characterization of the two tomato RILs derived from an interspecific cross, the F_1 generation (SCH) between them and the segregating F_2 generation allowed the identification of genetic variability for several fruit quality traits. Different polymorphism levels were detected by polypeptide profiles, SRAP and AFLP markers, showing that the observed phenotypic differences are associated with gene variation. Only SRAP and AFLP markers were able to detect QTLs for fruit quality traits of agronomic interest. This study demonstrated the presence of new gene blocks by chromosome rearrangement and recombination. These results suggest that it is possible to go on with a tomato breeding plan to generate phenotypes with higher performance for fruit quality in the studied F_2 generation.

ACKNOWLEDGEMENTS

The authors thank the Agencia Nacional de PromociónCientífica y Tecnológica (ANPCyT) for funding the project (PICT 2004 (SECyT-FONCyT) (Nº 08 25481)) and the Consejo Nacional de Investigaciones Científicas y Técnicas (CONICET) for scholarships.

REFERENCES

Aharoni A and O'Connell AP (2002) Gene expression analysis of strawberry achene andreceptacle maturation using DNA microarrays. **Journal of Experimental Botany 53**: 2073-2087.

Bai Y and Lindhout P (2007) Domestication and breeding of tomatoes: What have we gained and what can we gain in the future? **Annals of Botany 100**: 1085-1094.

Castro HA, Galvez MJ, González SR and Villamil CB (2006) Protein composition of *Cucurbita maxima* and *C. moschata*seeds. **Biologia Plantarum 50**: 251-256.

Cravero VP, Martín EA and Cointry E (2007) Genetic diversity in *Cynara cardunculus* determined by SRAP markers. **Journal of American Society of Horticultural Science 132**: 208-212.

Di Rienzo JA, Casanoves F, Balzarini MG, Gonzalez L, Tablada M and Robledo CW (2001) **Infostat**. Versión 2001, Grupo Infostat, FCA, Universidad Nacional de Córdoba, Argentina.

Falconer DS and Mackay TFC (1996) **Introduction to quantitative genetics**. Longmans Green, Harlow, 494p.

Figueiredo AST, Resende JTV, Faria MV, Paula JT, Schwarz K and Zanin DS (2015) Combining ability and heterosis of relevant fruit traits of tomato genotypes for industrial processing. **Crop Breeding and Applied Biotechnology** 15: 154-161.

Garg N, Cheema DS and Dhatt AS (2008) Genetics of yield, quality and shelf life characteristics in tomato under normal and late planting conditions. **Euphytica 159**: 275-288.

Giovannoni JJ (2004) Genetic regulation of fruit development and ripening. **Plant Cell 16**: 170-180.

Grandillo S, Zamir D and Tanksley SD (1999) Genetic improvement of processing tomatoes: A 20 years perspective. **Euphytica 110**: 85-97.

Grimplet J, Romieu C, Audergon JM, Marty I, Albagnac G, Lambert P, Bouchet JP and Terrier N (2005) Transcriptomic study of apricot fruit (*Prunus armeniaca*) ripening among 13,006 expressed sequence tags. Physiology Plantarum 125: 281-292.

Hennig L, Gruissem W, Grossniklaus U and Köhler C (2004) Transcriptional programs of early stages of plant reproduction. **Plant Physiology 35**: 1765-1775.

Hills J, Lethenborg P, Li PW, Rahman RH, Sorensen H and Sorensen JC (2003) Inheritance of progoitrin and total aliphatic glosinolates in oilseeds rape (*Brassica napus* L.). **Euphytica 134**: 179-187.

Ipsilandis CG, Tokatlidis IS, Vafias B and Stefanis D (2006) Criteria for developing second-cycle hybrids in maize. **Asian Journal of Plant Sciences**: 680-685.

Kearsey MJ and Pooni HS (1996) **The genetical analysis of quantitative traits**. Chapman and Hall, London, 396p.

Khialparast F, Abdemishani S, Yazdisamadi B, Naghavi MR and Foolad M (2013) Identification and characterization of quantitative trait loci related to chemical traits in tomato (*Lycopersicon esculentum* Mill.). **Crop Breeding Journal 3**: 13-18.

Kosambi DD (1943) The estimation of map distances from recombination values. **Annals of Eugenics 12**: 172-175.

Lecomte L, Duffé P, Buret M, Servin B, Hospital F and Causse M (2004) Marker-Assisted introgression of 5 QTL controlling fruit quality traits into three tomato lines revealed interactions between QTLs and genetic background. **Theoretical and Applied Genetics 109**: 658-668.

Li G and Quiros CF (2001) Sequence-related amplified polymorphism (SRAP), a new marker system based on a simple PCR reaction: its application to mapping and gene tagging in *Brassica*. **Theoretical and Applied Genetics 103**: 455-461.

Liberatti DR, Rodríguez GR, Zorzoli R and Pratta GR (2013) Tomato second hybrids differ from parents at three levels of genetic variation. **International Journal of Plant Breeding 7**: 1-6.

Mahuad SL, Pratta GR, Rodríguez GR, Zorzoli R and Picardi LA (2013) Preservation of *Solanum pimpinellifolium* genomic fragments in recombinant genotypes improved the fruit quality of tomato. **Journal of Genetics 92**: 195-203.

Marchionni Basté E, Liberatti DR, Mahuad SL, Rodríguez GR, Pratta GR, Zorzoli R and Picardi LA (2010) Diallel analysis for fruit traits among tomato recombinant inbred lines derived from an interspecific cross *Solanum lycopersicum* × *S. pimpinellifolium*. **Journal of Applied Horticulture 12**: 21-25.

Michael TP and Alba R (2012) The tomato genome flashed out. **Nature Biotechnology 30**: 765-767.

Moyle R, Fairbairn DJ, Ripi J, Crowe M and Botella JR (2005) Developing pineapple fruit has a small transcriptome dominated by metallothionein. **Journal of Experimental Botany 56**: 101:112.

Pereira da Costa JH, Rodríguez GR, Pratta GR and Zorzoli R (2009) Influencia de genes exóticos sobre la vida en estantería y el peso del fruto de tomate. **Agriscientia 16**: 7-13.

Pereira da Costa JH, Rodríguez GR, Pratta GR, Picardi LA and Zorzoli R (2014) Pericarp polypeptides and SRAP markers associated with fruit quality traits in interspecific tomato backcross. **Genetics and Molecular Research 13**: 2539-2547.

Pratta GR, Zorzoli R and Picardi LA (2003) Diallel analysis of production traits among domestic, exotic and mutant germplasms of *Lycopersicon*. **Genetics and Molecular Research 2:** 206-213.

Pratta GR, Rodríguez GR, Zorzoli R, Valle EM and Picardi LA (2011) Phenotypic and molecular characterization of selected tomato recombinant inbred lines derived from a cross *Solanum lycopersicum* x *S. pimpinellifolium*. **Journal ofGenetics 90**: 229-237.

Rick CM (1978) The tomato. **Scientific American 239**: 76-87.

Rodríguez GR, Pratta GR, Zorzoli R and Picardi LA (2006) Evaluación de caracteres de planta y fruto en líneas recombinantes autofecundadas de tomate obtenidas por cruzamiento entre *Lycopersicon esculentum* y *L. pimpinellifolium*. **Ciencia e Investigación Agraria 33**: 133-141.

Rodríguez GR, Sequin L, Pratta GR and Zorzoli R (2008) Protein profiling in F_1 and F_2 generations of two tomato genotypes differing in ripening time. **Biologia Plantarum 52**: 548-552.

Rodríguez GR, Pereira da Costa JH, Tomat DD, Pratta GR, Zorzoli R and Picardi LA (2011) Pericarp total protein profiles as molecular markers of tomato fruit quality traits in two segregating populations. **Scientia Horticulturae-Amsterdam 130**: 60-66.

Ruiz JJ, García-Martínez S, Picó B and Gao M (2005) Genetic variability and relationship of closely related Spanish traditional cultivars of tomato as detected by SRAP and SSR markers. **Journal of American Society of Horticultural Science 130**: 88-94.

Saliba-Colombani V, Causse M, Langlois D, Philouze Jand Buret M (2001) Genetic analysis of organoleptic quality in fresh market tomato. 1. Mapping QTL for physical and chemical traits. **Theoretical and Applied Genetics 102**: 259-272.

Shapiro SS and Wilk MB (1965) An analysis of variance test for normality (complete samples). **Biometrika 52**: 591-611.

Snedecor G (1964) **Métodos estadísticos**. Compañía Editorial, México, 593p.

Stevens ME, Kader AA and Albright M (1979) Potential for increasing tomato flavor via increased sugar and acid content. **Journal of American Society of Horticultural Science 104**: 40-42.

Tanksley SD, Ganal MW, Prince JP, de Vicente MC, Bonierbale MW, Broun P, Fulton TM, Giovannoni JJ, Grandillo S, Martin GB, Messeguer R, Miller JC, Miller L, Paterson AH, Pineda O, Riider MS, Wing RA, Wu W and Young ND (1992) High density molecular linkage map of the tomato and potato genomes. **Genetics 132**: 1141-1160.

Tanksley SD (1993) Mapping polygenes. **Annual Review of Genetics 27**: 205-233.

Terrier N, Glissant D, Grimplet J, Barrieu F, Abbal P, Couture C, Ageorges A, Atanassova R, León C, Renaudin JP, Dédaldéchamp F, Romieu C, Delrot S and Hamd S (2005) Isogene specific oligo arrays reveal multifaceted changes in geneexpression during grape berry (*Vitis vinifera* L.) development. **Planta 222**: 832-847.

Tomato Genome Consortium (2012) The tomato genome sequence provides insights into fleshy fruit evolution. **Nature 485**: 635-641.

Xu J, Ranc N, Muños S, Rolland S, Bouchet JP, Desplat N, Le Paslier MC, Liang Y, Brunel D and Causse M (2013) Phenotypic diversity and association mapping for fruit quality traits in cultivated tomato and related species. **Theoretical and Applied Genetics 126**: 567-581.

Zhang Y and Stommel JR (2001) Development of SCAR and CAPS Markers linked to the *Beta* Gene in Tomato. **Crop Science 41**:1602-1608.

16

Selection of cotton genotypes for greater length of fibers

Luiz Paulo de Carvalho[1], Francisco José Correia Farias[1], Camilo de Lellis Morello[1] and Paulo Eduardo Teodoro[2*]

*Corresponding author:
E-mail: eduteodoro@hotmail.com

[1] Embrapa Algodão, Centro Nacional de Pesquisa de Algodão, 58.428-095, Campina Grande, PB, Brazil
[2] Universidade Federal de Viçosa, Departamento de Biologia Geral, 36.571-000, Viçosa, MG, Brazil

Abstract: *In cotton breeding programs, it is necessary to identify genotypes with predictable behavior on the length of fibers, and which are responsive to environmental variations, in specific or broad conditions. The aim of this study was to employ the methodology of mixed models for simultaneous selection of cotton genotypes with greater length of fibers, adaptability and stability. It was evaluated 36 lines in three trials located in Apodi (2013 and 2014) and Santa Helena (2013,) in a randomized block design with two replications. Genetic parameters were estimated by the restricted maximum likelihood/best linear unbiased predictor method (REML/BLUP) and the selection was based on the method of harmonic mean of the relative performance of genetic values. The genotypes CNPA 2012-55, CNPA 2012-58, CNPA 2012-62 and CNPA 2012-64 can be grown in all tested environments, since they gather greater length of fibers, stability and adaptability.*

Key words: *BLUP/REML, genetic parameters, Gossypium hirsutum L.r. latifolium Hutch.*

INTRODUCTION

Upland cotton (*Gossypium hirsutum* L.r. latifolium Hutch.) produces one of the most important textile fiber of the world, and offers various utility products with great relevance in the Brazilian and the world's economy (Carvalho et al. 2015a, Menezes et al. 2015, Morello et al. 2015). However, changes in spinning technology, competition with synthetic fibers, and the globalization of cotton production and textile products increase the demand for higher quality fiber, making the length of fiber one of the main traits to be improved (Smith et al. 2008, Carvalho et al. 2016).

Genotype x environment interaction (G x E) is one of the greatest challenges in plant breeding, both in the selection procedures and in the recommendation of cultivars, and plant breeders usually look for stable genotypes with improved performance in relation to a particular trait (Ramalho et al. 2012, Cruz et al. 2014). Studies on G x E interaction in cotton are scarce in Brazil, and in total, these studies evaluate fiber yield, not the length (Suinaga et al. 2006, Souza et al. 2006, Silva Filho et al. 2008, Farias et al. 2016 a, b).

Thus, adaptability and phenotypic stability analysis are necessary, by which it is possible to identify genotypes with predictable behavior on the length of fibers, and which are responsive to environmental variations, in specific or broad conditions (Farias et al. 2016b). Therefore, selection methods that incorporate stability and adaptability in a single statistics can be considered superior, when compared with those which use only yield as selection criteria (Resende 2007).

The analysis carried out by the method of mixed models was proposed to perennial crops, such as pine (Resende et al. 1996), and its application has been spread in the analysis of information on several crops, both perennial and annual. REML (Restricted Maximum Likelihood) estimates variance components required by the model, and BLUP (Best Linear Unbiased Prediction) estimates the genotypic value (Resende 2007).

The use of mixed models allows ranking the genotypes taking into consideration, simultaneously, their genetic values (estimated trait) and stability (Rosado et al. 2012). The lower the standard deviation of genotypic behavior across sites, the greater is the harmonic mean of their genotypic values (HMGV) across sites. Thus, selection by the greatest HMGV values implies simultaneously in the selection for the evaluated trait and stability. In terms of adaptability, a statistics used in the context of mixed models refers to the relative performance of genotypic values (RPGV) across the environments. In this case, the predicted genotypic values (or the original data) are expressed as a proportion of overall mean of each location, and then, the mean value of this proportion across sites is obtained (Carbonell et al. 2007, Resende 2007).

The simultaneous selection of cotton genotypes with greater length of fibers, stability and adaptability, under the approach of mixed models, can be accomplished by the method of the harmonic mean of the relative performance of genetic values (HMRPGV) predicted. Carbonell et al. (2007) report the main advantages of this method: to provide estimates of adaptability and genotypic stability, whose breeding values are penalized by instability; to generate results in their own magnitude or scale of the evaluated trait, facilitating the recommendation of genotypes; to analyze unbalanced and not orthogonal designs; to analyze heterogeneous variances; to allow errors correlated within sites.

Thus, HMGV, RPGV and HMRPGV have been used as measures for the interpretation of genotypic stability and adaptability of crops, such as common bean (Carbonell et al. 2007), sugarcane (Bastos et al. 2007), cashew (Maia et al. 2009), eucalyptus (Rosado et al. 2012), rice (Regitano Neto et al. 2013), and cowpea (Torres et al. 2015, Torres et al. 2016). However, they have not been used for breeding purposes in cotton. The aim of this study was to employ the mixed models methodology for simultaneous selection of cotton genotypes with greater length of fibers, adaptability and stability.

MATERIAL AND METHODS

It was evaluated lines derivative from cross between Guazuncho2 with Acala SJ4 of long fiber. Seeds were advanced from F_2 to F_4. The genealogical selection procedure was used from the F_4 generation. In the latest generation, it was obtained 271 F_4 plants, and of these, 51 $F_{4:5}$ plants were selected with length ≥ 31 mm. Subsequently, in a greenhouse, it was selected 34 F_5:F_6 lines with length ≥ 32 mm, which were evaluated and analyzed by Carvalho et al. (2015b). In the remaining lines, it was selected only those with length ≥ 32 mm. These lines were evaluated in three trials located in Apodi-RN (under irrigation), in 2013 and 2014, and in Santa Helena-GO (without irrigation), in 2013, in a randomized block design with two replications.

The plot consisted of two 5m rows, spaced 0.80 m, totaling 60 plants per row. Cultural practices were normal for cotton crop, with the use of herbicide for weed control. Pest control was carried out according to the integrated management of pests recommended for the crop in the region. Length of fibers was evaluated in samples with 20 bolls per plot, using the *High volume instrument* device (HVI) of the Fibers Laboratory of Embrapa Algodão.

To evaluate the effect of G x E interaction, it was used the statistical model 54 (Model = elites cotton genotypes, multiple locations, multiple observations per plot) of the Selegen-Reml/Blup software (Resende 2007). This model corresponds to $y = Xb + Zg + Wc + e$, in which y, b, g, c, and e correspond respectively to the fixed effects of data vectors (blocks means by environments), genotype effects (random), effects of genotype x environment interaction (random), and random errors; and X, Z and W = incidence matrices for b, g, and c, respectively. The assumed distributions and mean structures (S) and variance (Var) were:

$$S\begin{bmatrix} y \\ g \\ c \\ e \end{bmatrix} = \begin{bmatrix} Xb \\ 0 \\ 0 \\ 0 \end{bmatrix}; \quad Var\begin{bmatrix} g \\ c \\ e \end{bmatrix} = \begin{bmatrix} I\sigma_g^2 & 0 & 0 \\ 0 & I\sigma_c^2 & 0 \\ 0 & 0 & I\sigma_e^2 \end{bmatrix}$$

The adjusted model was obtained from the mixed model equations: $\begin{bmatrix} X'X & X'Z & X'W \\ Z'X & Z'Z+I\lambda_1 & Z'W \\ W'X & W'Z & W'W+I\lambda_2 \end{bmatrix} \times \begin{bmatrix} \hat{b} \\ \hat{g} \\ \hat{c} \end{bmatrix} = \begin{bmatrix} X'y \\ Z'y \\ W'y \end{bmatrix}$, in which $\lambda_1 = \frac{\sigma_e^2}{\sigma_g^2} = \frac{1-h_g^2-c^2}{h_g^2}$; $\lambda_2 = \frac{\sigma_e^2}{\sigma_c^2} = \frac{1-h_g^2-c^2}{c_g^2}$, in which: $h_g^2 = \frac{\sigma_g^2}{\sigma_g^2+\sigma_c^2+\sigma_e^2}$ corresponds to broad-sense individual heritability in the block; $c^2 = \frac{\sigma_c^2}{\sigma_g^2+\sigma_c^2+\sigma_e^2}$ is the coefficient of determination of the effects of genotype x environment interaction; $\hat{\sigma}_g^2$ is the genotypic variance; $\hat{\sigma}_c^2$ is the variance of genotype x environment interaction; $\hat{\sigma}_e^2$ is the residual variance between plots; $r_{gloc} = \frac{\sigma_c^2}{\sigma_g^2+\sigma_c^2} = \frac{h_g^2}{h_g^2+c^2}$ is the genotypic correlation of genotypes across the environments.

Estimators of the variance components obtained by REML procedure via Expectation-Maximization algorithm (EM) are: $\hat{\sigma}_e^2 = \frac{\lfloor y'y-\hat{b}'X'y-\hat{g}'Z'y-\hat{c}W'y \rfloor}{[N-r(x)]}$; $\hat{\sigma}_g^2 = \frac{\lfloor \hat{g}'\hat{g}+\hat{\sigma}_e^2 trC^{22} \rfloor}{q}$; $\hat{\sigma}_c^2 = \frac{\lfloor \hat{c}'c+\hat{\sigma}_e^2 trC^{33} \rfloor}{s}$, in which: C^{22} and C^{33} come from

$$C^{-1} = \begin{bmatrix} C_{11} & C_{12} & C_{13} \\ C_{21} & C_{22} & C_{23} \\ C_{31} & C_{32} & C_{33} \end{bmatrix}^{-1} = \begin{bmatrix} C^{11} & C^{12} & C^{13} \\ C^{21} & C^{22} & C^{23} \\ C^{31} & C^{32} & C^{33} \end{bmatrix}$$

, in which C is the coefficient matrix of the mixed model equations; tr is the operator matrix trace; r(x) is the rank of the matrix X; N, q, s are the total number of data, the number of genotypes, and the number of genotype x environment combinations, respectively.

By using this model, it was obtained empirical BLUP predictors (eBLUP or REML/BLUP) of the genotypic values free of interaction, given by $\hat{\mu} + \hat{g}_i$, in which $\hat{\mu}$ is the mean of all environments, and $\hat{g}_i$ is the genotypic effect free of genotype x environment interaction. For each environment j, genotypic values (*Vg*) are predicted by $\hat{\mu} + \hat{g}_i + (\hat{g}e)_{ij}$, in which $\hat{\mu}_j$ is the mean of the environment j; $\hat{g}_i$ is the genotypic effect of genotype i on the environment j; and $(\hat{g}e)_{ij}$ is the effect of genotype x environment interaction in relation to the genotype i.

Prediction of genotypic values that capitalize the mean interaction in different environments is given by $\hat{\mu}_j + \hat{g}_i + \hat{g}e_m$, and is calculated by: $\hat{\mu} + \frac{\left(\frac{\hat{\sigma}_g^2+\hat{\sigma}_c^2}{n}\right)}{\hat{\sigma}_g^2}\hat{g}_i$, in which $\hat{\mu}$ is the overall mean of all environments; n is the number of environments; and $\hat{g}_i$ is the genotypic effect of the genotype i.

Harmonic mean values of the genotypic values (HMGV) for stability evaluation were obtained by the equation: $HMGH_i = \frac{n}{\sum_{j=1}^{n}\frac{1}{Vg_{ij}}}$, in which: n is the number of environments (n = 3) in which the genotype i was evaluated; Vg_{ij} is the genotypic value of the genotype i on the environment j, expressed by the ratio of the mean in this environment. The values of the relative performance of the genotypic values (RPGV) for adaptability were obtained according to the expression: $RPGV_i = \frac{1}{n}\frac{\sum_{j=1}^{n} Vg_{ij}}{M_j}$, in which M_j is the mean of the fiber length in the environment j. Joint selection simultaneously considering seed length of fibers, stability and adaptability is given by the statistical harmonic mean of the relative performance of genotypic values predicted (HMRPGV): $HMRPGV_i = \frac{n}{\sum_{j=1}^{n}\frac{1}{PRVg_{ij}}}$ (Resende 2007). Thus, the genotypes with greater HMRPGV are those that simultaneously present greater length of fibers, adaptability and genotypic stability in the environments evaluated in this study.

RESULTS AND DISCUSSION

Coefficient of experimental variation (CV_e) was 3.20% (Table 1), which is lower than those of other studies carried out with the cotton crop (Carvalho et al. 2015b, Carvalho et al. 2015c, Hoogerheide et al. 2007), which indicates excellent experimental precision. In another interpretation, the value obtained for CV_g (4.76%) indicates that a considerable fraction of MSgenotypes was extracted from MStotal. Joint evaluation of CV_g and CV_e is reflected in the statistics $\hat{r}_{g\hat{g}}$ (Resende and Duarte 2007). The accuracy obtained (96%) shows excellent experimental quality, and thus safety and credibility in the selection of superior genotypes for the trait length of fibers.

Heritability of the genotypes means ($\hat{h}^2_{mg}$) is estimated when using means of the blocks as evaluation and/or selection criteria (Resende 2007). Thus, in the face of the values obtained (0.91), there is reliability in the selection of elite cotton genotypes, based on the predicted genotypic values. In the estimate of the broad-sense individual heritability ($\hat{h}^2_g$), the total genetic dispersion is considered, which is relevant since in this research it was sought to explore all the genetic variance between cotton progenies. The variance of the genotypic effects ($\hat{\sigma}^2_g$) presented magnitude highly greater than the variance of the effects of G x E interaction ($\hat{\sigma}^2_c$), constituting 66% of the total phenotypic variability, represented by the heritability of individual plots ($\hat{h}^2_g$). Results with similar magnitude were observed by Maia et al. (2009), Rosado et al. (2012) and Laviola et al. (2012), when estimating genetic parameters via REML/BLUP in cashew clones, eucalyptus and jatropha, respectively.

Due to the greater or lesser degree of adaptability/genetic stability of individuals, the variance of the G x E interaction ($\hat{\sigma}^2_c$) can inflate the phenotypic expression of a trait (Bastos et al. 2007). This measure quantifies the fraction of the total variation due to the G x E interaction. Small magnitude estimates of $\hat{\sigma}^2_c$ indicates that the G x E interaction little influences on phenotypic value (Maia et al. 2009). In this context, a genotype with greater length of fibers in an environment tends to maintain the same level in different environments, since this genotype responds favorably to environmental influences (high correlation between genotypic values across the sites), in addition to have high predictability in the face of environmental changes. Estimates of $\hat{\sigma}^2_c$ can be considered low, corresponding to 3.61% of the total phenotypic variability, which favors the achievement of extremely high phenotypic correlation ($\hat{r}_{gloc}$=0.95) among the environments.

Phenotypic value corresponds to the values obtained in the field evaluations, which are influenced by genotypic effect and effect of G x E interaction. In the latter, the sum of the genotypic variance, of the residual variance between plots, and of the variance of the G x E interaction results in

Table 1. Estimates of variance components (individual REML) for length of fiber in 36 cotton genotypes evaluated in three environments

Parameter	Estimate
$\hat{\sigma}^2_g$	2.02 ± 0.12
$\hat{\sigma}^2_c$	0.11 ± 0.01
$\hat{\sigma}^2_e$	0.91 ± 0.26
$\hat{\sigma}^2_p$	3.04 ± 0.75
$\hat{h}^2_g$	0.66 ± 0.16
$\hat{h}^2_{mg}$	0.91 ± 0.18
$\hat{r}_{g\hat{g}}$	0.96 ± 0.09
c^2	0.04 ± 0.01
$\hat{r}_{gloc}$	0.95 ± 0.07
CV_g (%)	4.76 ± 0.27
CV_e (%)	0.91 ± 0.40
μ (mm)	29.81 ± 1.26

$\hat{\sigma}^2_g$: genotypic variance; $\hat{\sigma}^2_c$: variance of genotype x environment interaction; $\hat{\sigma}^2_e$: residual variance between plots; $\hat{\sigma}^2_p$: individual phenotypic variance; $\hat{h}^2_g$: broad-sense individual heritability; $\hat{h}^2_{mg}$: mean of genotype heritability; $\hat{r}_{g\hat{g}}$: accuracy in the selection of genotypes; c^2: coefficient of determination of effects of the genotype x environment interaction; $\hat{r}_{gloc}$: genotypic correlation across the environments; CV_g: coefficient of genetic variation; CV_e: coefficient of experimental variation; μ: overall mean.

individual phenotypic variance ($\hat{\sigma}_f^2$). Although the residual dispersion between plots ($\hat{\sigma}_e^2$) represents 30.13% of this value, significant genetic progress was achieved. Results in similar magnitude were observed by Maia et al. (2009), Rosado et al. (2012), in which it was evaluated the effects of G x E interaction on cashew and eucalyptus clones, respectively, via mixed models, it was found that the residual variance between plots represented 66.58% of the individual phenotypic variance (the two works had the same ratio between the variances).

The mean genotypic correlation of the genetic material performance across the environment ($\hat{r}_{gloc}$) indicates the ranking reliability of the best genotypes in the tested environments. In general, small changes were observed in the ranking of genotypes, due to the high magnitude of $\hat{r}_{gloc}$ (0.95), and due to the high $\hat{r}_{g\hat{g}}$ (0.96). This indicates the occurrence of simple fraction of G x E interaction, favoring the selection of genotypes with broader adaptation.

The genotypes CNPA 2012-55, CNPA 2012-58, CNPA 2012-62 and CNPA 2012-64 showed the best gain estimates with

Table 2. Estimates of genotypic values for length of fiber in 36 cotton genotypes evaluated in three environments and for average environment

Genotype	Environment 1 $\hat{\mu}_1 + \hat{g} + \hat{g}e$	Environment 2 $\hat{\mu}_2 + \hat{g} + \hat{g}e$	Environment 3 $\hat{\mu}_3 + \hat{g} + \hat{g}e$	Environment Mean $\hat{\mu} + \hat{g}$
CNPA 2012-55	33.92	30.94	32.88	32.58
CNPA 2012-56	33.81	30.63	32.72	27.61
CNPA 2012-57	33.65	30.43	32.54	26.29
CNPA 2012-58	33.45	30.26	32.35	32.19
CNPA 2012-59	33.29	30.15	32.21	31.28
CNPA 2012-60	33.16	30.05	32.11	30.14
CNPA 2012-61	33.06	29.97	32.02	29.30
CNPA 2012-62	32.99	29.87	31.92	31.40
CNPA 2012-63	32.92	29.77	31.83	28.64
CNPA 2012-64	32.84	29.69	31.77	31.85
CNPA 2012-65	32.77	29.61	31.70	29.23
CNPA 2012-66	32.71	29.54	31.64	29.86
CNPA 2012-67	32.65	29.47	31.59	29.55
CNPA 2012-68	32.57	29.40	31.53	31.28
CNPA 2012-69	32.51	29.33	31.46	30.62
CNPA 2012-70	32.44	29.26	31.38	27.81
CNPA 2012-71	32.37	29.20	31.32	28.64
CNPA 2012-72	32.31	29.14	31.25	29.61
CNPA 2012-73	32.25	29.08	31.19	30.38
CNPA 2012-74	32.20	29.02	31.13	30.79
CNPA 2012-75	32.14	28.97	31.07	29.00
CNPA 2012-76	32.08	28.92	31.01	29.85
CNPA 2012-77	32.02	28.86	30.96	28.79
CNPA 2012-78	31.97	28.80	30.91	30.78
CNPA 2012-79	31.92	28.75	30.86	30.79
CNPA 2012-80	31.87	28.69	30.80	30.44
CNPA 2012-82	31.82	28.63	30.74	29.51
CNPA 2012-83	31.77	28.58	30.68	29.94
CNPA 2012-84	31.71	28.52	30.62	29.13
CNPA 2012-85	31.67	28.47	30.56	29.41
CNPA 2012-86	31.62	28.41	30.51	27.61
CNPA 2012-87	31.56	28.35	30.45	31.23
CNPA 2012-88	31.49	28.28	30.38	29.88
CNPA 2012-89	31.42	28.21	30.31	30.73
CNPA 2012-90	31.35	28.14	30.23	28.71
CNPA 2012-91	31.26	28.04	30.13	28.31
Mean	32.38	29.21	31.30	29.81

Table 3. Stability of genetic values (HMGV), adaptability of genetic values (RPGV and RPGVμ), stability and adaptability of genetic values (HM-RPGV and HMRPGVμ) for length of fibers (mm) of 36 cotton genotypes predicted by BLUP analysis

Genotype	HMGV	RPGV	RPGVμ*	HMRPGV	HMRPGVμ*
CNPA 2012-55	32.53	1.09	32.59	1.10	32.60
CNPA 2012-56	27.55	0.93	27.61	0.95	27.65
CNPA 2012-57	26.22	0.88	26.28	0.90	26.30
CNPA 2012-58	32.13	1.08	32.19	1.10	32.20
CNPA 2012-59	31.22	1.05	31.28	1.05	31.30
CNPA 2012-60	30.08	1.01	30.14	1.01	30.15
CNPA 2012-61	29.24	0.98	29.30	1.00	29.50
CNPA 2012-62	31.35	1.05	31.41	1.05	31.40
CNPA 2012-63	28.57	0.96	28.63	0.97	28.65
CNPA 2012-64	31.79	1.07	31.85	1.07	31.85
CNPA 2012-65	29.14	0.98	29.22	0.99	29.27
CNPA 2012-66	29.81	1.00	29.86	1.00	29.87
CNPA 2012-67	29.50	0.99	29.56	0.99	29.57
CNPA 2012-68	31.23	1.05	31.28	1.05	31.30
CNPA 2012-69	30.57	1.03	30.63	1.05	30.65
CNPA 2012-70	27.74	0.93	27.81	0.95	27.85
CNPA 2012-71	28.57	0.96	28.63	0.97	28.65
CNPA 2012-72	29.56	0.99	29.62	1.00	29.65
CNPA 2012-73	30.32	1.02	30.38	1.02	30.40
CNPA 2012-74	30.73	1.03	30.79	1.03	30.80
CNPA 2012-75	28.95	0.97	29.00	0.98	29.00
CNPA 2012-76	29.80	1.00	29.85	1.00	29.85
CNPA 2012-77	28.72	0.97	28.79	0.98	28.80
CNPA 2012-78	30.70	1.03	30.77	1.05	30.77
CNPA 2012-79	30.74	1.03	30.80	1.05	30.80
CNPA 2012-80	30.37	1.02	30.44	1.02	30.45
CNPA 2012-82	29.45	0.99	29.51	1.00	29.51
CNPA 2012-83	29.88	1.00	29.94	1.00	29.95
CNPA 2012-84	29.06	0.98	29.13	0.99	29.15
CNPA 2012-85	29.35	0.99	29.41	1.00	29.41
CNPA 2012-86	27.55	0.93	27.61	0.95	27.61
CNPA 2012-87	31.18	1.05	31.24	1.05	31.25
CNPA 2012-88	29.81	1.00	29.88	1.00	29.90
CNPA 2012-89	30.68	1.03	30.74	1.05	30.75
CNPA 2012-90	28.65	0.96	28.71	0.97	28.75
CNPA 2012-91	28.25	0.95	28.31	0.95	28.35

* values of RPGVμ and HMRPGVμ were obtained by multiplying the genotypic value of the length of fibers from genotype i with the statistics RPGV and HMRPGV, respectively.

selection in all environments (Table 2). The genetic gains obtained with the selection of the genotypes were respectively: 19.29, 17.96, 15.78 and 14.98%. These genotypic values can also be considered for recommendation of these genotypes, which were selected in other environments with standard G x E interaction similar to those observed in this study. According to Maia et al. (2009), this is because the method of mixed models penalizes predicted genotypic values. Thus, the same behavior of the genetic means ($\hat{\mu} + \hat{g}$) of the length of fiber is expected, when the aforementioned genotypes are subjected to environments with characteristics similar to those of this study.

Table 2 shows that the genotypic value for the mean of environments ($\hat{\mu} + \hat{g}$) generated results similar to the methods in which adaptability (RPGV), adaptability and stability (HMRPGV) are simultaneously capitalized (Table 3). According to Maia et al. (2009), the capitalization of the G x E interaction depends on the selection of genotypes for greater adaptability and stability to the evaluated environments. This criterion is repeated in the genotypes CNPA 2012-55, CNPA 2012-58, CNPA 2012-62 and CNPA 2012-64, selected by the genetic means free of interaction ($\hat{\mu} + \hat{g}$).This fact indicates that these genotypes presented adaptability and genotypic stability between the three environments analyzed, in addition to having high length of fibers, i.e., the maintenance of this trait across the different environments. These results corroborate those obtained by Maia et al. (2009), Regitano Neto et al. (2013) and Torres et al. (2016), who verified the maintenance of the ranking of cashew clones, rice genotypes and cowpea, respectively, by genotypic value of methods for the mean of years ($\hat{\mu} + \hat{g}$), HMGV, RPGV and HMRPGV, attributing these results to genotypic correlation across sites ($\hat{r}_{gloc}$), which was positive and of similar magnitude to those of study.

REFERENCES

Bastos IT, Barbosa MHP, Resende MDV, Peternelii LA, Silveira LCI, Donda LR, Fortunato AA, Costa PMA and Figueiredo ICR (2007) Avaliação da interação genótipo x ambiente em cana-de-açúcar via modelos mistos. **Pesquisa Agropecuária Tropical 37**: 195-203.

Carbonell SAM, Chioratto AF, Resende MDV, Dias LAS, Beraldo ALA and Perina EF (2007) Estabilidade de cultivares e linhagens de feijoeiro em diferentes ambientes no Estado de São Paulo. **Bragantia 66**: 193-201.

Carvalho LP, Farias FJC and Rodrigues JIS (2015a) Selection for increased fiber length in cotton progenies from Acala and Non-Acala types. **Crop Science 55**: 1-7.

Carvalho LP, Salgado CC, Farias FJC and Carneiro VQ (2015b) Estabilidade e adaptabilidade de genótipos de algodão de fibra colorida quanto aos caracteres de fibra. **Ciência Rural 45**: 598-605.

Carvalho LP, Farias FJC, Morelo CL, Rodrigues JIS and Teodoro PE (2015c) Agronomic and technical fibers traits in elite genotypes of cotton herbaceous. **African Journal of Agricultural Research 10**: 4882-4887.

Carvalho LP, Farias FJC, Morelo CL and Teodoro PE (2016) Uso da metodologia REML/BLUP para seleção de genótipos de algodoeiro com maior adaptabilidade e estabilidade produtiva. **Bragantia 75**: 314-321.

Cruz CD, Carneiro PCS and Regazzi AJ (2014) **Modelos biométricos aplicados ao melhoramento genético**. 3rd edn, Editora UFV, Viçosa, 668p.

Farias FJC, Carvalho LP, Silva Filho JL and Teodoro PE (2016a) Biplot analysis of phenotypic stability in upland cotton genotypes in Mato Grosso. **Genetics and Molecular Research 15**: 1-8.

Farias FJC, Carvalho LP, Silva Filho JL and Teodoro PE (2016b) Usefulness of the HMRPGV method for simultaneous selection of upland cotton genotypes with greater fiber length and high yield stability. **Genetics and Molecular Research 15**: 1-8.

Hoogerheide ESS, Vencovsky R, Farias FJC, Freire EC and Arantes EM (2007) Correlações e análise de trilha de caracteres tecnológicos e produtividade de fibra de algodão. **Pesquisa Agropecuaria Brasileira 42**: 1401-1405.

Laviola BG, Alves AA, Gurgel FD, Rosado TB, Rocha RB and Albrecht JC (2012) Estimates of genetic parameters for physic nut traits based in the germplasm two years evaluation. **Ciência Rural 42**: 429-435.

Maia MCC, Resende MDV, Paiva JR, Cavalcanti JJV and Barros LM (2009) Seleção simultânea para produção, adaptabilidade e estabilidade genotípicas em clones de cajueiro, via modelos misto. **Pesquisa Agropecuária Tropical 39**: 43-50.

Menezes IPP, Hoffmann LV and Barroso PAV (2015) Genetic characterization of cotton landraces found in the Paraíba and Rio Grande do Norte states. **Crop Breeding and Applied Biotechnology 15**: 26-32.

Morello CL, Suassuna ND, Barroso PAV, Silva Filho JL, Ferreira ACB, Lamas FL, Pedrosa MB, Chitarra LG, Ribeiro JL, Godinho VPC and Lanza MA (2015) BRS 369RF and BRS 370RF: Glyphosate tolerant, high-yielding upland cotton cultivars for central Brazilian savanna. **Crop Breeding and Applied Biotechnology 15**: 290-294.

Ramalho MAP, Abreu AFB, Santos JB and Nunes JAR (2012) **Aplicações da genética quantitativa no melhoramento de plantas autógamas**. UFLA, Lavras, 328p.

Regitano Neto A, Ramos Júnior EA, Gallo PB, Freitas JG and Azzini LE (2013) Comportamento de genótipos de arroz de terras altas no estado de São Paulo. **Revista Ciência Agronômica 44**: 512-519.

Resende MDV (2007) **SELEGEN–REML/BLUP:** sistema estatístico e seleção genética computadorizada via modelos lineares mistos. Embrapa Florestas, Colombo, 359p.

Resende MDV, Prates DF, Yamada CK and Jesus A (1996) Estimação de componentes de variância e predição de valores genéticos pelo método da máxima verossimilhança restrita (REML) e melhor predição linear não viciada (BLUP) em pinus. **Boletim de Pesquisa Florestal 32-33**: 23-42.

Resende MDV and Duarte JB (2007) Precisão e controle de qualidade em experimentos de avaliação de cultivares. **Pesquisa Agropecuária Tropical 37**: 182-194.

Rosado AM, Rosado TB, Alves AA, Laviola BG and Bhering LL (2012) Seleção simultânea de clones de eucalipto de acordo com produtividade, estabilidade e adaptabilidade. **Pesquisa Agropecuária Brasileira 47**: 964-971.

Silva Filho J.L, Morello CL, Farias FJC, Lamas FM, Pedrosa MB and Ribeiro JL (2008) Comparação de métodos para avaliar a adaptabilidade e estabilidade produtiva em algodoeiro. **Pesquisa Agropecuária Brasileira 43**: 349-355.

Smith CW, Hague S, Hequet E, Thaxton S and Brown N (2008) Development of extra-long staple upland cotton. **Crop Science 48**: 1823-1832.

Souza AA, Freire EC, Bruno RLA, Carvalho LP, Silva Filho LP and Pereira WE (2006) Estabilidade e adaptabilidade do algodoeiro herbáceo no cerrado do Mato Grosso e Mato Grosso do Sul. **Pesquisa Agropecuária Brasileira 41**: 1125-1131.

Suinaga FA, Bastos CS and Rangel LEP (2006) Phenotipic adaptability and stability of cotton cultivars in Mato Grosso State, Brazil. **Pesquisa Agropecuária Tropical 36**: 145-150.

Torres FE, Teodoro PE, Rodrigues EV, Santos A, Corrêa AM and Ceccon G (2016) Simultaneous selection for cowpea (*Vigna unguiculata* L.) genotypes with adaptability and yield stability using mixed models. **Genetics and Molecular Research 15**: 1-11.

Torres FE, Teodoro PE, Sagrilo E, Corrêa AM and Ceccon G (2015) Interação genótipo x ambiente em genótipos de feijão-caupi semiprostrado via modelos mistos. **Bragantia 74**: 255-260.

17

BRSMG Uai: common bean cultivar with carioca grain type and upright plant architecture

Magno Antonio Patto Ramalho[1], Ângela de Fátima Barbosa Abreu[2*], José Eustáquio de Souza Carneiro[3], Leonardo Cunha Melo[2],Trazilbo José de Paula Júnior[4], Helton Santos Pereira[2], Maria José Del Peloso[2], Israel Alexandre Pereira Filho[5], Maurício Martins[6], Marcos Paiva Del Giúdice[3] and Rogério Faria Vieira[4]

***Corresponding author:**
E-mail: angela.abreu@embrapa.br

[1] Universidade Federal de Lavras (UFLA), Departamento de Biologia, C.P. 3037, 37.200-000, Lavras, MG, Brazil
[2] Embrapa Arroz e Feijão, Rod. GO-462, km 12, Zona Rural, C.P. 179, 75.375-000, Santo Antônio de Goiás, GO, Brazil
[3] Universidade Federal de Viçosa (UFV), Departamento de Fitotecnia, 36.570-000, Viçosa, MG, Brazil
[4] Empresa de Pesquisa Agropecuária de Minas Gerais (Epamig), Av. José Cândido da Silveira, 1647, União, 31.170-495, Belo Horizonte, MG, Brazil
[5] Embrapa Milho e Sorgo, Rod. MG-424, km 45, Sete Lagoas, C.P. 285, 35.701-970, Sete Lagoas, MG, Brazil
[6] Universidade Federal de Uberlândia (UFU), 38.400-902, Uberlândia, MG, Brazil

Abstract: *The common bean cultivar with carioca grain type, BRSMG Uai, is recommended for cultivation in Minas Gerais and stands out for its upright plant architecture, which facilitates cultivation and mechanical harvesting. This cultivar has high yield potential and is resistant to the major races of anthracnose that occur in region.*

Key words: *Breeding,Phaseolus vulgarisL.,grain yield, disease resistance.*

INTRODUCTION

Beans are among the main foods of the Brazilian population, with a consumption of 16 kg/inhabitant/year. Beans are important in human nutrition because they are important sources of protein and are rich in several minerals. Brazil is also a leading producer of beans. From 1993 to 2012, bean production increased by 13%, despite a reduction in planting area. This increase was due to a 62% increase in the average yield (Wander 2014). Part of this increase in yield was due to the genetic breeding of this crop.

Common beans with carioca grain (beige with a brown stripe) are preferred by consumers in Brazil. Therefore, breeding programs in Brazil have developed new lines with this grain type that combine good yield and resistance to stresses, including biotic stresses (Carvalho et al. 2013, Carbonell et al. 2014). However, there is an increasing demand related to plant architecture. The desired plants are erect, which facilitates cultivation and mechanical harvesting. This architecture also decreases disease severity by providing greater air circulation between plants and provides a better-quality grain because the pods are not in contact with the ground.

To meet the needs of bean producers, institutions working in partnership in the genetic improvement of common bean in Minas Gerais (Embrapa Rice and Beans, Agricultural Research Company of Minas Gerais (Epamig), Federal University of Lavras (UFLA) and Federal University of Viçosa (UFV)) have registered BRSMG Uai, which is a new common bean cultivar with a carioca grain type and upright plant architecture that is suitable for cultivation in the state of Minas Gerais.

BREEDING METHODS

The cultivar BRSMG Uai was obtained by hybridization in a recurrent selection program that aimed to obtain progenies with the carioca grain type, in association with upright plant architecture and high yields. In 2001, we performed a diallel cross among ten common bean inbred lines with upright plant architecture (Carioca MG, CNFC 9454, CNFC 9455, CNFC 9458, CNFC 9466, CNFC 9471, CNFC 9484, IAPAR 81, 9876 LP and IPR Uirapuru). The seeds of the 42 hybrids of the F_1 generation were sown to obtain the F_2 generation.The 42 F_2 populations were assessed for grain yield, with sowing in February 2002 in Lavras, MG, Brazil. Of these 42 populations, the 11 most promising were chosen to form the base population (S_0) of the first cycle (C_0). From these populations, 190 individual plants were selected, which gave rise to 190 progenies. These progenies were evaluated in the $S_{0:1}$, $S_{0:2}$, $S_{0:3}$, $S_{0:4}$ and $S_{0:5}$ generations from July 2002 to May 2004 and were selected for grain yield, grain type and plant architecture. In all generations, the progenies were assessed by the bulk method. In generation $S_{0:5}$, the selected progeny received the designations RP-1 and, subsequently, BRSMG Uai.

Line RP-1 was assessed in the experiments of elite lines of the Common Bean Breeding Program of UFLA, along with other 31 lines and five control cultivars (Carioca, Carioca MG, Pérola, BRSMG Talismã and Ouro Negro) in crop seasons with sowing in November 2004, February 2005, July 2005, November 2005, February 2006 and July 2006. Experiments

Table 1. Mean grain yield (kg ha^{-1}) of the cultivar BRSMG Uai and controls (BRSMG Talismã and Pérola) per location, sowing month and year of assessment in the state of Minas Gerais.

Location	Sowing month	Year	BRSMG Uai	Controls		% of the controls mean
				BRSMG Talismã	Pérola	
Lavras	July	2007	2650	2467	2016	118.2
Lambari	July	2007	1340	1385	1541	91.6
Patos de Minas	July	2007	2158	2058	1675	115.6
Sete Lagoas	July	2007	3200	3200	2933	104.4
Coimbra	July	2007	3187	2451	3170	113.4
Formoso	May	2007	1170	1053	1139	106.8
Lavras	November	2007	2400	2650	2238	98.2
Lambari	November	2007	1912	1467	1800	117.0
Coimbra	November	2007	3418	3218	3014	109.7
Lavras	February	2008	2554	2250	2562	106.2
Lambari	February	2008	2575	1883	2300	123.1
Patos Minas	February	2008	2004	1867	1717	111.8
Uberlândia	February	2008	1603	1802	1738	90.6
Florestal	February	2008	3514	3105	3514	106.2
Oratórios	February	2008	2229	1460	1581	146.6
Formoso	May	2008	1399	1591	1605	87.5
Sete Lagoas	July	2008	3575	3900	3330	98.9
Uberlândia	May	2008	2219	1964	2464	100.2
Coimbra	July	2008	3654	1681	2904	159.4
Lavras	November	2008	1345	1105	1358	109.2
Lambari	November	2008	2100	1392	1475	146.5
Patos de Minas	November	2008	2725	2332	2770	106.8
Uberlândia	November	2008	1504	1553	1410	101.5
Patos Minas	February	2009	2095	2192	2638	86.7
Viçosa	February	2009	2307	2272	2500	96.7
Uberlândia	February	2009	2439	2962	2769	85.1
Uberlândia	May	2009	2362	3322	3070	73.9
Coimbra	July	2009	2477	2243	2361	107.6
Mean sowing February			2369	2199	2369	103.7
Mean sowing May-July			2449	2276	2362	105.6
Mean sowing November			2201	1960	2009	110.9
General mean			2361	2172	2271	106.3

were carried out in Lavras and Lambari, southern Minas Gerais, and in Patos de Minas, Alto Paranaiba Region. A joint analysis of grain yield and other agronomic traits suggested that line RP-1 should be chosen to participate inValues for Cultivation and Use (VCU) trials in Minas Gerais.

RP-1 was assessed in VCU trials that were carried out from May 2007 until November 2009 in 28 environments in the state (Table 1). This line was evaluated along with 23 others and with the control cultivars BRSMG Talismã and Pérola. The experiments were arranged in a randomized complete block design, with three replications and plots consisting of four 4-m rows. The following traits were evaluated: grain yield (kg ha^{-1}); severity of angular leaf spot on a scale from 1-9, where 1 denotes a plant without disease symptoms and 9 indicates a completely infected plant (Schoonhoven and Pastor-Corrales 1987), considering resistant lines to be those with an average score up to 3 and susceptible lines to be those with an average score above 3; plant architecture, on a scale from 1-9, where 1 denotes upright plants and 9 indicates completely prostrate plants (Melo 2009); and degree of lodging, on a scale from 1-9, where 1 indicates an absence of lodging and 9 indicates all plants lodged (Melo 2009).

The reaction to races 65, 73, 81 and 89 of *Colletotrichum lindemuthianum*, which is the fungus that causes anthracnose, was evaluated in the laboratory according to the methodology described by Rava et al. (1994), as were the reaction to the common mosaic virus; the reaction to the fungus *Fusarium oxysporum*, which is the causal agent of fusarium wilt; the cooking time (in minutes) in a 25-seed Mattson cooker (Proctor and Watts 1987); and the protein content based on the total nitrogen grain content, as determined by the micro Kjedahl method using the factor 6.25 to convert the total N to protein (Horwitz 1980).

CULTIVAR CHARACTERISTICS

Plant architecture and lodging resistance

The cultivar BRSMG Uai has an indeterminate, type II growth habit. In the evaluations of plant architecture and lodging tolerance on a rating scale, the performance of BRSMG Uai was superior to that of BRS Estilo, which also has an erect architecture and is already recommended and widely accepted for cultivation in Brazil (Table 2) (Melo et al. 2010). Because it has taller plants and the tallest aboveground pod, 'BRSMG Uai' is adapted for direct mechanical harvesting, with fewer losses during this process.

Disease response

In assessments by artificial inoculation in a greenhouse, BRSMG Uai showed resistance to the races 65, 73, 81 and 89 of anthracnose, which are most frequently found in common bean crops in Brazil (Table 2).In addition, under controlled conditions, BRSMG Uai showed resistance to the common mosaic virus. In evaluations performed in the field, BRSMG Uai was susceptible to angular leaf spot and, in areas with a highly controlled infestation of *Fusarium oxysporum*, showed moderate resistance to fusarium wilt.

Crop cycle

The flowering of the cultivar BRSMG Uai occurs, on average,46 days after sowing. The crop cycle is similar to that of BRS Estilo and can be considered normal, ranging from 85 to 95 days from sowing to harvest, depending on the season and region.

Grain yield

The cultivar BRSMG Uai presented a grain yield that was 6.3% greater than the average of the controls Pérola

Table 2. Some traits of the cultivar BRSMG Uai and control BRS Estilo obtained in the VCU trials conducted in the state of Minas Gerais in the years 2007-2009 and under controlled conditions carried out at Embrapa Rice and Beans.

Trait	BRSMG Uai	BRS Estilo
Plant architecture[1]	2.0	3.4
Lodging[2]	2.6	4.0
Days until maturity	85	87
Angular leaf spot[3]	S	S
Anthracnose[3]		
Pathotype 65	R	S
Pathotype 73	R	S
Pathotype 81	R	S
Pathotype 89	R	R
Fusarium wilt[3]	MR	S
Common mosaic virus[3]	R	R
100 seed weight(g)	24	26
Cooking time (minutes)	32	26
Protein (%)	20	23

[1] Scale from 1-9, where 1 denotes upright plants and 9 indicates completely prostrate plants; [2] Scale from 1-9, where 1 indicates an absence of lodging and 9 indicates all plants lodged; [3] Reactiontodisease: R - resistant; MR - moderately resistant; S - susceptible.

and BRSMG Talismã when considering the average of all locations and seasons (Table 1).This value was higher than the average of the controls in the three sowing seasons but was most prominent in sowing performed in November, with a superiority of approximately 11%. Combined with the higher yield during this crop season, it the advantage that BRSMG Uai presents by having erect plants should once again be emphasized. When sowing is carried out in November, the harvest occurs in January, at which usually the rainfall is intense in the region. In prostrate plants, the pods can come into contact with moist soil and thus affect grain productivity and quality, which does not occur in BRSMG Uai.

Industrial and technical grain quality

Cultivar BRSMG Uai has a carioca grain type, with an average 100-seed weight of 24 g. The cooking time is 32 minutes, and the protein content is 20%, which are comparable to those of most currently recommended bean cultivars (Table 2).

BASIC SEED PRODUCTION

Cultivar BRSMG Uai was registered by Embrapa, Epamig, UFLA, and UFV in the National Register of Cultivars (RNC) of the Brazilian Ministry of Agriculture, Livestock and Supply (MAPA) on July 31, 2015, under number 33555, and the request for protection is being analyzed by the National Plant Varieties Protection Service (SNPC). Embrapa Products and Market is in charge of the seed production.

REFERENCES

Carbonell SAM, Chiorato AF, Bolonhezi D, Barros VLNP, Borges WLB, Ticelli M, Gallo PB, Finoto EL and Santos NCB (2014) 'IAC Milênio' - Common bean cultivar with high grain quality. **Crop Breeding and Applied Biotechnology 14**: 273-276.

Carvalho RSB, Lima IA, Alves FC and Santos JB (2013) Selection of carioca common bean progenies resistant to white mold. **Crop Breeding and Applied Biotechnology 13**:172-177.

Horwitz W (1980) **Official methods of analysis of the Association of Official Analytical Chemists (AOAC)**. 13th edn, Association of Official Analytical Chemists, Washington, 1018p.

Melo LC (2009) **Procedimentos para condução de ensaios de valor de cultivo e uso em feijoeiro-comum**. Embrapa Arroz e Feijão, Santo Antônio de Goiás, 104p.

Melo LC, Del Peloso MJ, Pereira HS, Faria LC, Costa JGC,Diaz JLC,Rava CA, Wendland A and Abreu AFB (2010) BRS Estilo - Common bean cultivar with carioca grain, upright growth and high yield potential. **Crop Breeding and Applied Biotechnology 10**: 377-379.

Proctor JR and Watts BM (1987) Development of a modified Mattson bean cooker produce based on sensory panel cookability evaluation. **Canadian Institute of Food Science andTecnology 20**: 9-14.

Rava CA, Purchio AF andSartorato A (1994) Caracterização de patótipos de *Colletotrichumlindemuthianum*que ocorrem em algumas regiões produtoras de feijoeiro comum. **Fitopatologia Brasileira 19**: 167-172.

Schoonhoven A and Pastor-Corrales MA (1987) **Standard system for the evaluation of bean germoplasm.** CIAT, Cali. 54p.

Wander AE (2014) Socioeconomia. In Ramalho MAP, Abreu AFB and Guilherme SR (Eds) **Informações técnicas para o cultivo do feijoeiro-comum na Região Central-Brasileira: 2015-2017**. UFLA, Lavras, p. 15-35.

18

Selection in energy cane families

Luís Cláudio Inácio da Silveira[1], Bruno Portela Brasileiro[2*], Volmir Kist[3], Heroldo Weber[2], Edelclaiton Daros[2], Luiz Alexandre Peternelli[3] and Márcio Henrique Pereira Barbosa[1]

***Corresponding author:**
E-mail: brunobiogene@hotmail.com

[1] Universidade Federal de Viçosa (UFV), Departamento de Fitotecnia, 36.570-900, Viçosa, MG, Brazil
[2] Universidade Federal do Paraná (UFPR), Departamento de Fitotecnia e Fitossanitarismo, 80.035-050, Curitiba, PR, Brazil
[3] Instituto Federal Catarinense, 89.703-720, Concórdia, SC, Brazil

Abstract: *The objective of this study was to identify superior parents for crosses and the best families to breed new clones of energy cane. The best-performing parents were RB867515, RB93509, KRAKATAU, IM76-228, IM76-229, and US85-1008. The heritability (0.59 – 0.85)and accuracy values (0.76 – 0.92) for the traits tons of cane per hectare, fiber, and sucrose content indicate high correlation between the predicted genotypic means and observed values, enabling efficient selection of the best cane energy families. The extensive genetic variation detected and the presence of promising seedlings in the selected segregating population indicate the possibility of using new crosses in some clones and even in future commercial plantings. It was possible to identify the best parents involved in the evaluated crosses, and to select, in the best families, seedlings with a high fiber and good sucrose content, seedlings with high fiber and low sugar content, as well as seedlings with the same fiber and sucrose contents as the current cultivars.*

Key words: *Saccharum spp., bioenergy, fiber, crop breeding.*

INTRODUCTION

Sugarcane (*Saccharum* spp.) is one of the most efficient crops in converting solar into chemical energy (Tew and Cobill 2008). The crop is an important food and bioenergy source and a significant component of the economy in many countries in the tropics and subtropics. With the feasibility of using cane residue and bagasse in ethanol production (second generation ethanol) and with the need to increase electricity production (thermoelectricity), sugarcane became one of the main options for the energy sector, mainly because of the high biomass yield and low production cost (Dias et al. 2013).

In recent years, sugarcane breeding focused on the development of cultivars with higher sucrose yield for sugar and ethanol production (Dal-Bianco et al. 2012, Santchurn et al. 2014, Iaia et al. 2014, Barbosa et al. 2015, Carneiro et al. 2015). The current demand for a higher biomass volume to produce second generation ethanol and electricity generates new research lines initiated for the development of cultivars with higher fiber content, the so-called cane energy (Santchurn et al. 2014, Silveira et al. 2015a, Silveira et al. 2015b).

However, since the metabolic pathways for sucrose and fiber synthesis are antagonistic (Ming et al. 2006), cane energy cultivar should be bred from crosses between genotypes of *Saccharum spontaneum* and *Saccharum officinarum*. The reason is that *S. spontaneum* has a higher fiber content than the other species of the genus *Saccharum*. However, their sucrose yield is low (Tew and Cobill 2008). On the other hand, the current sugarcane (*Saccharum* spp.) cultivars have high sucrose (>13%) and medium fiber contents (~13%). Thus, a combination of *Saccharum* spp. and *Saccharum spontaneum* seems ideal for the development of cane energy cultivars.

After hybridization with parents to obtain cane energy, families must be evaluated to then select individual plants within the best families, because selection in families with high genotypic values increases the probability of finding superior clones among the progenies (Barbosa et al. 2005). Based on this premise, family selection has been routinely applied prior to clone development in several breeding sugarcane programs (Stringer et al. 2011, Barbosa et al. 2012, Brasileiro et al. 2015, Silva et al. 2015).

Recently, the Rede Interuniversitária para o Desenvolvimento do Setor Sucroenergético (RIDESA - Inter-University Network for the Development of the Sugarcane Industry) initiated a hybridization program involving *S. spontaneum* and *S. robustum* accessions, cultivars of the Republic of Brazil (RB), and cultivars of other Brazilian breeding programs, for the development of clones with a fiber content above 17% and the current standard sucrose percentage of 13% in cultivars available for commercial cultivation (Silveira et al. 2015b).

As a contribution to the hybridization program and development of cane energy cultivars, the purpose of this study was to identify the best parents for future crosses, the best families for selection of individual plants and the best genotypes to be included in the next phase of the RIDESA breeding program.

MATERIAL AND METHODS

Family assessment

Fifty full-sib families from crosses performed in 2012 were evaluated at the Experimental Station of Serra do Ouro, of the Federal University of Alagoas, in the municipality of Murici (lat 9º 13' S, long 35º 50' W, and alt 450m asl), Alagoas, Brazil. After harvesting, seeds derived from biparental crosses were sent to the Experimental Station of Paranavaí of the Federal University of Paraná, in Paranavaí (lat 23º 05' S, long 52º 27' W, and alt 503 m asl), Paraná, Brazil, where they were sown in LVD soil. In October 2012, when grown to seedlings, the experiment was installed in a randomized block design with five replications. Each plot consisted of 20 seedlings, distributed in two 8m long rows, spaced 1.40m apart. The ratoon data were collected in July 2014, 11 months after the first cut.

The traits evaluated in families at the plot level were: mean number of stalks per plant (NS); mean stalk weight (MSW) estimated from the mean weight of 20 stalks per plot; fiber content in percent (FIB%) and sucrose content (SC%). The fiber and sucrose contents were determined in analyses of 500g samples of flour of 10 ground stalks, in 2 replications per family. In this way, we determined tons of cane per hectare, by weighing five replications (TCH.5r): TCH.5r = (NS × mSW × 10)/1.12, where: 1.12 = area occupied per plant in a plot in m^2 and 10 is the constant used to convert weight per plot into tons per hectare.

From the mean fiber content (FIB%), sucrose content (SC%) and tons of cane or stalks per hectare, in two experimental replications (TCH.2r), it was also possible to estimate the tons of fiber per hectare (TFH): TFH = (TCH.2r × FIB%)/100 and tons of sucrose per hectare (TSH):TSH = (TCH.2r × SC%)/100.

To analyze the data of TCH.5r, TCH.2r, FIB%, SC%, TFH, and TSH we used the following statistical model: y = Xr + Zg + e , where: y= data vector (y ~ N(Xr, V)); r is the vector of replication effects (assumed as fixed), added to the overall mean; g is the vector of genotypic effects,g ~ N(0, G), G = covariance matrix of genotypes ($G = I\sigma_g^2$); e is the error vector, where: e ~ N(0, R), R = residual covariance matrix ($R = I\sigma_e^2$). X and Z are the incidence matrices for these effects. The variance components σ_g^2 and σ_e^2 are the genetic and residual variance, respectively. Mixed model analysis REML/BLUP was performed using SELEGEN software (Resende 2007).

Clone selection and evaluation

The 10 best families were selected for the main yield-related traits in sugarcane (tons of cane and fiber and sucrose contents per hectare. Within each family, the target was to select at least 10% of the genotypes, i.e., 10 seedlings per family.

The selected seedlings were technically analyzed to establish estimates of the traits purity percentage (PUR%), percentage sucrose content (SC%), fiber content in percent (FIB%) and total recoverable sugar in kg of sugar per tons of cane (TRS).

RESULTS AND DISCUSSION

The genetic and environmental parameters resulting from the evaluation of 50 full-sib families by the REML/BLUP method are shown in Table 1. The significant effect of genetic variance indicates the existence of genetic variability in the evaluated families and reinforces the possibility of genetic gains by selection (Table 1).

The mean family heritability (h^2) was high for the traits FIB%, SC% and TCH.5r, which allowed selection of the best families, with accuracy higher than 76%, suggesting a high correlation between the predicted and actual mean genotypic values of the evaluated families, which indicates the possibility of genetic gain with selection of the best families (Table 1).

For TFH and TSH, h^2 was low due to the lower number of replications used in the estimates of the means of these traits (Table 1). With only two replications, heritability is greatly reduced and consequently, the accuracy in estimating the genotypic means of the trait tons of cane per hectare (TCH.2r). Since the technological analyses were carried out in only two replications, only the cane yield data of the same plots could be used to estimate the genetic parameters of the traits TFH and TSH, which is exactly what contributed to the low heritability of these traits (Table 1). However, in the estimates of genotypic means of FIB% and SC%, accuracies above 90% were obtained with only two replications, indicating that these traits were barely influenced by environmental effects.

In the evaluation of experimental traits, the accuracy of the coefficients of variation (CV%) was good (Table 1). The effects of genotypic variance were significant by Deviance analysis, indicating the presence of genetic variability among the evaluated families (Table 1).

Since the heritability for FIB%, SC% and TCH.5r was higher than heritability for tons of fiber per hectare (TFH) and tons of sucrose per hectare (TSH) (Table 1), the families were ranked for the traits TFH and TSH based on the genotypic means of FIB% and SC%, together with the genotypic mean of TCH.5r.

After ranking the families, the best 10 for TCH.5r, TFH and TSH were selected, followed by the selection of 3 to 17% of the seedlings in the best families, resulting in a total of 199 plants selected from 22 families (Table 2).

With the exploitation of the best families, possibly a higher number of high-yielding clones can be included in the subsequent steps of a sugarcane breeding programs (Barbosa et al. 2005, Brasileiro et al. 2015). Therefore, the identification of promising parents for crosses is essential to accelerate genetic gains and increase the likelihood of identification and selection of transgressive genotypes.

The families with highest means for tons of fiber per hectare (TFH) were 36 (IM76-228 × RB867515) and 46 (RB867515 × IM76-228). These families were outstanding for having a high fiber content (>14%) along with a high biomass yield, as well as the highest mean for tons of cane per hectare (TCH.5r) (Table 2 and Figure 1a).

Table 1. Estimates of genetic and environmental parameters of 50 full-sib families of the sugarcane

Parameters[a]	Traits					
	FIB%	SC%	TCH.5r	TCH.2r	TFH	TSH
$\hat{\sigma}^2_g$	1.83*	4.99*	431.33*	190.39*	10.41*	1.80*
h^2_g	0.75 +- 0.24	0.72 +- 0.24	0.32 +- 0.13	0.09 +- 0.08	0.23 +- 0.13	0.05 +- 0.06
h^2	0.85	0.84	0.59	0.16	0.38	0.10
Accuracy	0.92	0.91	0.76	0.40	0.62	0.32
r	2	2	5	2	2	2
CV%	5.95	12.00	31.86	32.74	33.30	36.72
Mean	12.99	11.35	121.43	132.72	17.33	14.85

[a]$\hat{\sigma}^2_g$: genotypic variance; h^2_g: heritability of individual plot in the broad sense, the effects total genotypic; h^2: heritability for mean families; r: number of replications; CV%: coefficient of experimental variation; *significant at 5% probability by the Deviance analysis; FIB% = fiber content (%); SC% = sucrose content (%); TCH.5r: tons of cane per hectare estimated in five replications; TCH.2r: tons of cane per hectare estimated in only two experimental replications; TFH: tons of fiber per hectare; and TSH: tons of sucrose per hectare.

Table 2. Genotypic means of the traits sucrose content (SC%), fiber content (FIB%), tons of cane per hectare in five replications (TCH.5r), tons of fiber per hectare (TFH), tons of sucrose per hectare (TSH), and number of seedlings selected (n_k) in the 22 best families

Family	Pedigree		Traits						n_k
	Mother	Father	SC%	FIB%	TCH.5r	TFH	TSH		
1	RB011941[b]	US85-1008[b]	12.37	12.54	125.75	15.77	15.56		10
3	B70710[b]	RB72910[a]	8.87	12.45	142.34	17.72	12.62		4
5	Co62175[b]	IANE48-21[b]	12.41	11.04	124.56	13.75	15.46		9
6	IAC87-3396[a]	US85-1008[b]	12.94	13.43	133.14	17.88	17.23		5
10	B70710[b]	RB93509[a]	8.75	11.85	139.96	16.58	12.25		11
11	RB83102[a]	IM76-229[c]	8.69	14.00	148.96	20.86	12.94		7
13	Co617[b]	KRAKATAU[b]	6.97	15.07	131.73	19.85	9.18		8
16	RB01649[a]	IN84-58[b]	9.23	16.04	115.62	18.55	10.68		9
22	RB867515[a]	US85-1008[b]	11.69	13.66	139.8	19.09	16.34		10
24	RB92579[a]	IM76-229[c]	9.47	14.24	142.04	20.22	13.45		9
27	RB93509[a]	Co285[b]	11.77	12.63	128.57	16.24	15.14		17
30	CTC5[a]	US85-1008[b]	13.23	13.28	121.54	16.15	16.08		3
34	RB93509[a]	KRAKATAU[b]	7.13	15.15	148.17	22.45	10.56		8
36	IM76-228[c]	RB867515[a]	9.38	14.53	158.03	22.96	14.83		10
39	Co453[b]	IAC50/134[a]	14.38	12.16	105.59	12.84	15.18		8
40	IM76-228[c]	US85-1008[b]	9.38	14.65	133.35	19.54	12.51		8
41	F150[b]	IN84-68[b]	9.46	15.39	122.45	18.84	11.59		8
43	RB867515[a]	US76-14[?]	12.31	12.97	141.77	18.39	17.45		10
44	RB93509[a]	B70710[b]	10.64	11.53	139.21	16.05	14.81		17
45	CTC9[a]	UM69/001[?]	13.03	12.34	133.4	16.46	17.38		8
46	RB867515[a]	IM76-228[c]	9.11	14.38	158.76	22.82	14.46		10
48	RB946022[a]	RB92579[a]	14.48	11.85	119.67	14.19	17.33		10
	Mean		10.71	13.41	134.29	18.08	14.22	Total	199

[a] *Saccharum* spp., [b] *Saccharum spontaneum*, [c] *Saccharum robustum*, [?] unknown parents.

The families 36 and 46 are reciprocal crosses between IM76-228 and RB867515. It shows the potential of combining these two parents to obtain energy cane cultivars, since their descendants have the vigor and productivity of RB867515, the cultivar most widely planted in the world (on nearly 3 million hectares in Brazil alone), associated with the high fiber content of IM76-228, a parent descendant of *Saccharum robustum*, whose great potential to obtain cane energy cultivars was observed in crosses.

Other families with high TFH means were 34 (RB93509 × KRAKATAU), 11 (RB83102 × IM76-229), 24 (RB92579 × IM76-229), 13 (Co617 × KRAKATAU), 40 (IM76-228 × US85- 1008), and 22 (RB867515 × US85-1008). This also demonstrated the potential of the parents KRAKATAU, IM76-229 and US85-1008 for breeding of cane energy cultivars, in view of their high fiber content and biomass production in more than one cross (Table 2). These parents should be explored in hybridizations with other parents of the breeding program of RIDESA (RB cultivars) as a way of optimizing the development process of cane energy cultivars. Furthermore, hybridization between these parents can contribute to intrapopulation improvement, since these genotypes belong to the same heterotic group.

Family 16 (RB01649 × IN84-58) had the highest mean for fiber content (FIB%) (Table 2 and Figure 1a), which may be a consequence of the genes transmitted by parent IN84-58. Therefore, new hybridizations between this parent and *Saccharum* spp. parents should be explored in order to assess the potential of this parent for the development of cane energy cultivars.

Outstanding performances were also observed in family 27 (RB93509 × Co285) and 44 (RB93509 × B70710). A large number of seedlings were selected (n_k = 17) from these families, exceeding the target (n_k = 10) during visual selection performed in the best families (Table 2).

In some families, the selection rate (n_k = 10) was not met, e.g., for family 30 (CTC5 × US85-1008), 6 (IAC87-3396 × US85-1008) and 3 (B70710 × RB72910) (Table 2). In these families, the high incidence of diseases such as smut

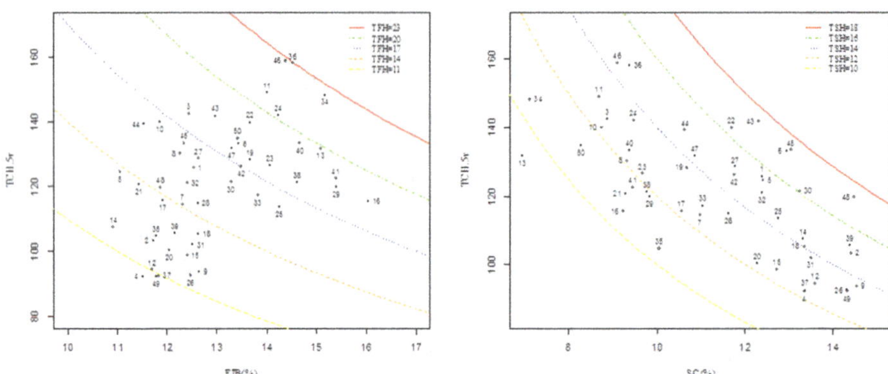

Figure 1. a) Isoquants of tons of fiber per hectare (TFH) inferred from the genotypic means of fiber content (FIB%) and tons of cane per hectare (TCH.5r) of 50 energy cane families and b) isoquants of tons of sucrose per hectare (TSH) obtained with the genotypic means of the sucrose content (SC%) and ton of cane per hectare (TCH.5r) of 50 energy cane families.

(*Sporisorium scitamineum*) and orange rust (*Puccinia kuehnii*), together with the reduced seedling vigor, resulted in the low selection rate.

The best performances in sucrose yield were observed in the families 43 (RB867515 × US76-14), 45 (CTC9 × UM69/001), 48 (RB946022 × RB92579), and 6 (IAC87-3396 × US85-1008), with sucrose yields of up to 17 tons per hectare (Table 2 and Figure 1b).

The SC was highest in family 48, which actually involved a cross between two parents, both descendants of *Saccharum* spp. This cross maintained the high sucrose content of conventional crosses but had one of the lowest fiber contents (Table 2). This result was expected, since both parents have low fiber content.

The reason to select families with high genotypic mean for TFH and TSH was mainly the maintenance of genetic variability in the second test phase (T2) of the breeding program. Thus, an advancement of clones with high fiber content during the next stages of selection is expected, in addition to the selection of clones with the industrial traits found in the current cultivars. On this basis, future parents for backcross programs can be selected, and in a more optimistic scenario, it is possible to expect the release of new cultivars of cane energy and/or conventional sugarcane, since there are promising clones among the selected.

Table 3 shows the fiber content (FIB%), sucrose content (SC%), purity (PUR%) and total recoverable sugar (TRS) of 35 of 199 of selected clones, which have industrial characteristics of energy cane.

The 35 most promising seedlings have a high fiber content and many of them also a high sucrose content. The description and agronomic evaluation of all clones during the following stages of the breeding program can confirm the potential of each selected genotype. Some of these clones may result in the first energy cane cultivars type 1, as described by Tew and Cobill (2008), with higher fiber and the current mean sucrose contents.

Among the 35 promising genotypes, 24 were selected from families with high mean FIB% and reduced mean SC%. These families were 11, 13, 16, 24, 36, 40, 41, and 46. The performance of these families for SC% and FIB% can be

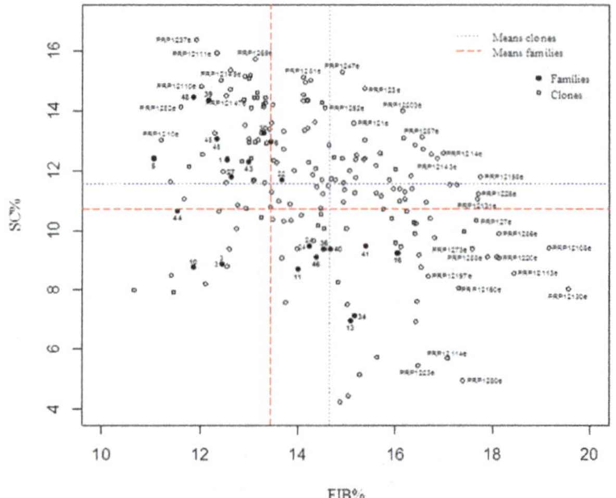

Figure 2. Dispersion of selected seedlings and of the 22 best families from the traits sucrose content (SC%) and fiber content (FIB%).

observed in Figure 1b. In family 36, with highest TFH (Table 2 and Figure 1b), 5 of 10 selected seedlings were among the 35 most promising genotypes, demonstrating once again the potential of the combination between the parents IM76-228 and RB867515.

As seen in Figure 2, there is broad genetic variability for SC% and FIB% both among families and among selected plants, demonstrating that the selection strategy was effective in maintaining the variability to be explored in the second selection stage.

By the selection strategy used here, it was possible to increase the probability of selecting seedlings with specific traits, such as the genotypes PRP1280e, PRP12108e, and PRP12130e, with fiber contents between 17.38% and 19.55% and low sucrose content, as well as genotypes with fiber content of over 15%, but also with high sucrose content,

Table 3. Means of fiber content (FIB%), sucrose content (SC%), purity (PUR%) and total recoverable sugar (TRS) and the origin (family) of 35 promising genotypes

Clone	Family	FIB%	SC%	PUR%	TRS (kg tons^{-1} of cane)
PRP1235e	1	16.17	13.08	89.29	128.75
PRP12135e	3	15.35	12.37	90.74	121.66
PRP12153e	6	16.00	12.04	86.41	119.49
PRP12108e	11	19.15	9.42	76.31	96.63
PRP1251e	11	15.36	13.00	87.67	128.42
PRP127e	11	17.65	10.32	79.76	104.61
PRP12113e	13	18.44	8.56	71.64	89.63
PRP12140e	13	16.56	12.71	91.34	124.68
PRP12187e	13	18.09	9.11	81.41	92.58
PRP1280e	13	17.38	4.96	59.70	58.38
PRP12130e	16	19.55	8.03	69.57	84.92
PRP12160e	16	17.29	8.04	67.53	85.82
PRP1288e	16	17.89	9.09	79.82	92.82
PRP12122e	22	15.19	12.49	80.66	125.30
PRP1214e	24	16.98	12.57	91.32	123.30
PRP12158e	24	17.75	11.80	81.72	118.18
PRP12114e	27	17.06	5.70	58.71	65.66
PRP1297e	27	15.67	12.57	88.75	123.99
PRP12143e	36	16.86	12.39	85.95	122.87
PRP121e	36	15.13	13.60	84.87	134.82
PRP1220e	36	18.12	9.07	74.73	93.80
PRP1286e	36	18.13	9.89	78.73	100.66
PRP1287e	36	16.54	13.12	87.96	129.38
PRP12203e	39	16.15	14.00	87.19	138.01
PRP12131e	40	17.68	11.05	80.44	111.38
PRP1217e	40	16.40	12.84	86.35	127.18
PRP1294e	40	17.25	11.52	85.22	114.74
PRP12107e	41	17.40	9.97	79.14	101.42
PRP1283e	41	17.12	11.52	80.24	115.92
PRP1226e	43	17.69	11.22	81.01	112.85
PRP1260e	43	16.73	12.55	84.66	124.72
PRP1273e	44	17.58	9.36	80.10	95.41
PRP12154e	45	15.36	12.39	84.06	123.54
PRP123e	46	15.36	14.74	86.79	145.19
PRP1257e	46	15.60	12.16	76.57	123.14
Mean		17.06	11.80	81.41	118.18
Standard deviation		1.12	2.26	8.11	19.84
Maximum		19.55	14.74	91.34	145.19
Minimum		15.19	4.96	58.71	58.38

similarly to PRP12158e, PRP1214e, PRP12203e, PRP1260e, and PRP123e, aside from other genotypes that also had higher means of SC% and/or FIB% than the selected clones (Figure 2).

In the breeding program of Barbados, clones with up to 42% fiber content and a Brix content of 23.5% were obtained by crossing parents originating from recurrent selection with *S. spontaneum* accessions to increase sucrose contents (Kennedy 2008). Santchurn et al. (2014) also obtained four types of high biomass genotypes in the Mauritius Sugarcane Industry Research Institute. The results obtained in Barbados, Mauritius, as well as in this study showed that the broad genetic variability in the *Saccharum* complex can be exploited for different purposes, in view of the numerous possibilities for new gene rearrangements in sugarcane, which will allow radical changes in the traits of future cultivar.

The selected genotypes have great potential to be exploited commercially and also in future crosses. Importantly, although these clones have interesting industrial traits, they must have high biomass productivity, be tolerant to the major pests and diseases, have longevity of the ratoon crop. Therefore, these traits should also be evaluated in the next selection stages.

The selection of sugarcane parents and hybrid combinations should be based on data of genetic diversity and the inbreeding coefficient to fully exploit the effects of dominance in hybridizations between parents of different heterotic groups (Brasileiro et al. 2014). The adequate choice of the parents and combinations between them has contributed to the development of new cultivars, such as the large number of cultivars developed from hybridizations of RB72454 and SP70-1143. The crosses involving these two parents resulted in 10 cultivars, including the most commonly cultivated and planted in Brazil (Daros et al. 2015). In fact, specific combinations are able to raise heterosis in segregating generations and increase the chances of identifying transgressive clones in their progenies (Silveira et al. 2015a).

Hybrid combinations should be indicated based on the potential *per se* of the parents and the magnitude of their dissimilarities. The findings of this study may be highly useful in the development of new crosses involving parents with higher fiber content, which are mainly descendants of *S. spontaneum* and *S. robustum* with current clones and cultivars (*Saccharum* spp.).

Breeders have thoroughly explored the high sucrose potential in the sugarcane breeding programs. However, due to difficulties in recent years in obtaining genetic gain for this trait (Dal-Bianco et al. 2012), apart from the growing worldwide demand for renewable energy, an alternative for the sugar-energy industry would be the development of cultivars with higher biomass production capacity and high fiber content. These new cultivars can significantly increase energy cogeneration, and optimize the exploitation of cellulose and hemicellulose, both in the production of second generation ethanol as of bio-oil and other products from biomass pyrolysis (Yang et al. 2006).

According to Tew and Cobill (2008), currently grown sugarcane contains around 12% fiber and 13% sugar. According to these authors, the target of breeding cane energy cultivars should be 30% fiber and 5% sugar. But the way to achieve genotypes with these characteristics is still very long. Nowadays, energy plants are not able to process sugarcane with more than 20% fiber content. In this case, to meet the present demand of the sugarcane industry, cultivars with contents of fiber around 17% and sucrose of 13% are needed. This increase in fiber content would be interesting from an industrial point of view, once the sugar yield would be maintained and quality of the bagasse used in power cogeneration would be better and the calorific value consequently higher, allowing higher electricity production per ton of processed waste.

Reciprocal recurrent selection (RRS) should be applied based on the pedigree information and genetic diversity in order to exploit the heterosis of future crosses as much as possible. The potential of RRS to obtain vigorous hybrids is considerable. Therefore, the populations of *Saccharum* spontaneum, *Saccharum* robustum and *Saccharum* spp. must be improved by intra and interspecific hybridizations of the best parents of each population and recombinant generation.

Increasing the fiber content via recurrent selection in *S. spontaneum* and *S. robustum* populations is essential for the success of the cane energy program, since the use of parents with fiber levels superior to those used in the 50 crosses evaluated in this study can generate plants with fiber levels higher than those selected to date.

The findings of this study are essential as orientation for new hybridizations and to develop new evaluation and selection strategies for cane energy families. Crosses between *Saccharum* spp. × *Saccharum spontaneum* and *Saccharum* spp. × *Saccharum robustum* should be heavily exploited, with a view to increasing the current fiber content in future cultivars

(Ming et al. 2006). In addition, due to the expansion of sugarcane cultivation in Brazil in recent years, especially in areas with soil and climatic constraints, crosses of current cultivars with descendants of *S. spontaneum* and *S. robustum* can also contribute to the introgression of alleles responsible for providing increased longevity of sugarcane with mechanical planting and harvesting, as well as alleles with higher tillering capacity and also for tolerance of the major pests and diseases, aside from other biotic and abiotic stresses occurring in sugarcane.

CONCLUSION

The best parents in the development of clones with high fiber productivity are RB867515, RB93509, RB92579, KRAKATAU, IM76-228, IM76-229, and US85-1008. It was also possible to identify and select seedlings of the best families with a high fiber content combined with good sucrose content, seedlings with fiber content close to 20% and a low sucrose content, as well as seedlings with the same fiber and sucrose contents as the currently planted cultivars.

ACKNOWLEDGEMENTS

The authors are indebted to the Coordination of Improvement of Higher Education Personnel (CAPES), and the National Research Council (CNPq) for financial support.

REFERENCES

Barbosa GVS, Oliveira RA, Cruz MM, Santos JM, Silva PP, Viveiros AJA, Sousa AJR, Ribeiro CAG, Soares L, Teodoro I, Sampaio Filho F, Diniz CA and Torres VLD (2015) RB99395: Sugarcane cultivar with high sucrose contente. **Crop Breeding and Applied Biotechnology 15**: 187-190.

Barbosa MHP, Resende MDV, Bressiani JA, Silveira LCI and Peternelli LA (2005) Selection of sugarcane families and parents by REML/BLUP. **Crop Breeding and Applied Biotechnology 5**: 443-450.

Barbosa MHP, Resende MDV, Dias LAS, Barbosa GVS, Oliveira RA, Peternelli LA and Daros E (2012) Genetic improvement of sugarcane for bioenergy: the Brazilian experience in network research with RIDESA. **Crop Breeding and Applied Biotechnology S12**: 87-98.

Brasileiro BP, Marinho CD, Costa PMA, Cruz CD, Peternelli LA and Barbosa MHP (2015) Selection in sugarcane families with artificial neural networks. **Crop Breeding and Applied Biotechnology 15**: 72-78.

Brasileiro BP, Marinho CD, Costa PMA, Moreira EFA, Peternelli LA and Barbosa MHP (2014) Genetic diversity in sugarcane varieties in Brazil based on the Ward-modified location model clustering strategy. **Genetics and Molecular Research 13**: 1650-1660.

Carneiro MS, Chapola RG, Fernandes Júnior AR, Cursi DE, Barreto FZ, Balsalobre TWA and Hoffmann HP (2015) RB975952 – Early maturing sugarcane cultivar. **Crop Breeding and Applied Biotechnology 15**: 193-196.

Dal-Bianco M, Carneiro MS, Hotta CT, Chapola RG, Hoffmann HP, Garcia AAF and Souza GM (2012) Sugarcane improvement: How far can we go? **Current Opinion in Biotechnology 23**: 265-270.

Daros E, Oliveira RA and Barbosa GVS (2015) **45 anos de variedades RB de cana-de-açúcar: 25 anos de Ridesa**. Graciosa, Curitiba, 156p.

Dias MOS, Junqueira TL, Cavalett O, Cunha MP, Jesus CDF, Mantelatto PE, Rossell CEV, Maciel Filho R and Bonomi A (2013) Cogeneration in integrated first and second generation ethanol from sugarcane. **Chemical Engineering Research and Design 91**: 411-1417.

Kennedy AJ (2008) Prospects for combining high sucrose content with increased fibre to generate multi-purpose cane varieties. http://www.jamaicasugar.org/wist/Proceedings/Prospects%20for%20combining%20high%20sucrose%20content.pdf

Iaia AM, Oliveira RA, Melo LJOT, Daros E, Simões Neto DE, Bastos GQ, Oliveira FJ, Chaves A and Melo TTAT (2014) RB002504 - New early-maturing sugarcane cultivar. **Crop Breeding and Applied Biotechnology 14**: 45-47.

Ming R, Moore PH, D`Hont A, Glaszmann JC, Tew TL, Mirkov TE, Silva J, Jifon J, Rai M, Schell RJ, Brumbley SM, Lakshmanan P, Comstock JC and Paterson AH (2006) Sugarcane Improvement through Breeding and Biotechnology. In Janick J (ed) **Plant breeding reviews**. John Wiley, New York, p. 15-118.

Resende MDV (2007) **Software SELEGEN-REML/BLUP: sistema estatístico e seleção genética computadorizada via modelos lineares mistos**. Embrapa Florestas, Colombo, 359p.

Santchurn D, Ramdoyal K, Badaloo MGH and Labuschagne M (2014) From sugar industry to cane industry: Evaluation and simultaneous selection of different types of high biomass canes. **Biomass and Bioenergy 61**: 82-92.

Silva FL, Barbosa MHP, Resende MDV, Peternelli LA and Pedrozo CA (2015) Efficiency of selection within sugarcane families via simulated individual BLUP. **Crop Breeding and Applied Biotechnology 15**: 1-9.

Silveira LCI, Brasileiro BP, Kist V, Daros E and Peternelli LA (2015a) Genetic diversity and coefficient of kinship among potential genitors for obtaining cultivars of energy cane. **Revista Ciência Agronômica 46**: 358-368.

Silveira LCI, Brasileiro BP, Kist V, Daros E, Peternelli LA and Barbosa MHP (2015b) Selection strategy in families of energy cane based on biomass production and quality traits. **Euphytica 201**: 443-455.

Stringer JK, Cox MC, Atkin FC, Wei X and Hogarth DM (2011) Family selection improves the efficiency and effectiveness of selecting original seedlings and parents. **Sugar Tech 13**: 36-41.

Tew TL and Cobill RM (2008) Genetic improvement of sugarcane *(Saccharum* spp.) as an energy crop. In Vermerris W (ed) **Genetic improvement of bioenergy crops**. Springer, New York, p. 273-294.

Yang H, Yan R, Chen H, Zheng C, Lee DH and Liang DT (2006) In-Depth investigation of biomass pyrolysis based on three major components: hemicellulose, cellulose and lignin. **Energy Fuels 20**: 388-393.

19

Genetic effects and potential parents in cowpea

Francisco Tiago Cunha Dias[1], Cândida Hermínia Campos de Magalhães Bertini[1*] and Francisco Rodrigues Freire Filho[2]

Abstract: *Six cowpea genotypes and their F_2 hybrid combinations were evaluated for general and specific combining ability. The Griffing's diallel cross design, Method 2, and mixed model B were used. The genotypes and hybrids differed statistically (p <0.01) for the 10 studied traits. With regard to the general and specific combining ability, there were statistical differences at 1% probability for all traits. The presence of additive and non-additive gene effects paves the way for breeding new hybrid cultivars. However, additive gene effects were predominant in the trait expression. Genotypes CE-542, CE-954 and CE-796 were identified as the most promising of the test group for inclusion in cowpea breeding programs.*

Key words: *Vigna unguiculata, combining ability, diallel analysis, yield components.*

***Corresponding author:**
E-mail: candida@ufc.br

[1] Universidade Federal do Ceará, Av. Mister Hull, 2977, Antônio Bezerra, 60.021-970, Fortaleza, CE, Brazil

[2] Embrapa Amazônia Oriental, Trav. Dr. Enéas Pinheiro, s/n, Bairro Marco, CP 48, 66.095-903, Belém, PA, Brazil

INTRODUCTION

Cowpea is one of the key food sources in the arid, semi-arid and tropical parts of Asia, Oceania, southern Europe, Africa, southern United States, and Central and South America (Singh et al. 2002). The variations in crop production under different environmental conditions stimulated the development of stable cultivars in breeding programs, with the selection of high-yielding lines (Romanus et al. 2008).

According to FAO (http://faostat3.fao.org/home/index.html), the world cowpea production in 2011 was 4.9 million tons, harvested from a cultivated area of more than 10 million hectares, mainly in Nigeria, Niger and Brazil. However, it is noteworthy that in Brazil, the IBGE does not distinguish data of common bean from those of cowpea, unlike several other producing countries. For Singh (2006), cowpea production is relevant in more than 65 countries. However, the lack of information about the crop in Brazil hampers data collection for FAO.

Historically, cowpea production in Brazil was concentrated in the Northeast (1.2 million hectares) and North (55,800 hectares). However, the crop area is expanding in the Midwest region, due to the development of cultivars with characteristics that favor mechanical cultivation (Saidi et al. 2010). Despite the expansion of cultivation into other regions of Brazil, cowpea yields in Ceará have been steadily declining, due to numerous factors, e.g., cultivars with low yield potential and cultivation as extensive secondary crop, associated with other crops, mainly maize, cassava and cotton.

One way to increase the species productivity is by the selection of plants in segregating populations obtained by hybridization of superior genotypes.

Sprague and Tatum (1942) proposed the use of diallel crosses to estimate the general and specific combining ability (GCA and SCA, respectively), in order to select superior parents for crosses. The GCA indicates the mean contribution of a parent or clone to hybrid combinations. The SCA is used to detect cases in which certain hybrid combinations perform better or worse than expected, based on the mean performance of the parents. The GCA is associated to genes with mainly additive effects, aside from the additive epistatic x additive effects. In turn, the SCA is basically defined by genes with dominant effects and various types of interactions (Cruz and Regazzi 2004).

Apart from providing important information about the GCA and SCA, the diallel analysis has been used in cowpea to estimate genetic variance and heritability and to detect the presence of maternal effects (Hazra et al. 1994). Therefore, the purpose of this research was to i) estimate the combining ability of early-maturing and upright genotypes and identify the most promising crosses, and ii) identify the dominant gene action in agronomic traits of cowpea.

MATERIAL AND METHODS

Genotypes and hybrid combinations

In the experiment, six cowpea genotypes from the genebank of the Universidade Federal do Ceará and the breeding program of Embrapa Meio Norte, Teresina, PI, were used. For the selection of the parents, the following criteria were used: earliness (cycle shorter than 70 days), upright growth (main and secondary short branches at right to acute angles in relation to the main stem), little or no common ancestry, and assessment of the agronomic value, based on the general appearance of each genotype. The genotypes CE-542, CE-796, CE-945, CE-954, F4RC1 (Plant J), and MNC03-737E-5-10 were selected.

The crosses between genotypes were made in a greenhouse of Embrapa Meio Norte, by two methods:

Methodology A

a) Collection of pollen from open flowers (pollen donor genotypes) in the morning, wrapping in paper bags and cold storage (4 °C);

b) Emasculation and pollination of developed flowers (maternal genotypes) that opened after 24 hours, or in the late afternoon of the day prior to natural anthesis.

Methodology B

a) Emasculation of developed flowers (maternal genotypes) prior to natural anthesis the next morning;

b) Collection of pollen from open flowers (pollen donor genotypes) used for pollination of emasculated flowers the day before.

The genotypes were sown in two steps. Four plants per meter of each genotype were sown in one 3-m row (spacing of 0.25 m between plants and 1.25 m between rows). The second sowing was carried out one week after the first.

After artificial pollination, the crosses were labeled with the cross code. The symbols used for the establishment of the cross order was recommended by Freire Filho (2005). Crosses were indicated by "x" and mathematical symbols (parentheses, brackets and keys) were used to establish the order of the crosses. The pods from the crosses were harvested separately, as they reached maturity.

Evaluation of hybrid combinations

The F_1 progenies were sown in trays with soil as substrate. After the emergence of the first trifoliate leaf, the seedlings were planted in a greenhouse in Acaraú (lat 3° 6′ N, long 40º 2′ W and alt 30 m asl), Ceará. To prevent the inclusion of selfings that could occur among the hybrids, the pods of each cross were maintained separate. Furthermore, in the F_1 plants of each plant pod the expression of morphological genetic markers was separately observed, e.g., for flower color, pod position, shape of the central leaflet, plant pigmentation, and growth habit. Thus, plants on which the markers could not be found were eliminated.

The F_2 generation was evaluated in a randomized complete block design with six replications. Each plot consisted of four 4.0 m rows, and with rows spaced 0.50 m and plants 0.20 m apart, corresponding to an area of 8 m^2. The two central rows, disregarding two plant spots on either end, were the evaluated portion (3.6 m^2). The traits evaluated in this experiment were days to flowering (DuF), insertion angle of the lateral branches (IALB), number of nodes on the main branch (NNMB), number of pods per plant (NPP), number of seeds per pod (NSP), pod length (PL), 100-seed weight (W100S), grain index (GI), grain yield (GY) in kg ha^{-1}, and days to maturity (DuM).

In the diallel analysis, we used the parents and F_2 progenies without reciprocal crosses. The diallel analysis was performed in the F_2 generation, since is rather difficult to obtain large amounts of F_1 seeds of cowpea and other pollinated plants. It is also interesting that in the F_2 generation, the contribution of dominance effects to the mean and variance was reduced to half and to a quarter, respectively, in comparison with the F_1 generation. Thus, the SCA in the F_2 generation is not estimated completely. However, the GCA is more accurately estimated because it depends mainly on the additive effects, which is particularly interesting for breeders of autogamous species.

This study was adopted for the analysis method 2 and the mixed model B of Griffing (1965). In this model, the effects of treatments are fixed and those of blocks and experimental error are random. The treatment effects being fixed means that any conclusions are only valid for the experimental material.

The chosen model enables the partitioning of the combinatorial sum of the partial diallels that include the parental genotypes. In this sense, the combining ability of parents can be compared when they are used as control. Moreover, the best hybrid combinations can be identified. All genetic statistical analyses were performed with software Genes (Cruz 2006).

RESULTS AND DISCUSSION

There were significant differences for all studied traits (Table 1). This presupposes genetic differences between cultivars in diallel crosses. This result is important with regard to the possibility of genetic gains in breeding. For Ramalho et al. (1993), the possibility of selection of superior genotypes depends strictly on the existing genetic diversity in the set of lines, which in turn is a function of the influence of the additive variance. For Krause et al. (2012), the existence of genetic variability in the cultivars used as progenitors is essential for the selection of promising genotypes.

In relation to GCA and SCA (Table 2), statistical differences were observed at 1% probability for all analyzed trait. These differences indicate variability, resulting from additive and non-additive gene effects, indicating the possibility of obtaining new cultivars or hybrid (Silva et al. 2004). Similar results were reported in cowpea by Carvalho et al. (2012), for the traits 100-seed weight, number of seeds per pod and yield; and by Kimani and Derera (2009) for the traits flowering, number of seeds per pod and 100-seed weight in common bean, indicating the possibility of gains with selection.

According to Medici et al. (2004), significant GCC effects indicate that at least one of the parental genotypes differs from the other in relation to the number of favorable genes with additive effects. A significant SCA indicates that the

Table 1. Estimates of mean squares, general (GCA) and specific (SCA) combining ability for 10 traits of cowpea genotypes in a 6 x 6 diallel analysis

Sources of Variation	df	Mean squares									
		GY[1]	NPP[2]	NSP[3]	PL[4]	W100S[5]	DuF[6]	IALB[7]	NNMB[8]	DuM[9]	GI[10]
Genotypes	20	79940.598*	69.00**	44.72**	53.47**	72.45**	107.90**	344.18**	48.07**	140.98**	0.014**
GCA	5	98797.24*	90.65**	104.35**	46.62**	46.57**	29.80**	482.03**	131.06**	192.95**	0.027**
SCA	15	73655.04*	61.78**	24.84**	55.76**	81.08**	123.87**	298.23**	20.41**	123.65**	0.010**
Error	100	44301.04	5.37	2.52	2.44	3.14	1.08	27.96	0.92	0.61	0.000741
Means		1187.51	13.79	12.01	17.58	15.19	36.06	34.84	9.35	65.38	0.78
CVe		17.72	16.81	13.22	8.90	11.67	2.88	15.18	10.29	1.20	3.48
Components											
GCA		4892.33	9.40	3.72	8.88	12.98	22.14	45.04	3.24	20.50	0.001569
SCA		1135.33	1.77	2.12	0.92	0.90	0.59	9.45	2.71	4.00	0.000567

[1] Grain yield (kg ha^{-1}); [2] Number of pods per plant; [3] Number of seeds per pod; [4] Pod length; [5] 100-seed weight; [6] Days until flowering; [7] Insertion angle of the lateral branches; [8] Number of nodes on the main branch; [9] Days until maturity; [10] Grain index. CVe: Coefficient of experimental variation.

hybrids have a higher or lower performance than expected, based on the GCA of the parents (Oliboni et al. 2013).

It was observed that the mean square effects of GCA and SCA were similar for all study traits (Table 1). However, according to Barelli et al. (2000), inferences based on mean squares are not recommended in autogamous plants, for not detecting the predominance of additive and non-additive gene action. Thus, the use of mean square effects for the genetic quadratic components for GCA and SCA is more appropriate and the assessment of the gene action involved in the trait expression. In this sense, predominant additive gene action was observed for most traits (Table 3). This is probably due to the fact that this analysis was conducted in the F_2 generation, where the additive gene action increases and the dominance deviations decrease.

The estimated GCA effects (g_i) inform about the potential of a genotype to generate good populations for breeding and about the concentration of alleles with predominantly additive effects. The higher the positive or negative g_i

Table 2. Estimates of the general combining ability (GCA) of six cowpea parents with upright growth and early cycle

Parents	Traits[1]									
	GY	NPP	NSP	PL	W100S	DuF	IALB	NNMB	DuM	GI
CE-542 (1)	43.1533	-2.3104	1.5441	1.3604	-1.1691	-1.5750	-4.9058	2.8466	2.7091	-0.0329
CE-796 (2)	13.8883	-0.6404	2.0091	0.3191	1.1595	0.0275	-2.9395	1.1429	-1.4795	0.0020
CE-945 (3)	29.7133	0.4170	-0.6970	0.2866	0.6545	0.4875	2.0954	-1.2995	0.9379	-0.0066
CE-954 (4)	84.2129	0.1308	-0.0545	-1.0595	-1.0616	0.2762	1.1666	-0.6595	-2.9995	-0.0016
F4RC1 (5)	10.0945	0.6458	-1.2670	-1.2558	-0.2904	0.4225	3.0816	-0.8458	-0.0008	0.0420
MNC03-737E-5-10 (6)	-12.6366	1.7570	-1.5345	0.3491	0.7070	0.3612	1.5016	-1.1845	0.8329	-0.0029
SD (gi)	27.7329	0.3054	0.2093	0.2061	0.2336	0.1370	0.6967	0.1268	0.1029	0.0035
SD ($g_i - g_j$)	42.9636	0.4731	0.3242	0.3193	0.3619	0.2122	1.0794	0.1965	0.1595	0.0055

[1] See abbreviations in Table 1.

Table 3. Estimates of the specific combining ability (SCA) of six cowpea parents with upright growth and early cycle

Parents/ Hybrids	Estimates of the specific combining ability (SCA)									
	GY[1]	NPP	NSP	PL	W100S	DuF	IALB	NNMB	DuM	GI
1 x 1	-56.09	3.94	-3.36	-3.80	-2.33	-10.24	10.86	-0.23	-1.62	0.01
1 x 2	63.46	-1.17	1.21	0.18	-1.58	-1.01	2.67	0.90	3.05	-0.02
1 x 3	-29.47	0.67	1.74	2.52	2.92	6.19	-1.83	-2.01	-4.85	-0.01
1 x 4	146.38	-1.61	3.84	1.93	0.26	5.90	-1.91	3.81	4.73	-0.02
1 x 5	-123.08	-3.10	-1.42	0.17	0.64	5.25	-12.54	-0.41	-0.58	0.01
1 x 6	54.88	-2.67	1.35	2.79	2.42	4.14	-8.10	-1.81	0.90	0.02
2 x 2	-20.55	-1.66	-1.49	-1.94	4.31	4.88	1.81	3.04	4.40	0.04
2 x 3	6.62	4.10	0.39	-3.61	-5.92	0.92	3.17	-1.94	-6.16	-0.02
2 x 4	-18.76	4.81	-1.21	-1.95	-5.38	-0.86	-4.62	-1.77	-3.57	0.04
2 x 5	-45.81	-3.41	1.51	2.30	-1.73	-4.51	-4.65	-2.04	-6.57	-0.07
2 x 6	25.59	-1.00	1.08	6.97	6.01	-4.28	-0.21	-1.23	4.43	-0.01
3 x 3	18.29	-1.28	1.36	0.53	-1.64	-3.36	-0.28	1.10	2.74	-0.02
3 x 4	-59.89	-3.29	-3.35	2.04	2.21	1.34	-1.27	-0.36	0.68	-0.03
3 x 5	2.07	1.09	-0.88	-0.20	2.53	-1.47	-3.09	1.18	2.51	0.05
3 x 6	44.08	-0.02	-0.61	-1.81	1.53	-0.24	3.59	0.93	2.34	0.06
4 x 4	8.91	2.90	0.69	0.70	0.58	0.21	8.29	-0.97	1.78	0.01
4 x 5	-113.27	-1.62	-1.40	0.95	1.81	-2.43	-6.83	-0.23	-6.21	-0.02
4 x 6	17.72	-4.08	0.73	-4.38	-0.07	-4.37	-1.94	0.51	0.78	0.02
5 x 5	80.73	1.65	1.15	-1.01	-2.25	0.59	13.41	0.75	4.12	0.02
5 x 6	18.61	3.74	-0.09	-1.20	1.26	1.98	0.29	0.01	2.61	-0.01
6 x 6	-230.45	2.01	-1.23	-1.18	-5.58	1.38	3.18	0.79	-5.54	-0.04
SD ($S_{ij} - S_{ik}$)	113.67	1.25	0.85	0.84	0.95	0.56	2.85	0.51	0.42	0.01
SD ($S_{ij} - S_{kl}$)	105.23	1.15	0.79	0.78	0.88	0.51	2.64	0.48	0.39	0.01

[1] See abbreviations in Table 1.

estimates, the greater is the superiority or not of the parent compared to the other genotypes in the diallel. On the other hand, values close to zero indicate that the performance does not differ from that of the general mean of the crosses (Bastos et al. 2003).

The GCA values of genotype CE-542 were high and positive for yield, NSP and PL and negative for the variables DuF and IRLA (Table 2). This indicates the possibility of using this parent to establish a base population to breed more productive progenies with more seeds, longer pods and more upright plants (negative g_i values).

The genotype that contributed most to earliness was CE-954 which, concomitantly, increased yield most (Table 2). Another genotype that can potentially reduce the plant and improve other studied traits was CE-796. Based on this information and the focus of the cowpea breeding program to obtain more productive, upright and early genotypes, crosses between the genotypes CE-542 and CE-954 and CE-796 were identified as the most promising to obtain gains for these traits.

The effect of SCA is interpreted as the deviation of a hybrid from what would be expected based on the GCA of its parents. It is therefore worth highlighting the importance of non-additive interactions resulting from genetic complementation between the parents. Thus, the study of SCA allows predict the possibility of genetic gain by the exploitation of heterosis (Bastos et al. 2003).

High positive or negative SCA (S_{ij}) estimates show that the performance of a cross is relatively better or worse than would be expected based on the GCA of the parents (Rainey and Griffiths 2005). Therefore, hybrid combinations with high means are expected when at least one parent has a more favorable effect of GCA (Tchiagam et al. 2011). Thus, the most promising crosses for higher yield were CE-542 x CE-954 and CE-542 x CE-796 (Table 3). The reason is that these combinations resulted in good yields and high SCA estimates, as were the genotypes with more favorable effects for GCA.

With regard to plant architecture, the insertion angle of lateral branches decreased most in cross CE-542 x F4RC1, probably because it involved more upright plants. For a short crop cycle, the cross with greatest reduction was CE-954 x F4RC1.

ACKNOWLEDGEMENTS

The authors thank Brazilian Federal Agency for Support and Evaluation of Graduate Education (CAPES) for funding this project and the Federal University of Ceará and Embrapa Meia Norte for their help in carrying out the research.

REFERENCES

Barelli MAA, Gonçalves-Vidigal MC, Amaral Júnior AT, Vidigal Filho PS, Scapim CA and Sagrilo E (2000) Diallel analysis for grain yield and yield components in *Phaseolus vulgaris* L. **Acta Scientiarum. Agronomy 22**: 883-887.

Bastos IT, Barbosa MHP, Cruz CD, Burnquist WL, Bressiani JA and Silva FL (2003) Análise dialélica em clones de cana-de-açúcar. **Bragantia 62**: 199-206.

Carvalho LCB, Silva KJD, Rocha MM, Sousa MB, Pires CJ and Nunes JAR (2012) Phenotypic correlations between combining abilities of F_2 cowpea populations. **Crop Breeding and Applied Biotechnology 12**: 211-214.

Cruz CD (2006) **Programa Genes**. 1st edn, Editora UFV, Viçosa, 382p.

Cruz CD and Regazzi AJ (2004) **Modelos biométricos aplicados ao melhoramento genético**. Editora UFV, Viçosa, 390p.

Freire Filho FR, Ribeiro VQ, Barreto PD and Santos AA (2005) melhoramento genético. In Freire Filho FR, Lima JAA and Ribeiro VQ (eds) **Feijão-caupi: avanços tecnológicos**. Embrapa Informação Tecnológica, Brasília, p. 29-92.

Griffing BAA (1956) Concept of general and specific combining ability in relation to diallel crossing systems. **Australian Journal of Biological Sciences 9**: 463-493.

Hazra P, Das PK and Som MG (1994) Analysis of heterosis for pod yield and its components in relation to genetic divergence of the parents and specific combining ability of the crosses in cowpea (*Vigna unguiculata* (L.) Walp). **Jornal Genetic Plant Breeding 53**: 418-423.

Kimani J and Derera J (2009) Combining ability analysis across environments for some traits in dry bean (*Phaseolus vulgaris* L.) under low and high soil phosphorus conditions. **Euphytica 166**: 1-13.

Krause w, Souza RS, Neves LG, Carvalho MLS, Viana AP and Faleiro FG (2012) Ganho de seleção no melhoramento genético intrapopulacional do maracujazeiro-amarelo. **Pesquisa Agropecuária Brasileira 47**: 51-57.

Medici LO, Pereira MB, Lea PJ and Azevedo RA (2004) Diallel analysis of maize lines with contrasting responses to applied nitrogen. **Journal of Agricultural Science 142**: 535-541.

Oliboni R, Faria MV, Neumann M, Resende JTV, Battistelli GM, Tegoni RF and Oliboni DF (2013) Análise dialélica na avaliação do potencial de híbridos de milho para a geração de populações-base para obtenção de linhagens. **Agronomia 34**: 7-18.

Rainey KM and Griffiths PD (2005) Diallel analysis of yield components of snap beans exposed to two temperature stress environments. **Euphytica 142**: 43-53.

Ramalho MAP, Santos JB and Zimmermann MJO (1993) **Genética quantitativa em plantas autógamas: aplicação ao melhoramento do feijoeiro**. Editora UFG, Goiânia, 271p.

Romanus KG, Hussein S and Mashela WP (2008) Combining ability analysis and association of yield and yield components among selected cowpea lines. **Euphytica 162**: 205-210.

Saidi M, Itulya FM, Aguyoh J and Ngouajio M (2010) Leaf harvesting time and frequency affect vegetative and grain yield of cowpea. **American Journal 102**: 827-833.

Scott AJ and Knott M (1974) Cluster analysis method for grouping means in the analysis of variance. **Biometrics 30**: 507-512.

Silva MP, Amaral Junior AT, Rodrigues R, Daher RF, Leal NR and Schuelter AR (2004) Análise dialélica da capacidade combinatória em feijão-de-vagem. **Horticultura Brasileira 22**: 277-280.

Singh BB, Ehlers JD, Sharma B and Freire Filho FR (2002) Recent progress in cowpea breeding. In Fatokun CA, Tarawali SA, Singh BB, Kormawa PM and Tamò M (eds) **Challenges and opportunities for enhancing sustainable cowpea production**. International Institute of Tropical Agriculture, Ibadan, p. 4-8.

Singh BB (2006) Cowpea breeding at IITA: Highlights of advances impacts. In **Anais do congresso nacional de feijão-caupi**. Embrapa Meio-Norte, Teresina, p. 1-4.

Sprague GF and Tatum LA (1942) General versus specific combining ability in single crosses of corn. **Journal of the American Society of Agronomy 34**: 923-932.

Tchiagam JN, Bell JM, Nassourou AM, Njintang NY and Youmbi E (2011) Genetic analysis of seed proteins contents in cowpea (*Vigna unguiculata* L. Walp.). **African Journal of Biotechnology 10**: 3077-3086.

Diallel analysis in cowpea aiming at selection for extra-earliness

Rosana Mendes de Moura Oliveira[1*], Francisco Rodrigues Freire Filho[2], Valdenir Queiroz Ribeiro[3], Ângela Celis de Almeida Lopes[1], Karla Annielle da Silva Bernardo[1] and Akemi Suzuki Cruzio[1]

Abstract: *In Brazil, the production of cowpea is concentrated in the Northeast and North; however, in recent years, its cultivation has expanded to the Cerrado biome of the Brazilian Midwest region, where it is incorporated into production arrangements in the form of off-season. In this work, it was carried out a study on the extra-earliness in cowpea in order to select parents and crosses with cycle shorter than 60 days. A test with an F_2 generation of a complete diallel cross involving five parents and ten crosses was carried out in a complete randomized block design with six replications. The parents IT82D-889 and AU94-MOB-816 stood out for extra-earliness flowering, while IT82D-60 and IT82D-889 stood out for extra-earliness maturity. The crosses IT82D-60 x AU94-MOB-816 and IT82D-60 x IT82D-889 were more promising in extra-earliness flowering, while IT82D-60 x IT82D-889 and IT82D-60 x MNC04-789B-119-2-3-1 were more promising for extra-earliness maturity.*

Key words: *Vigna unguiculata, combining ability, flowering and maturity.*

***Corresponding author:**
E-mail: rosanamendes.moura@gmail.com

[1] Universidade Federal do Piauí, Campus da Socopo, 64.049-550, Teresina, PI, Brazil
[2] Embrapa Amazônia Oriental, Travessa Doutor Enéas Pinheiro, s/n, Bairro Marco, 66.095-100, Belém, PA, Brazil
[3] Embrapa Meio-Norte, Av. Duque de Caxias, nº 5.650, Bairro Buenos Aires, 64.006-220, Teresina, PI, Brazil

INTRODUCTION

Cowpea (*Vigna unguiculata* (L.) Walp.) is an excellent protein source, and it presents all the essential amino acids, and is rich in carbohydrates, vitamins and minerals, besides having great fiber content and low fat content, constituting an important food component in several countries (Freire Filho et al. 2011).

It is grown by small and medium farmers in the North and Northeast of Brazil, and recently, it has been grown by large producers in these regions (Xavier et al. 2005). In the past years, its cultivation has expanded to the Cerrado biome of the Brazilian Midwest, where it is incorporated into production arrangements in the form of off-season, following soybeans, or rice and corn (Freire Filho et al. 2011). In these arrangements, cowpea is grown after the mid-rainy season, when there is no expectation of rainfall that would allow the cultivation of cornas second-crop. In this context, there is a short planting window for farmers interested in growing cowpea, to which cultivars of early and medium-early cycles fit perfectly, since their cycles last from 60 to 80 days. Thus, an alternative to enlarge the planting window for cowpea would be the use of extra-early cultivars with cycles shorter than 60 days.

Silva et al. (2007) mention that the main way to evaluate earliness is through the time between the sowing and the emergence of the first flowers. However,

Ribeiro et al. (2004) reported that early flowering genotypes do not necessarily present low cycle. According to Singh (1986), earliness is an essential factor in the process of adaptation of a culture to any agro-climatic zone, especially in the semiarid region, which may be associated with some stress factors that occur at the end of the development period. Adeyanju and Ishiyaku (2007) reported that earliness is an important agronomic trait, and it is important for cowpea in areas with relatively short rainfall cycle. Padi (2007) mentions that grain size and earliness are essential traits for the adoption of cowpea cultivars in the savannas of West Africa.

Based on maturity, Freire Filho et al. (2005) divided cowpea cycle into: extra-early cycle – maturity at up to 60 days after sowing; early cycle - maturity between 61 and 70 days after sowing; medium-early cycle - maturity between 71 and 80 days after sowing; medium-late cycle - maturity between 81 and 90 days after sowing; and late cycle - maturity from 91 days after sowing.

The knowledge of the potential combinations of genotypes that can be used in crosses serves as the basis for the identification of the parents and the crosses that will result in superior segregating populations. Among the methodologies used to evaluate parents and their combinations, diallels should be highlighted. The diallel scheme has been defined as a method, in which a group of homozygous lines is selected, which are crossed in pairs, providing a maximum of p^2 combinations (Hayman 1954).

Sprague e Tatum (1942) established and defined the terms of general combining ability (GCA) and specific combining ability (SCA), being GCA the average behavior of a parent in a series of hybrid combinations, and is associated with allele additive effects. On the other hand, SCA is used to denote hybrid combinations that are relatively superior or inferior than what is expected based on GCA (Cruz and Vencovsky 1989). Griffing (1956) proposed a new method of diallel analysis, in which GCA and the SCA are also estimated. This method is the most used due to its generality and application easiness.

This study aimed at evaluating parents regarding extra earliness and selecting potential crosses to produce extra early lines with cycle shorter than 60 days.

MATERIAL AND METHODS

Parents and diallel cross

As parents, it was used the lines IT82D-889 and IT82D-60, from the International Institute of Tropical Agriculture (IITA), Nigeria; AU94-MOB-816, from Auburn University, USA; and MNC04-789B-119-2-3-1 and MNC05-820B-240, from Embrapa Mid-North, Brazil. Seeds were provided by the Cowpea Breeding Program of Embrapa Mid-North, Teresina, PI (Table 1).

For the crosses, the five parents (n) were sown in a screen house, in order to avoid the occurrence of natural crosses by insects, and were crossed among each other in a complete diallel scheme excluding reciprocal $\left(\frac{(n-1)n}{2}\right)$, resulting in ten crosses. In the crosses, it was used the method presented by Rachie et al. (1975) and Zary and Muller Junior (1982).

Experimental design

The experiment was carried out at the experimental field of Embrapa Mid-North, Teresina, PI, in a randomized block design, with six replications, and 15 treatments, corresponding to five parents and ten F_2 crosses. The experimental plot

Table 1. Identification, origin, grain color, and number of days to flowering and maturity of five parents

Parents	Origin	Grain Color	NDF[3]	NDM[4]
IT82D-889	IITA, Nigéria[1]	Red	38.19	54.32
AU94-MOB-816	Auburn University, USA	Red/White	37.82	55.06
IT82D-60	IITA, Nigeria	Black eye	38.32	54.61
MNC04-789B-119-2-3-1	Embrapa, Brazil[2]	Black eye	39.82	55.06
MNC05-820B-240	Embrapa, Brazil	Black eye	39.82	56.10

[1] International Institute of Tropical Agriculture
[2] Brazilian Agricultural Research Corporation
[3] Number of days to flowering
[4] Number of days to maturity.

consisted of four 5.0 m rows, spaced 0.60 m between rows, and 0.25 m between plants within the row. Approximately 20 days after sowing, seven plants were labeled at random in the two central rows, corresponding to the useful area of the plot for the data collection, except for weight of 100 grains and grain yield, which were measured from the entire plot. The following traits were evaluated: number of days to flowering (NDF): number of days between sowing and the emergence of the first flower; number of days to maturity (NDM): number of days between sowing and the emergence of the first pod with color change, dry beans, ready for harvesting; plant height (PH): measured from the base of the plant to the top of the main branch, which was evaluated from the maturity of pods; pod length (PL): length of a pod randomly taken from each plant; number of grains per pod (NGP): number of grains obtained in the same pod used to measure the length; weight of 100 grains (W100G): weight of 100 grains sample randomly taken from the plot yield; and plot yield(PY): dry grains, obtained after harvesting all the plants of the plot, including plants individually harvested.

Statistical analysis

Initially, it was carried out analysis of variance in a complete randomized block for each trait. After that, by using the mean square error of the analyses of variance, it was carried out the analysis according to Method 2, model B of Griffing (1956), to estimate the effects of General Combining Ability (GCA) of the parent, and the Specific Combination Ability (SCA) of the cross, from a set of *p* parents and of the resulting generations, according to the statistical model:

$Y_{ij} = m + g_i + g_j + s_{ij} + e_{ij}$

in which,

Y_{ij}: is the mean value observed during the hybrid combination (i ≠ j) or the parent combination (i = j);

m: is the general mean;

g_i and g_j: are the effects of general combining ability of the i-th and j-th parent, respectively;

s_{ij}: is the effect of specific combining ability for crosses between parents i and j;

e_{ij}: is the experimental error.

The mean of NDF, NDM and NGP were transformed into $\sqrt{x}$ in order to obtain homoscedasticity of variances and error adjustment to an approximately normal distribution. The means of each trait were used for combining ability analysis through the unfolding of the treatment effect on GCA and SCA effects. The significance of the estimates of GCA and SCA effects were verified by the bilateral Student t test. It was carried out clustering analysis by the Scott-Knot test at 5% probability. Coefficient of variation was used for the evaluation of experimental precision. In all the analyses, it was used the Genes software (Cruz 2006).

RESULTS AND DISCUSSION

The effect of GCA was significant (P <0.01) for all traits, except for PH and PY, evidencing the importance of additive effect in inheritance. In this case, genotypes with greater effect for these traits had more favorable genes for cycle reduction. For SCA, in traits with significant effect (P <0.01), dominance proved to be relevant, and there may be complementarity between some parents that could generate superior lines (Table 2).

Table 2 also shows the coefficients of experimental variation. Most traits presented low values, ranging from 0.85 to 17.23% (Pimentel Gomes 2000). NDM had the lowest coefficient (0.85%), and PY had the highest coefficient (17.23%). The high coefficient of experimental variation obtained for PY can be explained by the fact that it is a quantitative trait, which is strongly influenced by the environment, and which can increase the estimate of the mean squares of the estimated errors. However, these values are expected, and evidence the good experimental precision of the study.

Table 3 shows GCA estimates. Lal et al. (1975) stated that for NDF it is more desirable that the parent presents negative estimate. The negative values obtained in parents for NDF and NDM indicate that a combination of these parents would result in a reduction in the expression of these traits. Since the main objective of the work is selection for extra-earliness, parents with GCA of high magnitude, but negative for NDF and NDM are the most important for reducing the expression of these traits. Thus, the parents AU94-MOB-816, IT82D-889 and IT82D-60 presented GCA

Table 2. Summary of the analysis of variance of traits related to earliness and grain yield in cowpea (*Vigna unguiculata* (L.) Walp.)

SV	df	Mean square						
		NDF (day)	**NDM** (day)	**PH** (cm)	**PL** (cm)	**NGP**	**W100G** (g)	**PY** (kg ha^{-1})
Treatments	14	0.053**	0.028**	75.104**	12.451**	0.415**	49.291**	103885.977**
GCA	4	0.139**	0.046**	36.979ns	38.167**	1.276**	149.110**	33006.113
SCA	10	0.019	0.021**	90.353**	2.165**	0.070**	9.363**	132237.922**
Error	70	0.017	0.004	19.499	0.410	0.015	2.175	14853.896
CV(%)		2.14	0.85	15.39	4.01	3.93	8.16	17.23

** Significant at 1% probability

Table 3. Estimates of the effects of general combining ability (GCA) for traits related to earliness and grain yield in cowpea (*Vigna unguiculata* (L.) Walp.)

Parents	General Combining Ability (GCA)						
	NDF (day)	**NDM** (day)	**PH** (cm)	**PL** (cm)	**NGP**	**W100G** (g)	**PY** (kg ha^{-1})
IT82D-889	-0.046**	-0.028**	-0.862	1.611**	0.142**	-2.519**	-21.389**
AU94-MOB-816	-0.052**	-0.010	1.053	-0.760**	0.188**	-0.829**	35.298**
IT82D-60	-0.025	-0.029**	0.864	-0.519**	-0.036**	0.089	-33.401
MNC04-789B-119-2-3-1	0.064**	0.022**	-0.953	0.095	-0.045**	0.670**	1.550**
MNC05-820B-240	0.059**	0.045**	-0.102	-0.427**	-0.249**	2.589**	17.941**

*, **significant p <0.05, and significant p <0.01, respectively, by the *t* test.

estimates more favorable for the reduction of NDF, indicating that they are the most promising for the reduction of the trait expression. These same parents also stood up for NDM, with negative values of GCA, and therefore were the most recommended for breeding programs for selection for extra-earliness in cowpea. All parents were significant for NDF, except for IT82D-60. For NDM, only AU94-MOB-816 presented no significance.

The values of GCA estimates for PH ranged from -0.953 (MNC04-789B-119-2-3-1) to 1.053 (AU94-MOB-816). The parents AU94-MOB-816 and IT82D-60 stood out, showing great potential for improving the expression of this trait. GCA estimates for PL ranged from -0.760 to 1.611. The most favorable estimates for this trait were obtained by IT82D-889 and MNC04-789B-119-2-3-1, with values of 1.611and 0.095, respectively. Among these parents, IT82D-889 stood out for presenting GCA estimate superior to the other parents. All parents were significant for this trait, except for MNC04-789B-119-2-3-1. For NGP, all parents showed significant effects. The highest estimates were 0.188 and 0.142 for AU94-MOB-816 and IT82D-889, respectively. The parent MNC05-820B-240 had the highest GCA estimate for W100G. For PY, all parents showed significant effects, except for IT82D-60. The parents AU94-MOB-816 and MNC05-820B-240 showed the highest GCA estimate for PY, proving to be promising for the use in programs aimed at yield increase. According to Lorencetti et al. (2005), the choice of parents for the formation of segregating populations is crucial for the success of a breeding program, and the combining ability with the presence of complementary genes is responsible for the success of the crosses.

Table 4 presents SCA estimates. There was no statistically difference for NDF, showing that all crosses behaved similarly. However, IT82D-60 x AU94-MOB-816, IT82D-889 x AU94-MOB-816 and IT82D-60 x IT82D-889 should be highlighted for presenting negative estimates for this trait. It was also observed that the three genotypes that participated in these crosses presented the best GCA estimates for the trait in question, which is in accordance with Cruz et al. (2004), who stated that a good cross must present significant SCA estimate, at the time that their parents require high GCA estimates for the trait under study.

For NDM, IT82D-60 x IT82D-889 and IT82D-60 x AU94-MOB-816 presented significant effect. The most promising crosses for NDM were IT82D-60 x AU94-MOB-816, IT82D-60 x IT82D-889 and IT82D-60 x MNC04-789B-119-2-3-1. The first two combinations also stood out regarding NDF estimates, proving to be potential crosses to produce extra early descendants. In PH, the cross IT82D-60 x AU94-MOB-816 stood out for presenting significant effect and also for showing high SCA estimate. For PL, six crosses had positive estimates, highlighting the crosses MNC05-820B-240 x AU94-MOB-816, and AU94-MOB-816 x MNC04-789B-119-2-3-1 with high estimates. The crosses MNC05-820B-240 x IT82D-60, AU94-

Table 4. Estimates of the effects of specific combining ability (SCA) for traits related to earliness and grain yield in cowpea (*Vigna unguiculata* (L.) Walp.)

Crosses	Specific Combining Ability (SCA)						
	NDF (day)	NDM (day)	PH (cm)	PL (cm)	NGP	W100G (g)	PY (kg ha^{-1})
IT82D-889 x AU94-MOB-816	-0.055	-0.016	2.406	0.531**	-0.183**	2.526**	-124.549**
IT82D-60 x IT82D-889	-0.052	-0.055*	2.478	-0.350	0.063	-0.466	101.858*
IT82D-889 x MNC04-789B-119-2-3-1	0.020	0.039	2.590	-0.481**	-0.001	-1.213*	-15.780
MNC05-820B-240 x IT82D-889	-0.014	-0.028	2.112	-0.126	-0.008	-0.466	68.776
IT82D-60 x AU94-MOB-816	-0.075	-0.123**	7.229**	0.004	0.049	-0.156	-146.638**
AU94-MOB-816 x MNC04-789B-119-2-3-1	0.013	0.021	-1.814	0.798**	0.182**	0.596	214.018**
MNC05-820B-240 x AU94-MOB-816	-0.036	-0.001	0.430	1.078**	0.109*	-0.156	43.027
IT82D-60 x MNC04-789B-119-2-3-1	0.024	-0.044	0.736	0.148	-0.033	1.510**	65.778
MNC05-820B-240 x IT82D-60	-0.023	-0.005	-0.614	0.337	0.617**	-1.068*	124.727**
MNC05-820B-240 x MNC04-789B-119-2-3-1	0.046	-0.005	1.366	-0.110	-0.106*	-0.489	-305.424**

*, ** Significant p <0.05, and significant p <0.01, respectively, by the t test.

MOB-816 x MNC04-789B-119-2-3-1, and MNC05-820B-240 x AU94-MOB-816 had the highest SCA estimates for NGP.

The cross IT82D-889 x AU94-MOB-816 was the most promising for W100G. However, both parents of this cross had negative GCA estimates for the same trait. This fact is evident since different genotypes may be more efficient if they are combined. Such result may have occurred due to the high complementarity between the parents, with respect to genes related to the trait. According to Silva et al. (2004), not always the parent with high GCA estimates generate the best cross when combined with each other. The cross MNC05-820B-240 x MNC04-789B-119-2-3-1 presented one of the lowest SCA estimates for W100G (-0.489). However, these same parents had the highest GCA estimates for the trait in question. Result shows the low complementarity between the parents for this trait.

Grain yield is one of the most important traits for breeders. Therefore, the best crosses for the trait are selected, i.e., those with high SCA estimates, involving parents with high GCA (Oliveira Júnior et al. 1999). Thus, the cross AU94-MOB-816 x MNC04-789B-119-2-3-1 presented genetic potential superior to the others, since the parents presented the highest GCA estimates for PY. The crosses MNC05-820B-240 x IT82D-60 and IT82D-60 x IT82D-889 also stood out for this trait, with high estimates.

According to the GCA and SCA estimates, the materials which showed the greatest potential for extra earliness selection were the parents IT82D-889, AU94-MOB-816 and IT82D-60, and the crosses IT82D-60 x AU94-MOB-816, IT82D-889 x AU94-MOB-816 and IT82D-60 x IT82D-889. According to SCA, the cross IT82D-60 x AU94-MOB-816 is the most recommended for extra earliness selection, since it showed significant estimates for NDF and NDM. For PY, this cross presented negative estimate, which is not satisfactory. However, when it is taken into account both extra earliness and yield, the most recommended cross is IT82D-889 x IT82D-60.

Table 5 shows the means and the result of the clustering by the Scott-Knott test. Genotypes were distributed into two to five groups. For NDF, the test clustered the genotypes in only two clusters. "Cluster A", which allocates the genotypes with the best mean, had only one parent (AU94-MOB-816). NDF general mean was 38.19 days, which was below the value found by Bertini et al. (2009), with mean of 47.17 days. For NDM, the crosses IT82D-889 x MNC05-820B-240 and AU94-MOB-816 x MNC05-820B-240 stood out and presented the best means, 52.85 days and 52.14 days, respectively. These two crosses also stood out for NDF. The general mean for NDM was 54.46 days, which was below the value obtained by Santos et al. (2012).

In cowpea, the ideal height of the plant for mechanical harvesting has not been determined yet. Simone et al. (1992), referring to the ideotype of common bean (*Phaseolus vulgaris* L.) upright cultivar for mechanical harvesting, suggest that the height is between 50 to 55 cm. PH general mean obtained in the experiment was 28.69 cm, well below the suggested value. The cross AU94-MOB-816 x MNC05-820B-240 showed the best mean for this trait (37.83 cm). The cross IT82D-889 x IT82D-60 stood out for PL, with mean of 19.62 cm, which had general mean of 15.98 cm, higher than that obtained by Teixeira et al. (2007). For NGP, three crosses obtained the best means and were allocated in "Cluster

Table 5. Means of the traits: number of days to flowering (NDF), number of days to maturity (NDM), plant height (PH), pod length (PL), number grains per pod (NGP), weight of 100 grains (W100G), and plot yield (PY) of cowpea

Genotypes	NDF[1] (day)	NDM[1] (day)	PH (cm)	PL (cm)	NGP[1]	W100G (g)	PY (kg ha^{-1})
MNC05-820B-240	39.82 B	56.10 C	26.83 C	14.5 D	6.81 D	24.33 A	648.05 B
IT82D-60	38.32 B	54.61 C	28.83 B	15.33 C	8.47 C	19.67 B	680.50 B
IT82D-889	38.19 B	54.32 C	29.83 B	17.00 B	9.12 C	17.67 C	643.88 B
AU94-MOB-816	37.82 A	55.06 C	30.07 B	15.83 C	10.18 B	19.67 B	669.67 B
MNC04-789B-119-2-3-1	39.82 B	55.5 C	29.00 B	15.50 C	7.51 D	20.83 B	351.17 D
IT82D-60 x MNC05-820B-240	38.32 B	54.61 C	25.50 C	14.83 D	9.00 C	18.33 B	473.06 C
IT82D-889 x MNC05-820B-240	36.72 A	52.85 A	31.17 B	16.68 B	10.96 B	15.17 D	628.67 B
AU94-MOB-816 x MNC05-820B-240	36.36 A	52.14 A	37.83 A	14.67 D	11.15 B	17.17 C	468.83 C
MNC04-789B-119-2-3-1 x MNC05-820B-240	38.94 B	55.06 C	29.33 B	15.67 C	9.12 C	20.33 B	617.72 B
IT82D-889 x IT82D-60	37.70 A	54.17 C	22.17 C	19.62 A	12.18 A	12.83 E	541.17 C
IT82D-889 x AU94-MOB-816	36.36 A	53.73 B	31.28 B	16.85 B	10.82 B	17.24 C	497.24 C
IT82D-60 x MNC04-789B-119-2-3-1	38.69 B	55.06 C	29.46 B	17.67 B	10.43 B	15.00 D	559.76 C
AU94-MOB-816 x IT82D-60	37.82 A	55.06 C	26.67 C	13.50 E	11.83 A	15.00 D	654.17 B
IT82D-889 x MNC04-789B-119-2-3-1	38.44 B	55.06 C	26.97 C	15.99 C	11.97 A	18.50 B	798.5 A
AU94-MOB-816 x MNC04-789B-118-2-3-1	39.19 B	54.64 C	25.34 C	16.00 C	9.18 C	19.20 B	609.28 B
Parents means	38.81	55.06	28.91	15.63	8.35	20.43	598.65
Crosses means	37.82	54.17	38.57	16.15	10.63	16.88	583.94
General mean	38.19	54.46	28.69	15.98	9.86	18.06	588.84

[1] analysis carried out with data transformed into $\sqrt{x}$

Means followed by the same letter belong to the same cluster by the Scott-Knott test at 5% probability.

A" IT82D-889 x IT82D-60 (12.18), AU94-MOB-816 x IT82D-60 (11.83), and IT82D-889 x MNC04-789B-119-2-3-1 (11.97). Five clusters were formed for W100G. Only the genotype MNC05-820B-240 was allocated in "Cluster A", with mean of 24.33g. The general mean (18.06 g) obtained for this trait was greater than the value obtained by Bertini et al. (2009), which was 15.84 g. The most productive cross was IT82D-889 x MNC04-789B-119-2-3-1, with mean of 789.5 kg ha^{-1}.

ACKNOWLEDGEMENTS

The authors thank the staff of Embrapa Mid-North, Manuel Gonçalves da Silva, Agripino Ferreira and Paulo Sérgio Monteiro, for their valuable collaboration in the execution of the experiment. The authors also thank the Coordination for the Improvement of Higher Education Personnel (CAPES) for granting scholarship.

REFERENCES

Adeyanju AO and Ishiyaku MF (2007) Genetic study of earliness in cowpea (*Vigna unguiculata* L. Walp) under screen house condition. **International Journal of Plant Breeding and Genetics1**:34-37.

Bertini CHCM, Teofilo EM and Dias TC (2009) Divergência genética entre acessos de feijão-caupi do banco de germoplasma da UFC. **Ciência Agronômica40**: 99-105.

Cruz CD (2006) **Programa Genes: Biometria**. Editora UFV, Viçosa, 382p.

Cruz CD, Regazzi AJ and Carneiro PCS (2004) **Modelos biométricos aplicados ao melhoramento genético**. Editora UFV, Viçosa, 480p.

Cruz CD and Vencovsky R (1989) Comparação de alguns métodos de análise dialélica. **Revista Brasileira de Genética12**:425-438.

Freire Filho FR, Ribeiro VQ, Barreto PD and Santos AA (2005) Melhoramento genético. In Freire Filho, FR, Ribeiro VQ and Lima JAA (eds) **Feijão-caupi: avanços tecnológicos.** Embrapa Informação Tecnológica, Brasília, p. 29-92.

Freire Filho FR, Ribeiro VQ, Rocha MM, Silva KJD, Nogueira MSR and Rodrigues EV (2011) **Feijão-caupi: produção, melhoramento genético, avanços e desafios**. Embrapa Informação Tecnológica, Brasília, 81p.

Pimentel Gomes F (2000) **Curso de estatística experimental.** Degaspari, Piracicaba, 477p.

Griffing B (1956) Aconcept of general and specific combining ability in relation to diallel crossing systems. **Australian Journal of Biological Sciences9**:463-493.

Hayman BI (1954) The theory and analysis of diallel crosses. **Genetics39**:789-809.

Lal S, Singh M and Pathak MM (1975) Combining ability in cowpea. **Indian Journal of Genetics and Plant Breeding35**:375-378.

Lorencetti C, Carvalho FIF, Benin G, Marchioro VS, Oliveira AC, Silva

JAG, Hartwig I, Schmidt DAM and Valério IP (2005) Capacidade combinatória e heterose em cruzamento dialélico de aveia (*Avena sativa* L.). **Revista Brasileira Agrociência11**:143-148.

Oliveira Júnior A, Miranda GV and Cruz CD (1999) Predição de populações F_3 a partir de dialelos desbalanceados. **Pesquisa Agropecuária Brasileira34**:781-787.

Padi FK (2007) Response to selection for grain yield and correlated response for grain size and earliness in cowpea based on early generation testing. **Annals of Applied Biology152**:361-368.

Rachie KO, Rawal KM and Franckowiak JD (1975) **A rapid method for hand crossing cowpea**. IITA, Ibadan, 5p. (IITA. Technical Bolletin, 2).

Ribeiro ND, Hoffmann Júnior L and Possebon SB (2004) Variabilidade genética para ciclo em feijão dos grupos preto e carioca. **Revista Brasileira Agrociência10**:19-29.

Santos A, Ceccon G, Correa AM, Durante LGY and Regis JAVB (2012) Análise genética e de desempenho de genótipos de feijão-caupi cultivados na transição do cerrado-pantanal. **Cultivando o saber5**: 87-102.

Silva FB, Ramalho MAP and Abreu AFB (2007) Seleção recorrente fenotípica para florescimento precoce de feijoeiro carioca. **Pesquisa Agropecuária Brasileira42**:1437-1442.

Silva MP, Amaral Júnior AT, Rodrigues R, Daher RF, Leal NR and Schuelter AR (2004) Análise dialélica da capacidade combinatória em feijão de vagem. **Horticultura Brasileira22**:277-280.

Simone M, Failde V, Garcia S and Panadero PC (1992) **Adaptación de variedades y líneas de judias secas (*Phaseolus vulgaris* L.) a la recolección mecanica directa.** INTA, Salta:, 5p.

Sprague GF and Tatum LA (1942) General vs. specific combining ability in single crosses of corn. **Journal of the American Society of Agronomic 34**: 923-932.

Teixeira NJP, Machado CF, Freire Filho FR, Rocha MM and Gomes RLF (2007) Produção, componentes de produção e suas inter-relações em genótipos de feijão-caupi (*Vigna unguiculata* (L.) Walp.) de porte ereto. **Ceres 54**: 374-382.

Xavier GR, Martins LMV, Rumjanek NG and Freire Filho FR (2005) Variabilidade genética em acessos de caupi analisada por meio de marcadores RAPD. **Pesquisa Agropecuária Brasileira40**:353-359.

Zary KW and Miller Junior JC (1982) Comparisson of two methods of hand-crossing *Vigna unguiculata* (L.) Walp. **HortScience17**: 246-248.

21

Genetic diversity and apple leaf spot disease resistance characterization assessed by SSR markers

Gustavo H.F. Klabunde[1], Camila F.O. Junkes[2], Sarah Z.A. Tenfen[1], Adriana C.M. Dantas[3], Carla R.C. Furlan[4], Adelar Mantovani[5], Frederico Denardi[6], José I. Boneti[7] and Rubens O. Nodari[1*]

***Corresponding author:**
E-mail: rubens.nodari@ufsc.br

[1] UFSC, Rodovia Admar Gonzaga, 1346, 88.034-000, Florianópolis, SC, Brazil
[2] UFPel, Rua Gomes Carneiro, 1, 96.010-610, Pelotas, RS, Brazil
[3] UERGS, Rua Júlio de Castilhos, 3947, 95.010-005, Caxias do Sul, RS, Brazil
[4] Faculdades Integradas FACVEST, Avenida Marechal Floriano, 947, 88.501-103, Lages, SC, Brazil
[5] UDESC, Avenida Luis de Camões, 2090, 88.520-000, Lages, SC, Brazil
[6] EPAGRI, Rua Abílio Franco, 1500, 89.500-000, Caçador, SC, Brazil
[7] EPAGRI, Rua Araújo Lima, 102, 88.600-000, São Joaquim SC, Brazil

Abstract: *Among the cultivation problems of apple production in Brazil, Apple Leaf Spot (ALS) disease represents one of the main breeding challenges. This study aims at analyzing the genetic diversity among 152 apple scion accessions available at the Apple Gene Bank of EPAGRI, located in Caçador, Santa Catarina/ Brazil. Eleven genomic SSR loci were analyzed to assess genetic diversity of ALS resistant and susceptible accessions. Results revealed high genetic diversity of the studied accessions, being 120 exclusive alleles (67 unique) from scion accessions resistant to ALS, and a mean PIC of 0.823. The locus Probability of Identity (I) ranged from 0.017 to 0.089. The combined I was 4.11 x 10^{-16}, and the Power of Exclusion was 99.99999259%. In addition, the DNA fingerprint patterns will contribute as additional descriptors to select parental for crosses and early identification of apple accessions for breeding purposes, and also for cultivar protection.*

Key words: *Colletotrichum gloeosporioides, DNA fingerprinting, Malus spp, genetic identity, microsatellite.*

INTRODUCTION

Brazil is the tenth largest apple producer country, with a total production of 1.335 million tons in 2012 (FAO 2014). However, the growers still face serious limiting problems, such as poor climatic adaptation of the current varieties due to insufficient chilling during the winter. High rainfall, associated with high temperatures, during the growing season are the main causes of development of many serious diseases problems in Southern Brazil. Breeding programs carried out by researchers and germplasm bank curators in this region focus on identifying promising germplasm to be used as parent for specific traits, such as short juvenility, high yield, fruit quality, disease-resistance rootstocks, and for low to moderate chilling requirement (Furlan et al. 2010). In addition, the development of scion cultivars resistant to Apple Leaf Spot disease (ALS) is one of the major challenges for apple breeders in southern states of Brazil. The disease is caused by the fungus *Colletotrichum gloeosporioides,* which triggers the appearance of several brown patches, followed by intense defoliation of trees, and consequently decrease in fruit production (Weir et al. 2012).

One of the most effective strategies to control ALS is to develop disease-resistant cultivars. In this context, the Apple Gene Bank plays an important role as a source of high genetics (Janick et al. 1996, Hokanson et al. 1997). *Malus* ×

domestica Borkh. is one of the most important domesticated fruit species, and its distinct varieties has been submitted to countless crossings and selections over the years. However, a limited number of *Malus* species has been used up to present for breeding disease resistance, except few breeding programs, like Purdue Rutgers, Illinois, USA, and some European apple breeding programs (Korban and Tartarini 2009). The majority of the parents used in breeding programs have explored a strait gene pool, involving crosses among widespread commercial cultivars (Kumar et al. 2010). For instance, the varieties Red Delicious, Golden Delicious and Jonathan, have been frequently utilized on apple breeding, producing several of the current cultivars (Noiton and Alspach 1996). In addition, the selection and release of sport mutants of widespread cultivars have also potentialized the trend towards genetic similarity in today's commercial cultivars (Brooks and Olmo 1994).

The accessibility to diverse apple germplasm is essential for pursuing successful breeding, since it increases the number of allele combinations and allows the development of new cultivars with improved and desirable traits. Consequently, broadening the germplasm used in breeding will increase the diversity in cultivation, avoid genetic vulnerability, and preserve their unique genetic characteristics available in this new genetic combinations for future generations.

In plant germplasm characterization, duplicates and mislabeling of accessions might occur, which are unwanted, costly and time consuming (Gustavsson et al. 2008). In these cases, molecular markers are very useful to detect identical, synonymous and homonymous accessions, and also to help breeders to form their core collections (Gross et al. 2012). In addition, proper identification and characterization of accessions will help better protect cultivars under intellectual property rights and identify parents carrying alleles of interest for apple breeding (Goulão et al. 2001).

The EPAGRI´s Apple Gene Bank (EAGB) contains 442 apple accessions introduced from more than 20 countries. Among them, there are several wild apple species, such as *M. aldenhamensis, M. atrosanguinea, M. baccata, M. eley, M. hillieri, M. platicarpa, M. robusta, M. prunifolia, M. pumila, M. floribunda,* and 19 *M. pumila* rootstock accessions. The referred Gene Bank can be then considered a potential source of germplasm for many genetic resistances. There are 33 accessions with vertical resistance to apple scab (*Venturia inaequalis*). Furlan et al. (2010) also identified 187 resistant and 58 susceptible accessions to *Colletotrichum gloeosporioides*.

The use of molecular markers has been valuable for assessing species and cultivars' genetic diversity, determining phylogenetic relationships and identifying alleles of interest (Han and Korban 2010). Among the molecular ones, microsatellite markers (SSR-*simple sequence repeats*) are highly polymorphic, multiallelic, co-dominant, reproducible and are distributed throughout the entire genome, making them ideal for revealing genetic diversity (Morgante et al. 2002). Therefore, the use of molecular markers to access the apple germplasm is a trustworthy tool (Gustavsson et al. 2008, Zhuang et al. 2011, Potts et al. 2012, Reim et al. 2013, Burak et al. 2014). Almost 400 microsatellite markers have already been identified and mapped in *Malus* and other fruit species (Guilford et al. 1997, Liebhard et al. 2002, Silfverberg-Dilworth et al. 2006, Han and Korban 2008). These molecular markers have been successfully used for assessing the gene pool and relatedness among distinct *Malus* germplasm (Pereira-Lorenzo et al. 2008, Gasi et al. 2010); and also for identifying loci associated with target alleles and for map-based cloning, regarding their co-dominant heritage and high polymorphism level (Hokanson et al. 1998, Liebhard et al. 2002, Oraguzie et al. 2005, Naik et al. 2006, Han and Korban 2010, Zhuang et al. 2011).

Thus, the primary goal of this work was to assess genetic diversity and relatedness of a large collection of apple germplasm maintained by EPAGRI´s *Malus* breeding program. The objective was to identify and establish DNA fingerprinting patterns based on SSR markers, and to associate specific alleles and genotypes to ALS resistant and susceptible phenotypic data.

MATERIAL AND METHODS

Plant material and DNA isolation

The 152 apple scion accessions (Table 1) selected for this study represent a wide genetic spectrum containing several target alleles of particular interest to apple breeding in Southern Brazil. The selected apple accessions used in this work are originated from 18 countries and are currently located at EPAGRI – Caçador Experimental Station (lat 26º 49' 07" S, long 50º 59' 07" W, and alt 960 m asl), Caçador, SC, Brazil. Total DNA was extracted from 100 mg of leaf material, based on CTAB method from Doyle and Doyle (1990). DNA quality and quantity were determined by spectrophotometry at 260 nm, using Nanodrop 1000 spectrophotometer (Thermo Scientific, Waltham, MA, USA).

Table 1. Names and country of origin of apple accessions genotypes used in this study

	Accessions	Origin		Accessions	Origin		Accessions	Origin
1	2137964	*	52	Greensleeves	UK	103	† Planaltina	Brazil
2	† Akagui	Japan	53	GrothRed	*	104	POME 10	Brazil
3	Angius	*	54	† Hame 6	*	105	POME 12	Brazil
4	Argentina 2	Argentina	55	Harrold Red	USA	106	† POME 13	Brazil
5	Arlet	Switzerland	56	† Hatsuaki	Japan	107	POME 14	Brazil
6	Baronesa	Brazil	57	Hokuto	Japan	108	POME 15	Brazil
7	Bem Davis	Brazil	58	Holland	USA	109	POME 16	Brazil
8	Black Jon	USA	59	Honey Gold	USA	110	POME 17	Brazil
9	Bonita[a]	Brazil	60	† Horey	Japan	111	† POME 19	Brazil
10	Bonita[b]	Brazil	61	Imperatore	Brazil	112	POME 20	Brazil
11	Braebum	New Zealand	62	Imperatriz	Brazil	113	† POME 22	Brazil
12	Carla	Brazil	63	Israel 8-3	Israel	114	POME 28	Brazil
13	Catarina	Brazil	64	† Ivette	N.lands	115	POME 03	Brazil
14	Centenária	Brazil	65	Jersey Mac	USA	116	Porporate	*
15	† Condessa	Brazil	66	† Jona Free	USA	117	Priam	USA
16	Comrade Red	England	67	† Jonagold	USA	118	† PX1032	France
17	Coop 14	USA	68	Jonared	USA	119	PX1033[a]	France
18	Coop 16	USA	69	Liberty	USA	120	PX1033[b]	France
19	Coop 24	USA	70	† Lisgala	Brazil	121	†PX216	France
20	Coop 06	USA	71	M.Aldenhamensis	China	122	† PX565	France
21	† D1R100T209	USA	72	M.Atrosanguinea	UK	123	Quinte	Canada
22	D1R102T116	USA	73	† M. Baccata	China	124	† Rainha	Brazil
23	D1R103T245	USA	74	M.Eleyi	China	125	Red Delicious	USA
24	D1R68T571	USA	75	M.Robusta	Canada	126	Red Free	USA
25	D1R73T94	USA	76	M6039	*	127	Redgold	USA
26	† D1R98T188	USA	77	Maayan	Israel	128	Reinette du Canada	Canada
27	D1R99T15	USA	78	Mac Free	USA	129	Reinette du Mans	France
28	D2R30T30	USA	79	† Marquesa	Brazil	130	René Reinetes	France
29	D2R31T237	USA	80	Mechinoku	Japan	131	Rome Beauty	USA
30	D2R38T126	USA	81	Melrose	USA	132	Sansa	Japan
31	D2R39T243	USA	82	Mere	*	133	† Senshu	Japan
32	D2R40T253[a]	USA	83	Monroe	USA	134	Shellred	USA
33	D2R40T253[b]	USA	84	† Mutsu	Japan	135	SM69-3	Brazil
34	Daiane	Brazil	85	Natsumidori	Japan	136	† Splendor	New Zealand
35	Delcon	USA	86	† Nero 26	USA	137	Stark J. Grimes	USA
36	† Discovery	UK	87	NewtonPippin	USA	138	Starkjonadel	New Zealand
37	Dorsett Golden	Bahamas	88	Niagara	USA	139	Stayman	USA
38	Ein Shemer	Israel	89	NJ36	USA	140	† Summered	Canada
39	† Elstar	Netherlands	90	NJ41	USA	141	Summerland	Canada
40	Erwin	Germany	91	NJ44	USA	142	Suntan	UK
41	Eva	Brazil	92	† NJ45	USA	143	Supper Kidd's	UK
42	Fiesta (Red Pippin)	UK	93	NJ49	USA	144	† Toukou	Japan
43	FR8	Brazil	94	NJ50	USA	145	Tropical Beauty	South Africa
44	Fuji Suprema	Japan	95	† NJ96	USA	146	Wealthy	USA
45	† Gala	New Zealand	96	NJR75	USA	147	Webster	USA
46	Galícia	Brazil	97	Nova Easygro	USA	148	Wemershock	*
47	Gloster 69	Germany	98	Ohrin	Japan	149	† Willie Sharp	Netherlands
48	† Golden Delicious	USA	99	Orankis Tem	*	150	† Wilmuta	Belgium
49	Goldjon	Italy	100	Pachacamac	Peru	151	Winter Gold	*
50	Gorden	*	101	Paulared	USA	152	Yoko	Japan
51	Grangille Red	*	102	Pilot	USA			

* Unknown origin; † Indicates cultivar susceptibility to ALS disease.

SSR genotyping

All accessions were genotyped using 11 perfect dinucleotide microsatellite loci from seven apple linkage groups (Table 2). PCR reactions were carried out in a 13.75 μL volume containing 40 ng of genomic DNA, 1X *Taq* buffer, 1 U *Taq* DNA polymerase (Invitrogen, Carlsbad, CA, USA), 0.25 μM of each dNTP (Fermentas, Vilnius, Lithuania), 1.5 mM magnesium chloride and 0.3 μM of each primer pair. A Mastercycler Gradient (Eppendorf, Hamburg, Germany) thermalcycler was used to amplify the eleven SSR loci with the following cycles: 94 ºC for 5 min; followed by 30 cycles at 95 ºC for 30 s; annealing temperature (according to the related literature of each primer) for 1 min; 72 ºC for 1 min; and a final extension of 5 min at 72 ºC. Each 5′ forward oligo was labeled with a fluorophore HEX or 6-FAM to enable the automated genotyping (Table 2). After amplification reactions, PCR products were diluted 10X in ultrapure water in order to be genotyped by capillary electrophoresis in a MegaBACE 1000 DNA Analysis System (GE Healthcare, Little Chalfont, UK).

Table 2. SSRs names, linkage groups, 5′fluorescent modifications, and sequences of the primers used in this study and the expected size range of the amplified fragments

SSR name	Linkage group*	5′ Labeling	Forward sequence	Reverse sequence	Size range (bp) Original publication
Ch03c02 [a]	12	Hex	tca cta ttt acg gga tca agc a	gtg cag agt ctt tga caa ggc	116-136
Ch02b10 [a]	02	6-FAM	caa gga aat cat caa aga ttc aag	caa gtg gct tcg gat agt tg	121-159
Ch05a05 [a]	06	Hex	tgt atc agt ggt ttg cat gaa c	gca act ccc aac tct tct ttc t	198-230
Ch05c06 [a]	16	Hex	att gga act ctc cgt att gtg c	atc aac agt agt ggt agc cgg t	104-126
Ch02c11 [a]	10	Hex	tga agg caa tca ctc tgt gc	ttc cga gaa tcc tct tcg ac	219-239
Ch02b03b [a]	10	Hex	ata agg ata caa aaa ccc tac acag	gac atg ttt ggt tga aaa ctt g	77-109
Ch01g12 [a]	12	6-FAM	ccc acc aat caa aaa tca cc	tga agt atg gtg gtg cgt tc	112-186
Nz02b01 [b]	15	Hex	ccg tga tga caa agt gca tga	atg agt ttg atg ccc ttg ga	212-238
Ch02g09 [a]	08	6-FAM	tca gac aga aga gga act gta ttt g	caa aca aac cag tac cgc aa	98-138
Ch03b06 [a]	15	Hex	gca tcc ttg aat gag gtt cac t	cca atc acc aaa tca atg tca c	111-131
Hi03g06 [c]	15	6-FAM	tgc caa tac tcc ctc att tac c	gtt taa aça gaa ctg cac cac atc c	182-204

[a] Liebhard et al. (2002), [b] Guilford et al. (1997), [c] Silfverberg-Dilworth et al. (2006)
*Fiesta x Discovery apple genetic reference linkage map (Liebhard et al. 2003)

Data analysis

Alleles peaks from raw data were scored by comparison with ET 550-R size standard (GE Healthcare) using Fragment Profiler analysis software version 1.2 (GE Healthcare). Observed and Expected heterozygosity (*Ho and He*); number of alleles per locus; allele frequencies; exclusive and unique alleles; Polymorphism Information Content $PIC = 1 - J = 1 - \Sigma(1 = J)p2ij$, where *pij* is the frequency of the *j*th allele for *i*th marker (Anderson et al. 1993); Probability of Identity (1; being1 = $\Sigma pi^1 + \Sigma(2pipi)^2$), where p_i and p_j are the frequencies of the *i*th and *j*th alleles and i ≠ j), and (1 unbiased = $n^3 (2a^22 - a_1) - 2n^22(a_3 + 2a_2) + n(9a_2 + 2) - 6/(n - 1)(n - 2)(n - 3)$) where *n* is the sample size, a_i equals Σpj^i and p_j is the frequency of the *j*th allele (Paetkau and Strobeck 1994) and Power of Exclusion (Q) (Vandeputte 2012) were calculated using the CERVUS 3.0 software (Kalinowski et al. 2007). Alleles were considered rare when their frequencies were lower than 0.05. A dendrogram was constructed with the use of unweighted pair-group method with arithmetic means - UPGMA (Sneath and Sokal 1973), using the software NTSYSpc version 2.02 (Rohlf 2000), and visualized using FigTree software (tree.bio.ed.ac.uk/*software/figtree/*).

RESULTS AND DISCUSSION

The eleven genomic SSR loci amplified a total of 242 alleles in the 152 apple accessions (Table 3), with a mean of 22 alleles per locus. Availability of marker data ranged from 99.98% (Ch02b03), 99.99% (Ch02g09 and Ch03c02) to 100% (remaining markers), with a total of 0.42% missing data. The joint analyses of all plants and all loci showed a proportion of heterozygotes to homozygotes of 3.38. When only ALS resistant plants were analyzed (118 accessions), the proportion of heterozygotes to homozygotes was 3.41. Similar proportion value (3.58) was observed for the 34 susceptible accessions. The number of alleles per locus ranged from 16 (Nz02b01) to 29 (Ch01g12) (Table 2). One hundred and twenty alleles (49.59%) were found to be exclusive to resistant accessions, and 13 alleles (5.37%) were

Table 3. Primers, observed size range (bp), number of alleles (A), Polymorphism information content (PIC), Observed heterozygosity (*Ho*), Expected heterozygosity (*He*), Probability of Identity (I) and Power of Exclusion (Q) for resistant, susceptible and total accessions

Primer	range (bp)	Resistant accession/cultivars (n=118)				Susceptible accessions/cultivars (n=34)				Total (n=152)					
		A (*)	PIC	*Ho*	*He*	A (*)	PIC	*Ho*	*He*	A	PIC	*Ho*	*He*	I	Q
Ch03c02	92-218	21(10)	0.786	0.508	0.813	13(2)	0.798	0.438	0.832	23	0.790	0.493	0.815	0.025	0.3174
Ch02b10	106-182	25(12)	0.907	0.746	0.917	14(1)	0.809	0.824	0.838	26	0.895	0.763	0.905	0.022	0.3105
Ch05a05	160-250	19(11)	0.788	0.805	0.812	11(3)	0.617	0.676	0.669	22	0.759	0.776	0.785	0.017	0.3029
Ch05c06	92-226	17(10)	0.761	0.881	0.792	7	0.553	0.824	0.619	17	0.729	0.868	0.762	0.086	0.3892
Ch02c11	200-250	17(6)	0.857	0.720	0.873	11	0.833	0.559	0.863	17	0.854	0.684	0.870	0.031	0.3243
Ch02b03b	76-124	20(11)	0.887	0.836	0.900	9	0.790	0.909	0.824	20	0.880	0.852	0.893	0.030	0.3244
Ch01g12	96-188	29(17)	0.858	0.856	0.873	12	0.850	0.971	0.877	29	0.868	0.882	0.881	0.058	0.3581
Nz02b01	190-260	14(7)	0.740	0.814	0.771	9(2)	0.631	0.765	0.692	16	0.734	0.803	0.767	0.089	0.3923
Ch02g09	90-212	26(15)	0.861	0.771	0.876	12(1)	0.806	0.875	0.838	27	0.854	0.793	0.869	0.071	0.3764
Ch03b06	90-154	17(9)	0.858	0.856	0.875	12(4)	0.767	0.706	0.807	21	0.847	0.822	0.864	0.048	0.3497
Hi03g06	164-240	24(12)	0.787	0.712	0.807	12	0.780	0.882	0.816	24	0.808	0.750	0.827	0.033	0.3277
Combined														**	***

(*) Exclusive alleles between parenthesis
** Combined probability of identity for the eight analyzed SSR loci = $4{,}11\times10^{-16}$
*** Power of exclusion for the eight analyzed SSR loci = 99.99999259%

exclusive to susceptible accessions (Tables 3 and 4). Exclusive alleles for ALS resistant accessions were found in all SSR markers, with a mean of 10.91 alleles per locus (Table 4). On the other hand, exclusive alleles were also found in six loci for susceptible ALS accessions. A large number of alleles (163 alleles, 67.36 %) were classified as rare alleles (frequency < 0.5%), ranging from 10 (Ch02c11) to 23 (Ch01g12). A total of 67 unique alleles (27.69%) were detected, out of which 30 were detected in ALS resistant, and 37 in susceptible accessions.

To measure the informativeness of these markers, the polymorphism information content (PIC) for each SSR locus was calculated. The PIC value varied from 0.740 to 0.907, with mean of 0.823 for resistant accession, and from 0.553 to 0.850 for susceptible accessions, with mean of 0.702. Total PIC values for all markers varied from 0.734 (Nz02b01) to 0.895 (Ch02b10), with mean of 0.820 (Table 3). Although Ch02b10 was the most informative locus, with the highest PIC value, Ch01g12 marker presented the highest number of alleles. Total observed heterozygosity varied from 0.493 (Ch03c02) to 0.868 (Ch05c06), with mean of 0.771. Total expected heterozygosity varied from 0.762 (Ch05c06) to 0.905 (Ch02b10), with mean value of 0.840 (Table 3).

In the present study, it was found 16 to 29 alleles per locus. Higher number of alleles, Ho, PIC (Table 3), and wider size ranges were observed for all SSR markers in this study, when compared to those reported on the original primer note studies (Guilford et al. 1997, Liebhard et al. 2002, Silfverberg-Dilworth et al. 2006). The higher genetic diversity values are mainly related to the great diversity of apple accessions in conjunction with the large sample size (152). Guilford et

Table 4. Alleles (bp) identified in 152 accessions in each of the 11 SSRs used in this work

Primer	Alleles
Ch03c02	92 104 106 108 110 116 118 120 122 124 126 128 130 132 134 136 144 146 154 164 166 172 218
Ch02b10	106 108 110 112 116 118 120 122 124 126 128 130 132 134 136 138 140 144 146 148 152 158 160 164 170 182
Ch05a05	160 162 194 198 200 202 204 206 208 210 212 214 216 218 220 222 226 228 232 234 250 264
Ch05c06	92 102 104 106 108 110 112 114 116 118 120 122 126 128 132 218 226
Ch02c11	200 204 208 210 218 220 222 226 228 230 232 234 236 238 240 244 250
Ch02b03b	76 78 80 82 84 86 88 90 92 94 96 98 100 102 104 106 110 116 122 124
Ch01g12	96 104 106 108 110 112 114 116 118 120 124 128 132 134 136 138 140 142 146 148 152 154 156 158 162 166 180 186 188
Nz02b01	190 198 208 218 220 222 228 232 234 236 240 242 244 246 250 260
Ch02g09	90 92 96 98 100 102 104 106 108 110 112 114 116 118 120 122 126 128 134 136 138 140 142 150 154 168 212
Ch03b06	90 96 100 102 104 106 108 110 112 114 116 118 120 122 124 128 130 138 140 152 154
Hi03g06	164 168 172 174 178 182 184 188 192 194 196 198 200 202 204 206 208 210 214 216 218 228 236 240

Dark gray background indicates exclusive alleles for susceptible accessions. Gray background indicates exclusive alleles for resistant accessions.

al. (1997) found 4.5 alleles per locus in 21 apple cultivars; Gianfranceschini et al. (1998) found 8.2 alleles per locus in 19 cultivars, and Coart et al. (2003) found 19.58 alleles per locus in 119 apple genotypes. Oraguzie et al. (2005) found 9.7 alleles per locus in 66 apple rootstocks cultivars; Hokanson et al. (1998) found 12.1 alleles per locus in 142 accessions of 23 *Malus* species; and Gharghani et al. (2009) found 17 alleles per locus in other 159 apple cultivars and wild species from several geographical origins and species.

High level of genetic diversity was expressed also in terms of total expected heterozygosity (He=0.84). The high level of heterozygosity of the cultivars is in accordance with former studies on *M. sylvestris* and *M. domestica* (Coart et al. 2003, Larsen et al. 2006, Koopman et al. 2007). Allelic frequencies showed wide variations, ranging from 0.017 to 0.828. The most frequent alleles were detected in three SSR loci, i.e., Hi02b07, Hi08g06, and CTG1066091, with frequencies higher than 0.8.

It was taken the most diverse cluster of accessions as one cluster. In fact, the accessions of this cluster are 92% genetically different from the remaining accessions. In addition, they are not used in breeding programs, the reason why they diverge so much from the other accessions. Thus, the dendrogram (Figure 1) based on genetic similarity among the 152 accessions showed two major clusters. Only Winter Gold, *M. eley* and *M. atrosanguinea* accessions form the first major cluster. *M. eley* and *M. atrosanguinea* accessions are wild apples, non-domesticated and share very few alleles with the other major cluster. In addition, these two species have not recently been used for breeding of modern cultivars; however they could be used, since all wild accessions are resistant to ALS. Other wild accessions, such as *M. robusta, M. baccata* and *M. andenhamensis,* clustered near the first major cluster. The second major cluster is sub-clustered in several minor ones, with similarity between pairs of accessions ranging from 17% to 86%. The highest similarity value

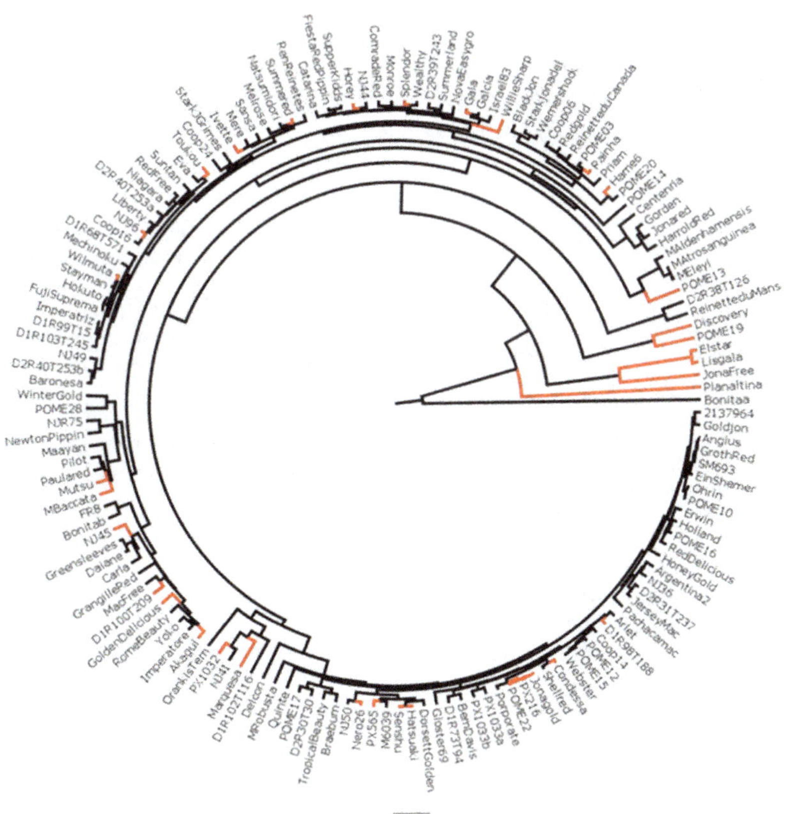

Figure 1. UPGMA dendrogram based on polymorphisms of 11 SSR loci in 152 apple accessions. The red branches indicate susceptibility to Apple Leaf Spot disease (*Colletotrichum gloeosporioides*), being all other accessions resistant to ALS. The scale represents the dissimilarity coefficient between accessions.

of 86% was found in two pairs of accessions; however, identical accessions were not found. The first pair is formed by Fuji Suprema and Hokuto, and the second pair, by Imperatriz and D1R99T15. This high similarity could be the result of a second backcross generation (BC_2). The dendrogram revealed also that the great majority of susceptible accessions were positioned far from wild resistant accessions. Pairs of accessions sharing 82% of alleles were found five times. In three of these cases, the pair is formed by a resistant and a susceptible accession, suggesting that one or more amplified alleles linked to resistance QTLs may be found at the 18% non-shared alleles (Table 4).

The Power of Exclusion (Q) of 99.99999259% from all combined data set provides reliability to the results as discussed above. In addition, the extremely low Probability of Identity (I) of 4.11×10^{-16} provides an unambiguous way to discriminate apple accessions (Table 3). These indices allow cultivar identification by analyzing a relatively smaller number of loci than the 11 that were used in the present work.

Brazilian apple breeding programs are in constant need of new resistant scion cultivars which include fruit quality traits. Therefore, combining knowledge of the genetic markers with potential link to ALS resistant phenotype could lead to the faster development of ALS resistant cultivars for use in crossbreeding. Marker assisted selection techniques could make use of such markers associated to ALS resistant phenotypes for further analysis of their genetic linkage and chromosomal location, either for a single locus or quantitative trait loci (QTLs). The development of new ALS resistant cultivars would benefit not only Brazilian and worldwide apple breeders, but also consumers due to the lesser use of pesticides. In addition, the selection of these markers was successful in characterizing a reasonable large number of apple accessions, which could be of great use to other apple germplasm curators worldwide. Nonetheless, further works should be carried out in order to characterize the interaction between resistant accessions and environmental conditions by performing phenotypic analysis in farm conditions.

ACKNOWLEDGEMENTS

To the Conselho Nacional de Desenvolvimento Científico e Tecnológico (CNPq) for the scholarships granted to GHFK and RON, to Coordenação de Aperfeiçoamento de Pessoal de Nível Superior (CAPES) for the scholarship granted to SZAT and AÇÃO TRANSV/ICTS/ EMPRESAS/Ed 2007/2013 of FINEP.

REFERENCES

Anderson JA, Churchill GA, Autrique JE, Sorrells ME, and Tanksley SD (1993) Optimising parental selection for genetic linkage maps. **Genome 36**: 181-186.

Brooks RM and Olmo HP (1994) Register of new fruit and nut varieties list 35. **Horticultural Science 29**: 942-969.

Burak M, Ergül A, Kazan K, Akcay ME, Yüksel C, Bakir M, Mutaf F, Akpinar AE, Yasasin AS and Ayanoglu H (2014) Genetic analysis of Anatolian apples (*Malus* sp.) by simple sequence repeats. **Journal of Systematics and Evolution 9999**: 1-9.

Coart E, Vekermans X, Smulders MJM, Wagner I, Huylenbroeck JV, Bockstaele EV and Roldán-Ruiz I (2003) Genetic variation in the endangered wild apple (*Malus sylvestris* (L.) Mill.) in Belgium as revealed by amplified fragment length polymorphism and microsatellite markers. **Molecular Ecollogy 12**: 845-857.

Doyle JJ and Doyle JL (1990) Isolation of plant DNA from fresh tissue. **Focus 12**: 13-15.

FAO (2014) Food and Agriculture Organization for the United Nations – FAOSTAT. Available at <http://faostat3.fao.org>. Accessed in May 2016.

Furlan CRC, Dantas ACM, Denardi F, Becker WF and Mantovani A (2010) Resistência genética dos acessos do banco de germoplasma de macieira da EPAGRI à mancha foliar de glomerella (*Colletotrichum gloeosporioides*). **Revista Brasileira de Fruticultura 32**: 507-514.

Gasi F, Simon S, Pojskic N, Kurtovic M and Pejic I (2010) Genetic assessment of apple germplasm in Bosnia and Herzegovina using microsatellite and morphologic markers. **Scientia Horticulturae 126**: 164-171.

Gustavsson LG, Brantestam AK, Sehic J and Nybom H (2008) Molecular characterization of indigenous Swedish apple cultivars based on SSR and S-allele analysis. **Hereditas 145**: 99112.

Gharghani A, Zamani Z, Talaie A, Oraguzie NC, Fatahi R, Hajnajari H, Wiedow C and Gardiner SE (2009) Genetic identity and relationships of Iranian apple (*Malus* x *domestica* Borkh.) cultivars and landraces, wild *Malus* species and representative old apple cultivars based on simple sequence repeat (SSR) marker analysis. **Genetic Resources and Crop Evolution 56**: 829-842.

Gianfranceschi L, Seglias N, Tarchini R, Komjanc M and Gessler C (1998) Simple sequence repeats for the genetic analysis of apple. **Theoretical and Applied Genetics 96**: 1069-1076.

Goulão L and Oliveira CM (2001) Molecular characterisation of cultivars of apple (*Malus* x *domestica* Borkh.) using microsatellite (ISSR and SSR) markers. **Euphytica 122**: 81-89.

Gross B, Volk GM, Richards CM, Forsline P, Fazio G and Chao CT (2012) Identification of "duplicate" accessions within the USDA-ARS national plant germplasm system *Malus* collection. **Journal of the American Society of Horticultural Science 137**: 333-342.

Guilford P, Prakash S, Zhu JM, Rikkerink E, Gardiner SE, Bassett H and Forster R (1997) Microsatellites in *Malus* x *domestica* (apple) abundance, polymorphism and cultivar identification. **Theoretical and Applied Genetics 94**: 249-254.

Han YP and Korban SS (2010) Strategies for map-based cloning in apple. **Critical Reviews in Plant Science 29:** 265-284.

Hokanson SC, McFerson JR, Forsline L and Lamboy WF (1997) Collecting and managing wild *Malus* germplasm in its center of diversity. **Horticultural Science 32**: 173-176.

Hokanson SC, Szewc-Mcfadden AK, Lamboy WF and Mcferson JR (1998) Microsatellite (SSR) markers reveal genetic identities, genetic diversity and relationships in a *Malus* x *domestica* Borkh. core subset collection. **Theoretical and Applied Genetics 97**: 671-683.

Janick J, Cummins JN, Brown SK and Hemmat H (1996) Apple. In Janick J and Moore JN (ed) **Fruit breeding: tree and tropical fruits**. John Wiley & Sons, New York, p. 1-78.

Kalinowski ST, Taper ML and Marshall TC (2007) Revising how the computer program CERVUS accommodates genotyping error increases success in paternity assignment. **Molecular Ecology 16**: 1099-1006.

Koopman WJM, Li Y, Coart E, Van de Weg WE, Vosman B, Roldán-Ruiz I and Smulders MJM (2007) Linked vs. unlinked markers: multilocus microsatellite haplotype-sharing as a tool to estimate gene flow and introgression. **Molecular Ecology 16**: 243-256.

Korban SS and Tartarini S (2009) Apple structural genomics. In Folta KM and Gardiner SE (eds) **Genetics and genomics of rosaceae**. Springer, Berlin, p. 85-119.

Kumar S, Volz RK, Alspach PA and Bus VGM (2010) Development of a recurrent apple breeding programme in New Zealand: a synthesis of results, and a proposed revised breeding strategy. **Euphytica 173**: 207-222.

Larsen AS, Asmussen CB, Coart E, Olrik DC and Kjaer ED (2006) Hybridization and genetic variation in Danish populations of European crab apple (*Malus sylvestris*). **Tree Genetics & Genomes 2**: 86-97.

Liebhard R, Gianfranceschini L, Koller B, Ryder CD, Tarchini R, Van De Weg E and Gessler C (2002) Development and characterization of 140 new microsatellites in apple (*Malus* x *domestica* Borkh.). **Molecular Breeding 10**: 217-241.

Liebhard R, Koller B, Gianfranceschi L and Gessler C (2003) Creating a saturated reference map for the apple (*Malus* x *domestica* Borkh.) genome. **Theoretical and Applied Genetics 106**: 1497-1508.

Morgante M, Hanafey M and Powell W (2002) Microsatellites are preferentially associated with nonrepetitive DNA in plant genomes. **Nature Genetics 30**: 194-200.

Naik S, Hampsom C, Gasic K, Bakkerem G and Korban SS (2006) Development and linkage mapping of E-STS and RGA markers for functional gene homologues in apple. **Genome 49**: 959-968.

Noiton DAM and Alspach PA (1996) Founding clones, inbreeding, coancestry and status number of modern apple cultivars. **Journal of American Society of Horticultural Science 121**: 773-782.

Oraguzie NC, Yamamoto T, Soejima J, Suzuki T and De Silva HN (2005) DNA fingerprinting of apple (*Malus* spp.) rootstocks using Simple Sequence Repeats. **Plant Breeding 124**: 197-202.

Paetkau D and Strobeck C (1994) Microsatellite analysis of genetic variation in black bear populations. **Molecular Ecology 3**: 489-495.

Pereira-Lorenzo S, Cabrer AMR, Díaz JG and Hernández MDB (2008) Genetic assessment of local apple cultivars from La Palma, Spain, using simple sequence repeats (SSRs). **Scientia Horticulturae 117**: 60-166.

Potts SM, Han Y, Khan MA, Kushad MM, Rayburn AL and Korban SS (2013) Genetic diversity and characterization of a core collection of *Malus* germplasm using simple sequence repeats (SSRs). **Plant Molecular Biology Reporter 30**: 827-837.

Reim S, Höltken A and Höfer M (2012) Diversity of the European indigenous wild apple (*Malus sylvestris* (L.) Mill.) in the East Ore Mountains (Osterzgebirge), Germany: II. Genetic characterization. **Genetic Resources and Crop Evolution 60**: 879-892.

Rohlf JF (2000) NTSYS-pc: Numerical taxonomy and multivariate analysis system. Exeter Software, Setauket, 44p.

Silfverberg-Dilworth E, Matasci CL, Van de Weg WE, Van Kaauwen MPW, Walser M, Kodde LP, Soglio V, Gianfranceschi L, Durel CE, Costa F, Yamamoto T, Koller B, Gessler C and Patocchi A (2006) Microsatellite markers spanning the apple (*Malus* x *domestica* Borkh) genome. **Tree Genetics & Genomes 2**: 202-224.

Sneath PHA and Sokal RR (1973) **Numeral taxonomy**. WH Freeman, San Francisco, 573p.

Weir BS, Johnston PR and Damm U (2012) The *Colletotrichum gloesporioides* complex species. **Studies in Mycology 73**: 115-180.

Vandeputte M (2012) An accurate formula to calculate exclusion power of marker sets in parentage assignment. **Genetics Selection Evolution 44**: 34-40.

Zhuang Y, Liu HT, Li CM, Wang Y, Zhao YB, Chen DM, Han ZH and Zang XZ (2011) Inheritance of and molecular markers for susceptibility *of Malus domestica* to fruit ring rot (*Botryosphaeria dothidea*). **Journal of Phytopathology 159**: 782-788.

22

Environmental effect on sunflower oil quality

Amadeu Regitano Neto[1*], Ana Maria Rauen de Oliveira Miguel[2], Anna Lúcia Mourad[2], Ercília Aparecida Henriques[2] and Rosa Maria Vercelino Alves[2]

***Corresponding author:**
E-mail: amadeu.regitano@embrapa.br

[1] Embrapa Semiárido, Rodovia BR-428, km 152, CP 23, 56.302-970, Petrolina, PE, Brazil
[2] Instituto de Tecnologia de Alimentos, ITAL, Av. Brasil, 2880, CP 139, 13.070-178, Campinas, SP, Brazil

Abstract: *Sunflower is one of the most important oilseed crops and produces a high-quality edible oil. Balance of fatty acids in standard sunflower oil shows preponderance of linoleic rather than oleic acid, and conditions during seed development, such as temperature, changes the oleic/linoleic ratio of the oil. This work aimed to evaluate the environmental effect on fatty acid profile in a group of standard and high oleic varieties and hybrids. Seeds were produced during regular season crop and during off-season crop featuring different temperatures from anthesis to maturity. Fatty acid composition was determined by gas chromatography. Levels of oleic acid, in standard oil genotypes, raised as the crop developed in warmer environment while levels of linoleic acid decreased, and the opposite was observed when the crop was grown under lower temperature. High oleic genotypes were less sensitive to environment switching and showed lower variation on fatty acid composition.*

Key words: *Helianthus annuus, high oleic, fatty acid, genotype x environment interaction.*

INTRODUCTION

Sunflower cultivation plays a key role in edible oil production worldwide, represents an important alternative for crop rotation and provides intercropping and succession in producing regions. In Brazil, sunflower production is concentrated in the second crop, during summer-fall period. The production in 2014/15 crop will reach 189.7 thousand tons and the state of Mato Grosso, the major producer, is responsible for 85% of Brazilian sunflower production (Conab 2015).

Fatty acid (FA) composition of sunflower seed lipids is determined by plant genotype and depending on it, this composition is more or less affected by environmental conditions such as light and temperature (Schulte et al. 2013).

Traditional sunflower genotypes produce high quality edible oil with low content of saturated fatty acids, about 100 g kg^{-1} of palmitic (16:0) and stearic acid (18:0) and elevated concentration (up to 90%) of unsaturated FA, mainly 18:1 and 18:2, with the predominance of linoleic acid (18:2) (Rodriguez et al. 2002). Breeding efforts have conducted to the release of genotypes with much higher concentration of oleic acid rather than linoleic acid. Stable high oleic (HO) genotypes were developed through induced mutation by Soldatov (1976) and are nowadays present in a wide range of hybrids (Fernández-Martínez et al. 2007).

Following this trend, other variants have been produced such as high stearic or high palmitic acid cultivars (Fernández-Moya et al. 2005, Skorić et al. 2008).

FA composition of the oil determines its use (Metzger and Bornscheuer 2006). Multiple applications have been identified for the new FA profiles such as pharmaceutical, cosmetic, industrial or edible use (Gupta 2014).

Although traditional sunflower genotypes produce high quality oil for cooking, the development of HO sunflower genotypes represents a very important breeding achievement. The largest advantage of this type of oil is its higher degree of oxidative stability than traditional sunflower oils with standard oleic acid concentration (Grompone 2005).

The effect of temperature during the plant cycle, mainly from anthesis to maturity, on the FA composition of sunflower oil has been reported to change the oleic/linoleic acid (O/L) ratio, known as unsaturation ratio, in the oil. Oil content and composition are modified by high constant temperature during sunflower grain development, and previous reports indicated a marked reduction in the percentage of linoleic acid in standard sunflower oil, apparently due to the effect of temperature on the activity of the enzyme oleate desaturase, which is responsible for the conversion of oleic to linoleic acid (Flagella et al. 2002, Grompone 2005). Garcés and Mancha (1991), who established that the oleyl-phosphatidyl-choline-desaturase activity, in sunflower, is inhibited at temperatures above 20 °C, specifically investigated this effect and Rolletschek et al. (2007), afterwards, demonstrated the mechanism by which temperature modifies the unsaturation degree of the sunflower oil through its effect on dissolved oxygen levels in the developing seed.

Substantial changes are reported for standard genotypes, while less significant changes in the unsaturation ratio have been observed in HO genotypes influenced by temperature (Flagella et al. 2002, Izquierdo and Aguirrezábal 2008, Grunvald et al. 2013, Piao et al. 2014).

In Brazil, mainly in southeastern and western states, it became an usual practice to take advantage of the mild climate and the still occurring rains to sow the off-season crop, which represents a second opportunity for sowing, just subsequent to the crop season. Crop season takes place during spring/summer and off-season crop in summer-fall period. These two periods present a very diverse environment in terms of temperature during crop development, with anthesis occurring in summer and fall, respectively. The State of Mato Grosso is the largest national sunflower producer and even presenting a tropical climate, the temperature during plant development is quite distinct between the two adopted sowing periods.

The aim of this work was to investigate the effect of sowing date and consequently the effect of temperature on oil yield and FA composition of seeds from standard and HO sunflower hybrids grown in southeastern Brazil.

MATERIAL AND METHODS

Sunflower seeds were supplied by seed producers and comprised four open pollinated varieties, six single cross hybrids and three three-way cross hybrids of standard sunflower and two HO three-way cross hybrids that are listed in Table 1.

Table 1. Description of genotypes

Cultivars	Genotype structure	Oil (%)	Oil type	Seed color
Catissol	open population	30-40%	standard	black
CIA	open population	38-44%	standard	black
Iarama	open population	38-42%	standard	gray
Uruguai	open population	28-35%	standard	gray stripped
AG 975	single cross hybrid	45-50%	standard	black
Aguará 6	single cross hybrid	45-50%	standard	black
Helio 250	single cross hybrid	44-48%	standard	black
Helio 251	single cross hybrid	40-44%	standard	gray stripped
Helio 253	single cross hybrid	42-46%	standard	gray stripped
Helio 358	single cross hybrid	44-55%	standard	black
AG 962	three-way cross hybrid	43-50%	standard	black
Charrua	three-way cross hybrid	45-50%	standard	black
Helio 360	three-way cross hybrid	43-47%	standard	gray stripped
Olisun 3	three-way cross hybrid	45-50%	high oleic	black
Olisun 5	three-way cross hybrid	45-50%	high oleic	black

In order to obtain the grain samples for analysis, the commercial seeds were sown in two consecutive years during off-seasons (2011 and 2012) and during regular seasons (2011/12 and 2012/13) at the IAC Experimental Center (Agronomic Institute, Farm Santa Elisa) in Campinas, SP, Brazil (lat 22° 52′ S, long 47° 05′ W, and alt 660 m asl). Cultivation practices were the same over the four field assays and comprised liming, sowing fertilization (300 kg ha^{-1} of 8-28-16), stand (40000 plants ha^{-1}) and top dressing (40 kg ha^{-1} of N) with 1 kg ha^{-1} of boron at 30 days from germination. Weeds were hoed when necessary and no agrochemicals were needed in these fields.

The plants were harvested and aquenes were detached from sunflower heads after drying. Grain samples were processed for total lipids and FA composition analysis at the Institute of Food Technology in Campinas, SP, Brazil. The samples were ventilated and grains were homogenized in a Tecnal, model TE-631/2 multipurpose mill with water circulation. A weighted sample of 5 ± 0.5 g was taken of each genotype, placed in a double paper filter cartridge and submitted to an organic solvent-based extraction, with petroleum ether (80 mL) as the extraction solvent in a Butt-type extraction equipment (8 hours). The cartridge was discarded and the miscela (oil + solvent) was separated by rotary evaporation (40-50 °C). The residual solvent was eliminated by current of nitrogen and the balloon containing the lipids was maintained in a drying oven at 100 ± 5 °C for 1 hour, cold and weighted. The percentage of oil in the seeds was then determined by gravimetric measurement of the collected oil and expressed as a weight percent relative to the initial weight of the oilseeds (Firestone 2008). All analyses were performed in duplicate.

Compositions in FA were determined by the method of gas chromatography with capillary column and detection by flame ionization. The samples were prepared following the obtainment of FAME (fatty acid methyl ester) according to Hartman and Lago (1973), and the identification of FA was performed by a gas chromatograph Varian, model 3900, and capillary column by a Chrompack CP-Sil 88. The FA was identified by its time of retention, and chromatograms were compared to well-known commercial standards and quantification was accomplished as a relative percentage of area, as suggested by Firestone (2008).

Data set from 15 genotypes (G), sowed for two years (Y) in two sowing dates (SD) each, were analyzed through analysis of variance (ANOVA) in a completely randomized block design with two replications. Joint ANOVAs were performed for each FA and for oil content considering mathematical model Yijk = m + Gi + Yj + SDk + GYij + GSDik + YSDjk + GYSDijk + Eijk, with Y of random effect. Treatment means were compared by Tukey's test ($p<0.05$). Coefficients of determination (R^2) were calculated to access proportional contribution of genotype interactions (G x Y, G x SD, G x Y x SD) to total genotype x environment interaction using related sum of squares. Coefficients of correlation were estimated with respective *p*-values, indicating the probability of observing the correlation coefficient under the null hypothesis or no correlation, with n-2 degrees of freedom. Analyses were performed using SAS software (2016).

RESULTS AND DISCUSSION

Mean lipid concentrations largely varied among genotypes and sowing dates (Table 2). The cultivar Uruguai, developed by IAC, was not designed for oil production, but bird food or silage, and presented the lowest oil concentration for all sowing dates (25.6 and 26.4%) in relation to any other genotype. Oil concentration from conventional oil hybrids ranged from 31.9% (Helio 251) to 47.0% (Helio 358) in summer-fall plantings and from 28.5% (Helio 253) to 45.3% (Helio 358) in spring-summer crop. HO three-way hybrids produced more oil (42.85 and 44.48%) during season crop than during off-season crop (38.3 and 39.4%). Lipid concentration mean squares (MS) (Table 4) showed significant differences ($p<0.05$) for all sources of variation, except for Y x SD interaction. Considering all genotype x environment interactions for lipid content, estimated coefficient of determination (R^2) indicated the largest contribution of G x Y (45.0%) interaction along with G x SD (28.8%) and G x Y x SD (26.0%) to estimated sum of squares.

These results are consistent and expected, since genotypes have diverse genetic background on hybrid formation and were selected and meant for planting on different locations (Table 1). Moreover, environment conditions during cultivation, considering Y and SD were distinct, regarding temperature and water availability, although no water deficit occurred during these periods (Table 5). Lipid concentration showed no significant coefficient of correlation with FA and temperatures (Table 6) considering standard oil genotypes. Lipids concentration of HO hybrids showed significant negative correlation with stearic acid while all other coefficients were not significant. In a different environmental condition, Onemli (2012), evaluating conventional hybrid, reported strong correlation of oil content and oleic (-0.84), linoleic (0.84)

Table 2. Concentration of lipids (LIP %) and fatty acids (means %): oleic (C 18:1, %) and linoleic (C 18:2, %) in sunflower seeds, evaluated during 2011 and 2012 Off Season Crop and 2011/12 and 2012/13 Season Crop in Campinas, SP, Brazil

Genotypes	LIP %				C 18:1				C 18:2			
	Off Season		Season		Off Season		Season		Off Season		Season	
	2011	2012	2011/12	2012/13	2011	2012	2011/12	2012/13	2011	2012	2011/12	2012/13
AG 962	45.84 aA1*	40.76 abA1	42.71 aA1	37.39 abA1	20.16 bB2	25.84 bB2	47.68 bcA1	47.05 bcA1	65.43 aA1	61.04 aA1	40.38 bB2	41.47 abB2
AG 975	45.55 aA1	38.94 abA1	42.59 abA1	38.75 abA1	17.76 bB2	21.55 bB2	43.46 bcA1	48.78 bcA1	68.35 aA1	66.31 aA1	46.47 abB2	40.64 bB2
Aguará 06	40.72 abA1	34.79 abAB1	43.72 abA1	30.87 bB1	19.87 bB2	17.81 bB2	50.25 bcA1	46.01 bcA1	66.87 aA1	70.98 aA1	40.76 bB2	44.71 abB2
Catissol	36.36 bA1	38.20 abA1	30.48 bB1	39.70 abA1	25.30 bB2	23.90 bB2	50.37 bcA1	45.85 bcA1	61.63 aA1	64.95 aA1	39.73 bB2	44.30 abB2
Charrua	42.33 abA1	36.09 abA1	43.92 abA1	41.98 aA1	17.05 bB2	17.27 bB2	53.02 bA1	61.07 bA1	70.28 aA1	71.45 aA1	38.03 bB2	29.76 bB2
CIA	46.81 aA1	38.67 abA1	43.97 aA1	37.84 abA1	20.47 bB2	27.51 bB2	49.14 bcA1	53.94 bcA1	65.25 aA1	60.71 aA1	39.27 bB2	36.54 bB2
Helio 250	42.48 abA1	36.98 abA1	43.74 abA1	45.07 aA1	13.71 bB2	18.23 bB2	35.47 bcA1	38.48.cA1	73.41 aA1	70.17 aA1	54.53 abB2	51.86 abB2
Helio 251	35.84 bA1	31.97 bA1	38.98 abA1	37.37 abA1	23.25 bA1	22.41 bA1	34.83.cA1	34.09.cA1	64.54 aA1	66.22 aA1	55.50 aA1	56.06 aA1
Helio 253	42.46 abA1	37.60 abAB1	42.46 abA1	28.58 bB2	27.21 bB1	21.71 bB2	42.18 bcB1	61.00 bA1	60.78 aA1	66.90 aA1	47.82 abA1	27.43 bB2
Helio 358	47.08 aA1	39.88 aA1	45.34 aA1	41.94 aA1	16.87 bB2	20.88 bB2	43.52 bcA1	45.11.cA1	71.53 aA1	68.08 aA1	47.46 abB2	44.68 abB2
Helio 360	39.22 abA1	41.17 aA1	42.60 abA1	41.84 aA1	20.70 bA1	23.20 bA1	35.46.cA1	35.88.cA1	67.25 aA1	65.68 aA1	54.87 abA1	54.53 abA1
Iarama	39.84 abA1	34.24 abA1	38.82 abA1	39.09 abA1	17.12 bB2	25.23 bB2	45.51 bcA1	56.12 bcA1	70.35 aA1	63.57 aA1	43.83 abB2	34.35 bB2
Olisun 03	39.48 abA1	38.38 abA1	44.48 abA1	42.85 aA1	78.06 aA1	68.43 aA2	86.82 aA1	88.53 aA1	13.58 aA1	22.93 aA1	5.72.cA1	3.52.cB2
Olisun 05	38.59 abA1	36.21 abA1	44.41 abA1	37.19 abA1	79.61 aA1	76.19 aA1	89.03 aA1	88.16 aA1	11.09 bA1	14.92 bA1	3.71.cA1	3.61.cB2
Uruguai	26.41.cA1	29.90 bA1	33.08 bA1	25.62 bA1	24.72 bB2	25.53 bB2	51.29 bA1	53.94 bcA1	63.00 bA1	63.45 bA1	40.37 bB2	36.52 bB2
Mean Standard	40.84	36.86	40.95	37.39	20.32	22.39	44.78	48.26	66.82	66.12	45.31	41.76
Mean HO	39.03	37.30	44.45	40.02	78.83	72.31	87.93	88.35	12.34	18.93	4.72	3.57

* Means in columns followed by the same lowercase letters are not significantly different; Means in rows followed by the same uppercase letters are not significantly different; Means in rows followed by the same number are not significantly different, within years, according to Tukey's test at $p<0.05$. Mean HO comprises Olisun 03 and Olisun 05 means, exclusively.

and palmitic acid (0.93) concentration.

Means of total saturated FA (TSFA) comprising concentrations of myristic (C14:0), palmitic (C16:0), stearic (C18:0), arachidic (C20:0), behenic (C22:0) and lignoceric (C24:0) FA summed (Table 3), ranged, in season crop, from 8.09 to 11.59% in standard genotypes and from 6.86 to 7.84% in HO hybrids. In off-season crop TSFA concentrations were between 10.71 and 14.14% in conventional genotypes and between 7.97 and 8.92% in HO hybrids. Major contribution to TSFA content were given by palmitic and stearic acids, and although standard oil genotypes had higher TSFA means when compared to HO oils, concentrations of TSFA in both types showed a tendency to be slightly lower when sunflower were cultivated in warmer environments.

Significant coefficients of correlation (Table 6) between palmitic acid and stearic acid were low and positive (0.296), while between palmitic and oleic were moderate and negative (-0.610) and moderate positive (0.609) between palmitic and linoleic acid. For stearic acid correlations with oleic (-0.459) and linoleic acid (0.413) significant coefficients were found in standard genotypes. Correlations of palmitic and stearic acids in HO hybrids showed not significant coefficients for all traits. Such effect could be attribute to the small sample size (N=8) used for the estimates. Although not significant, such correlations with oleic and linoleic acids showed the same magnitude and direction observed for conventional oil genotypes.

ANOVA of TSFA percentage (Table 4) showed significant differences for all sources of variation, except for a non-significant triple interaction, indicating that the effect of G x Y and G x SD were responsible for 32.31 and 27.14%, respectively, of total G x E interactions.

The effect of the environment on sunflower saturated FA concentration has been reported (Izquierdo and Aguirrezábal 2008, Piao et al. 2014) and attributed to planting date (Zheljazkov et. al. 2009), night temperature (Izquierdo et al. 2002) and hybrids (Zheljazkov et al. 2011). This indicates the possibility of producing sunflower oil with improved dietary characteristics through cultivar selection and management of the environment in which the crop will develop by choosing more adequate SD.

Content of total unsaturated fatty acids (TUFA) in all studied sunflower oils revealed broad range of variation among genotypes (Table 3). Oleic and linoleic acids summed, from standard genotypes, showed concentrations ranging from 85.78% (AG 962) to 89.26% (Uruguai) in off-season plantings and from 88.27% (AG 962) to 91.92% (Uruguai) in regular season plantings. Considering HO hybrids, TUFA levels were 91.08% (Olisun 5) and 92.03% (Olisun 3) in off-season crop and from 92.16% to 93.12 in Olisun 5, in season crop. In average, all studied genotypes had higher TUFA content when plant cycle occurred during the warmer environment.

Oleic acid contents were different among standard but were not among HO genotypes (Table 2). Observed levels of oleic acid for standard genotypes were from 13.71% (Helio 250) to 27.51% (CIA) during off-season crop and from 34.09% (Helio 251) to 61.07% (Charrua) in season crop. For HO genotypes, concentrations of oleic acid were between 68.43% (Olisun 3) and 79.61% (Olisun 5) in off-season crop, and from 86.82% (Olisun 3) to 89.03% (Olisun 5) in season crop.

Observed levels of linoleic acid (Table 2), in its turn, varied in off-season crop from 60.78% (Helio 253) to 73.41% (Helio 250) and in season crop from 27.43% (Helio 253) to 56.06% (Helio 251) among standard genotypes, while in HO hybrids were 11.09% (Olisun 5) and 22.93% (Olisun 3) in off-season crop, and 3.71% (Olisun 5) and 5.72% (Olisun 3) in season crop.

Table 3. Concentration means of total saturated fatty acids (TSFA, mean %) and total unsaturated fatty acids (TUFA, mean %) evaluated during 2011 and 2012 Off Season Crop and 2011/12 and 2012/13 Season Crop in Campinas, SP, Brazil

Genotypes	TSFA		TUFA	
	Off Season	Season	Off Season	Season
AG 962	13.53 aA*	11.36 aB	86.43 cB	88.57 dA
AG 975	12.82 abA	10.12 bcB	87.18 cB	89.88 cA
Aguará 06	11.93 bA	8.79 cB	88.03 bcB	91.18 bA
Catissol	12.80 abA	10.22 bcB	87.20 cB	89.67 cdA
Charrua	11.80 bA	9.44 bcB	88.20 bcB	90.46 bcA
CIA	11.69 bA	8.63 cB	88.31 bcB	91.22 bA
Helio 250	12.00 bA	9.49 bcB	88.00 bcB	90.43 bcA
Helio 251	11.50 bA	9.46 bcB	88.49 bB	90.54 bcA
Helio 253	11.50 bA	10.35 bB	88.53 bA	89.53 cdA
Helio 358	11.04 bA	9.12 cB	88.96 bB	90.66 bcA
Helio 360	11.30 bA	9.36 bcB	88.70 bB	90.64 bcA
Iarama	11.57 bA	9.62 bcB	88.40 bcB	90.21 bcA
Olisun 03	8.10 cA	7.05 dA	91.90 aA	92.70 aA
Olisun 05	8.69 cA	7.36 dB	91.31 abB	92.64 abA
Uruguai	11.36 bA	8.61 cB	88.64 bB	91.34 bA
Mean Standard	11.91	9.58	88.08	90.33
Mean HO	8.39	7.21	91.61	92.67

* Means in columns followed by the same lowercase letters are not significantly different; Means in rows followed by the same uppercase letters are not significantly different, according to Tukey's test at $p<0.05$. Mean HO comprises Olisun 03 and Olisun 05 means, exclusively.

Table 4. Mean Squares (MS) and *p*-values from joint ANOVA of oil yield (Oil, %), oleic acid (C 18:1, %) and linoleic acid (C 18:2, %), total saturated FA (TSFA, %) and total unsaturated FA (TUFA, %) in Campinas, SP, Brazil

Sources	df	Oil		C18:1		C18:2		TSFA		TUFA	
		MS	*p*	MS	*p*	MS	*p*	MS	*p*	MS	*p*
Genotypes (G)	14	109.59	0.00	2387.48	0.00	2137.50	0.00	11.97	0.00	11.83	0.00
Years (Y)	1	406.42	0.00	118.94	0.00	66.81	0.02	9.15	0.00	9.70	0.00
Sowing Date (SD)	1	20.20	0.04	16546.56	0.00	13737.97	0.00	141.91	0.00	131.00	0.00
G x Y	14	28.76	0.00	31.18	0.00	28,51	0.01	1.00	0.00	1.11	0.00
G x SD	14	18.42	0.00	126.82	0.02	115.80	0.00	0.84	0.00	0.93	0.00
Y x SD	1	0.00	1.00	34.79	0.10	90.68	0.01	13.93	0.00	13.74	0.00
G x Y x SD	14	16.62	0.00	28.68	0.01	28.83	0.00	0.26	0.10	0.31	0.10
CV%		5.42		8.77		6.96		3.88		0,48	
Mean		39.17		40.32		48.98		10.35		89.60	

ANOVA of oleic and linoleic acid (Table 4), presented significant (p<0.05) MS for all sources of variation, except for a not significant Y x SD interaction for oleic acid. Proportional contribution to G x E sum of squares showed the preponderance of the genotype x sowing date interaction, with coefficients of determination (R^2) of 67.04 and 64.47% for oleic and linoleic acids levels. G x Y interaction had R^2 of 16.48 and 15.87% while triple interaction contributed with 15.16 and 16.05% for total interaction, for unsaturated acids in the same order. Oil content and FA composition have been shown to be significantly influenced by the environment and significant G x E interaction has been reported (Qadir et al. 2006, Van Der Merwe et al. 2013).

As expected, significant and negative coefficient of correlation (Table 6) was found between oleic and linoleic acid concentrations in standard (-0.997) and HO (-0.997) genotypes, as reported by Onemli (2012), who found a strong negative correlation estimate (-0.99) between oleic and linoleic acids percentage.

Table 5. Environmental conditions during grain filling* (60 to 115 days after sowing), in Campinas, SP, Brazil

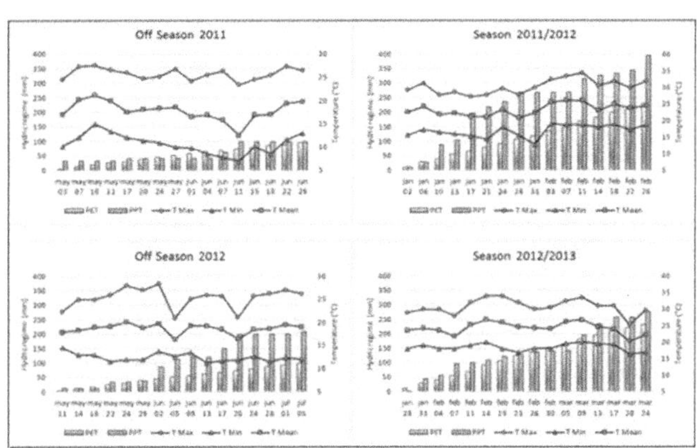

*- Average daily maximum (TMax), mean (TMean) and minimum (TMin) temperatures, precipitation (PPT) and potential evapotranspiration (PET). Source: CIIAGRO on line

Presented data corroborates with data from literature reports. Levels of oleic and linoleic acids in sunflower oil have been shown to vary according to temperature after flowering and variation in the oleic acid content to be inversely proportional to variation in the linoleic content (Ungaro et al. 1997, Qadir et al. 2006, Roche et al. 2006, Izquierdo and Aguirrezábal 2008, Turhan et al. 2010, Grunvald et al. 2013, Van Der Merwe et al. 2013, Piao et al. 2014). All the authors found the highest oleic acid content when crops were cultivated in warmer environments with higher temperature during seed development, and conversely, linoleic acid content increased when the crop developed under colder environments. The observed effect of temperature on the oleic/linoleic ratio in sunflower oil has been attributed to the synthesis or activity of the oleate desaturase enzyme, which is stimulated by low temperature and repressed by high temperature (Sarmiento et al. 1998, Flagella et al. 2002).

The more expressive differences between oleic and linoleic acid concentration due to temperature were observed in standard oil genotypes. Oleic acid content varied, in average (Table 2), from 21.36 to 46.52% and inversely linoleic acid content ranged from 66.47 to 43.53% from colder to warmer environments. HO genotype oils, in turn, showed less

Table 6. Mean Pearson's coefficient of correlation and respective p-values for lipids concentration, fatty acids* and temperatures for standard genotypes (above diagonal, N=52) and high oleic hybrids (below diagonal, N=8)

	LIP %	C16:0	C18:0	C18:1	C18:2	TMax	TMean	TMin
LIP %		-0.067 0.637	0.028 0.844	-0.132 0.351	0.126 0.375	0.027 0.848	-0.001 0.996	-0.031 0.829
C16:0	-0.119 0.779		0.296 0.033	-0.610 <.0001	0.609 <.0001	-0.523 <.0001	-0.539 <.0001	-0.549 <.0001
C18:0	-0.800 0.017	-0.166 0.695		-0.459 0.001	0.413 0.002	-0.533 <.0001	-0.560 <.0001	-0.583 <.0001
C18:1	0.606 0.111	-0.657 0.077	-0.560 0.149		-0.997 <.0001	0.900 <.0001	0.910 <.0001	0.908 <.0001
C18:2	-0.562 0.147	0.690 0.058	0.504 0.203	-0.997 <.0001		-0.887 <.0001	-0.893 <.0001	-0.888 <.0001
TMax	0.617 0.103	-0.352 0.392	-0.754 0.031	0.920 0.001	-0.904 0.002		0.994 <.0001	0.977 <.0001
TMean	0.601 0.115	-0.306 0.460	-0.782 0.022	0.889 0.003	-0.870 0.005			0.994 <.0001
TMin	0.576 0.135	-0.254 0.544	-0.803 0.016	0.846 0.008	-0.825 0.012			

*- Fatty acids: C16:0 = palmitic acid, C18:0 = stearic acid, C18:1 = oleic acid, C18:2 = linoleic acid; TMax, TMean and TMin – average of maximum, mean and minimum temperatures during grain filling.

expressive change in unsaturated acid percentage, and oleic acid, in average, increased from 75.57 to 88.14% and linoleic acid diminished from 15.63 to 4.14%, when cultivation occurred in colder and warmer environments, respectively. The less expressive effect of temperature on oleic and linoleic acids concentration of HO hybrids has been reported (Flagella et al. 2002, Izquierdo and Aguirrezábal 2008, Grunvald et al. 2013, Van Der Merwe et al. 2013). This effect has been attributed to the activity of the oleate desaturase, limited to the early stage of the embryo development (Garcés and Mancha 1991) and also that its transcript is not accumulated during grain filling (Lagravère et al. 2004).

Most expressive changes in unsaturation ratio among standard oil genotypes due to moving sowing dates from colder to warmer environment, were observed in Charrua, which increased O/L ratio 6.95 times, from 0.24 to 1.68, followed by Aguará 06 (4.12 times) and Iarama (4.11 times) changing from 0.27 and 0.32 in off season crop to 1.13 and 1.30 in season crop, respectively. Less expressive changes in O/L ratio were observed in Helio 251 (1.77 times) and Helio 360 (1.99 times), switching from 0.35 and 0.33 to 0.62 and 0.65, in the same order. Changes in HO genotypes were from 4.01 to 18.98 in Olisun 03 (4.73 times) and from 5.99 to 24.19 in Olisun 05 (4.04 times).

Taken into account the sunflower crop productivity in Brazilian largest producer state, about 1500 kg ha^{-1} (Conab 2015) along with industry preferences and possible rewarding, using these results, one must consider that conventional hybrids planted in the season crop (spring-summer) will produce 377.33 kg ha^{-1} more oleic acid than the one planted in the off-season crop, while in opposite direction, the production of linoleic acid would be 344.04 kg ha^{-1} greater when planted in the off-season crop.

In these same conditions, HO hybrids would produce 1322.05 kg ha^{-1} of oleic acid during season crop, decreasing 188.46 kg ha^{-1} with a raise of 172.35 kg ha^{-1} of linoleic acid when sowed in off-season crop.

High levels of oleic acid in oils are related to longer shelf life due to their higher stability and resistance to oxidation. Oil industry and sunflower producers can take advantage of the temperature effects on FA synthesis by managing genotypes and sowing dates in order to have the best response for their needs.

ACKNOWLEDGEMENTS

Authors acknowledge the financial support of the National Counsel of Technological and Scientific Development (CNPq 555877/2010/8).

REFERENCES

Ciiagro (2015) Centro integrado de informações agrometeorológicas [homepage on the internet]. São Paulo: Instituto Agronômico; 1988-2003. Available at <http://www.ciiagro.sp.gov.br/ciiagroonline> Accessed on April 20, 2015.

Conab - Companhia Nacional de Abastecimento (2015) Acompanhamento da safra brasileira de grãos v.2 – Safra 2014/15, n.8 – Oitavo Levantamento. Conab, Brasília, 118p.

Fernández-Martínez JM, Pérez-Vich B, Velasco L and Domínguez J (2007) Breeding for specialty oil types in sunflower. **Helia 30**: 75-84.

Fernández-Moya V, Martínez-Force E and Garcés R (2005) Oils from improved high stearic acid sunflower seeds. **Journal of Agricultural and Food Chemistry 53**: 5326-5330.

Firestone D (2008) **Official methods and recommended practices of the American Oil Chemists Society**. AOCS Press, Champaign, 1200p.

Flagella Z, Rotunno T, Tarantino E, Caterina RD and Caro AD (2002) Changes in seed yield and oil fatty acid composition of high oleic sunflower (*Helianthus annuus* L.) hybrids in relation to the sowing date and the water regime. **European Journal of Agronomy 17**: 221-230.

Garcés R and Mancha M (1991) *In vitro* oleate desaturase in developing sunflower seeds. **Phytochemistry 30**: 2127-2130.

Grompone MA (2005) Sunflower oil. In Shahidi F (ed) **Bailey's industrial oil & fat products-edible oils and fat products: edible oils**. John Wiley & Sons, Hoboken, p. 655-730.

Grunvald AK, Carvalho CGP, Leite RS, Mandarino JMG, Andrade CAD, Amabile RF and Godinho VDC (2013) Influence of temperature on the fatty acid composition of the oil from sunflower genotypes grown in tropical regions. **Journal of the American Oil Chemists Society 90**: 545-553.

Gupta MK (2014) Sunflower oil: history, applications and trends. **Lipid Technology 26**: 260-263.

Hartman L and Lago RCA (1973) Rapid preparation of fatty acid methyl from lipids. **Laboratory Practice 22**: 475-473.

Izquierdo NG, Aguirrezábal LAN, Andrade F and Pereyra V (2002) Night temperature affects fatty acid composition in sunflower oil depending on the hybrid and the phenological stage. **Field Crops Research 77**: 115-126.

Izquierdo NG and Aguirrezábal LAN (2008) Genetic variability in the response of fatty acid composition to minimum night temperature during grain filling in sunflower. **Field Crops Research 106**: 116-125.

Lagravère T, Kleiber D, Surel O, Calmon A, Bervillé A and Dayde J (2004) Comparison of fatty acid metabolism of two oleic and one conventional sunflower hybrids: A new hypothesis. **Journal of Agronomy and Crop Science 190**: 223-229.

Metzger JO and Bornscheuer U (2006) Lipids as renewable resources: current state of chemical and biotechnological conversion and diversification. **Applied Microbiology and Biotechnology 71**: 13-22.

Onemli F (2012) Impact of climate changes and correlations on oil fatty acids in sunflower. **Pakistan Journal of Agricultural Science 49**: 455-458.

Piao X, Choi SY, Jang YS, So YS, Chung JW, Lee S, Jong J and Kim HS (2014) Effect of genotype, growing year and planting date on agronomic traits and chemical composition in sunflower (*Helianthus annuus* L.) germplasm. **Plant Breeding and Biotechnology 2**: 35-47.

Qadir G, Ahmad S, Hassan F and Cheema MA (2006) Oil and fatty acid accumulation in sunflower as influenced by temperature variation. **Pakistan Journal of Botany 38**: 1137-1147.

Rodriguez DJ, Philips DBS, Rodriguez-Garcia R and Angulo-Sanchez JL (2002) Grain yield and fatty acid composition of sunflower seed for cultivars developed under dry land conditions. In Janick J and Whipkey A (eds) **Trends in new crops and new uses**. American Society for Horticultural Science Press, Alexandria, p. 139-142.

Roche J, Bouniols A, Mouloungui Z, Barranco T and Cerny M (2006) Management of environmental crop conditions to produce useful sunflower oil components. **European Journal of Lipid Science and Technology 108**: 287-297.

Rolletschek H, Borisjuk L, Sánchez-García A, Gotor C, Romero LC, Martínez-Rivas JM and Mancha M (2007) Temperature-dependent endogenous oxygen concentration regulates microsomal oleate desaturase in developing sunflower seeds. **Journal of Experimental Botany 58**: 3171-3181.

Sarmiento C, Garcés F and Mancha M (1998) Oleate desaturation and acyl turnover in sunflower (*Helianthus annuus* L.) seed lipids during rapid temperature adaptation. **Planta 205**: 595-600.

SAS (2016) **SAS university edition: installation guide for Windows**. SAS Institute Inc., Cary, 24p.

Schulte LR, Ballard T, Samarakoon T, Yao L, Vadlani P, Staggenborg S and Rezac M (2013) Increased growing temperature reduces content of polyunsaturated fatty acids in four oilseed crops. **Industrial Crops and Products 51**: 212-219.

Skorić D, Jocić S, Sakac Z and Lecić N (2008) Genetic possibilities for altering sunflower oil quality to obtain novel oils. **Canadian Journal of Physiology and Pharmacology 86**: 215-221.

Soldatov KI (1976) Chemical mutagenesis in sunflower breeding. In **Proceedings of the 7th International sunflower conference**. International Sunflower Association, Krasnodar, p. 352-357.

Turhan H, Citak N, Pehlivanoglu H and Mengul Z (2010) Effects of ecological and topographic conditions on oil content and fatty acid composition in sunflower. **Bulgarian Journal of Agricultural Science 16**: 553-558.

Ungaro MRG, Sentelhas PC, Turatti JM and Soave D (1997) Influência da temperatura do ar na composição de aquênios de girassol. **Pesquisa Agropecuária Brasileira 32**: 351-356.

Van Der Merwe R, Labuschagne MT, Herselman L and Hugo A (2013) Stability of seed oil quality traits in high and mid-oleic acid sunflower hybrids. **Euphytica 193**: 157-168.

Zheljazkov VD, Vick BA, Baldwin BS, Buehring N, Coker C, Astatkie T and Johnson B (2011) Oil productivity and composition of sunflower as a function of hybrid and planting date. **Industrial Crops and Products 33**: 537-543.

Zheljazkov VD, Vick BA, Baldwin BS, Buehring N, Astatkie T and Johnson B (2009) Oil content and saturated fatty acids in sunflower as a function of planting date, nitrogen rate and hybrid. **Agronomy Journal 101**: 1003-1011.

23

Genetic and cytological diversity in cherry tree accessions (Eugenia involucrata DC) in Rio Grande do Sul

Divanilde Guerra[1,3*], Paulo Vitor Dutra de Souza[1], Sérgio Francisco Schwarz[1], Maria Teresa Schifino-Wittmann[2], Claudio André Werlang[1] and Pedro Augusto Veit[1]

Abstract: *This study aimed to evaluate the genetic and cytological diversity and stability of 35 cherry tree accessions collected in Rio Grande do Sul. We used 15 RAPD (Random Amplified Polymorphic DNA) molecular markers and performed cytological analysis and number count of anthers. Analyses of genetic diversity allowed the separation of accessions into four groups, resulting in an average of 8.93 bands per primer amplified, 7.89 polymorphic bands, 88.08% of polymorphism and 86% of genetic similarity. Cytological analyses of gametic cells allowed for the characterization of accessions as diploids with n=11. In these, the average of meiotic cells considered normal was 82.12%; average pollen viability was 92.44% and in vitro germination was 40.26%; the average number of anthers was 161.85 anthers/flowers. Therefore, the accessions evaluated showed high genetic similarity and cytological stability and can be used in commercial plantations or hybridizations.*

Key words: *Breeding, hybridizations, molecular analysis, cytogenetic analysis.*

***Corresponding author:**
E-mail: divanildeguerra@yahoo.com.br

[1] Universidade Federal do Rio Grande do Sul (UFRGS), Faculdade de Agronomia, Departamento de Horticultura e Silvicultura, Avenida Bento Gonçalves, 7712, 91.501-970, Porto Alegre, RS, Brazil

[2] UFRGS, Faculdade de Agronomia, Departamento de Plantas Forrageiras e Agrometeorologia

[3] Universidade Estadual do Rio Grande do Sul (UERGS), Curso de Agronomia, Rua Cipriano Barata, 47, 98.600-000, Três Passos, RS, Brazil

INTRODUCTION

The Myrtaceae Family is one of the largest plant families with around 3600 species in 150 genera and is distributed in different environments throughout tropical, subtropical and temperate regions, highlighting its wide adaptive power (Landrum and Kawasaki 1997, Romagnolo and Souza 2004). In this family, the genus *Eugenia* is one of the most representatives since it consists of about 1000 species distributed mainly throughout Central and South America, with more than 350 native species in Brazil (Landrum and Kawasaki 1997).

The cherry tree (*Eugenia involucrata* DC.) is a native species of southern Brazil, belonging to the genus *Eugenia* and can be found from Minas Gerais to Rio Grande do Sul in forest formations of the Atlantic complex and in the forests and savannas of the Paraná basin (Donadio et al. 2004). It also occurs in Mato Grosso do Sul, Minas Gerais and Goiás and in other countries of South America (Rodrigues and Carvalho 2001). This species is commonly known as cherry tree, *cerejeira-do-mato*, *cereja-do-rio-grande* and black cherry (Lorenzi 2002) and is of great ecological importance since it is a species dispersed by animals and appropriate for the recovery of degraded areas, besides its landscape value due to the beauty of its flowers and stems (Lorenzi 2002, Carvalho 2008). The fruit have commercial potential and can be consumed fresh or processed in the form of jams, jellies, liqueurs and juices (Oliveira 2007), in addition to the medicinal properties of the fruit and leaves (Rodrigues and Carvalho 2001).

Although having enormous potential to integrate production systems, the cherry tree is seldom used for this purpose due to lack of agronomic information, absence of improved cultivars, large need for labor to harvest the fruit (and their perishable nature), as well as having a preferential sexual system of propagation, which results in plants with high variability of agronomic traits, complicating the implementation and management of orchards (Carvalho 2008). Furthermore, the basic characteristics of the species are still unknown, so cytological analyses and technologies using molecular markers can assist in breeding programs, identifying and selecting plants with greater genetic variability and cytological stability that can be used as parents in directed crosses or in vegetative propagation (Wunch and Hormaza 2007). Thus, the aim of this study was to: a) evaluate the genetic variability of cherry tree accessions through RAPD molecular markers; b) evaluate the cytological stability of the accessions.

MATERIAL AND METHODS

Thirty-five cherry free accessions from natural populations, as well as pre-selected and cultivated populations in domestic environments were selected for the evaluation of genetic diversity and cytological analyses. The cities in the state of Rio Grande do Sul where the accessions were collected and their identification in the paper are: Venâncio Aires (1); Porto Alegre (2); Gravataí (3); Gravataí (4); Cachoeirinha (5); Porto Alegre (6); Porto Alegre (7); Porto Alegre (8); Santa Cruz do Sul (9); Charqueadas (10); Eldorado do Sul (11); Eldorado do Sul (12); Eldorado do Sul (13); Eldorado do Sul (14); Eldorado do Sul (15); Eldorado do Sul (16); Porto Alegre (17); Porto Alegre (18); Canoas (19); Canoas (20); Esteio (21); Porto Alegre (22); Estrela (23); Campina das Missões (24); Venâncio Aires (25); Guaporé (26); Santa Cruz do Sul (27); Porto Alegre (28); General Câmara (29); Eldorado do Sul (30); Três Passos (31); Bom Progresso (32); São Luiz Gonzaga (33); São Borja (34) and Tenente Portela (35).

Young leaves were collected for DNA extraction and genetic diversity evaluation through the protocol proposed by Ferreira and Grattapaglia (1998). The quantity and quality of DNA were evaluated by applying the samples to 1% agarose gel, which was stained with ethidium bromide (0.5 ng mL^{-1}) and submitted to electrophoresis for an hour at 110 V. The quantification was performed by comparison with λ50, λ100, λ200 and λ500 Lambdas standards and the DNA quality was evaluated by the absence of DNA traces (Guerra et al. 2016).

The genetic diversity was obtained using RAPD ("Random Amplified Polymorphic DNA") molecular markers. The reaction had a total volume of 25 uL containing: 10 mM of Tris-HCl, pH 8.3; 50 mM of KCl; 2.0 mM of MgCl2; 0.4 mM of dNTPs; 0.25 uM of primer; 5.0 ng of DNA; 1 unit of Taq DNA Polymerase and ultrapure water to complete the volume. The amplification of specific regions was performed using PCR (Polymerase Chain Reaction) technique in a thermocycler, being performed for 48 cycles: 92 °C for 30 seconds; 37 °C for 1 minute and 30 seconds; and 72 °C for 1 minute and 30 seconds. Preceding the cycle, an initial denaturation was performed at 94 °C for 5 minutes and, afterwards, a final extension at 72 °C for 5 minutes. After the amplification, the samples were applied to agarose gel (1.8%) and compared with the standard of known size (Gibco BRL with 100 base pairs). The time for the electrophoretic separation was of two hours at 110 volts. The gels were photographed and the fragments determined by comparison with the standard 100pb, using Kodak EDAS 290 (Electrophoresis Documentation and Analysis System) software. The results were analyzed by observing the gels and generating a binary matrix in which the individuals were genotyped in regards to the presence (1) or absence (0) of bands, creating a matrix to calculate the genetic similarity and the formation of the dendrogram using the R program (Pillar 1999). These were also evaluated in regards to the total number of amplified bands, number of bands and percentage of polymorphic bands of each primer.

For the cytological analysis, flowers were collected and fixed for 24 hours in an ethanol and acetic acid solution/ in a ratio of 3:1, and were then transferred to 70% ethanol and stored in a freezer. For the evaluations, slides were prepared with the anthers of each flower, which are macerated and stained with propionic carmine dye (2%) (Guerra et al. 2013). For the analysis of meiotic behavior, ten flowers per accessions and at least ten meiotic cells with good chromosome spreading per slide and at all stages of meiosis were evaluated. For the estimation of pollen viability ten slides were analyzed, randomly counting 1,000 pollen grains, making a total of 10,000 grains per accession. Viability was estimated according to coloring capacity; the grains were considered viable when stained and non-viable when empty or colorless, and this was determined by dividing the number of stained pollen grains by the total number of grains observed and multiplying the result by 100 (Simioni and Valle 2011).

For the analysis of in vitro pollen germination, flowers at maturity stage and with closed petals were collected. The culture medium used was that described by Sahar and Spiegel-Roy (1984), with 1% agar, 15% sucrose, 100ppm H3BO3, 1000ppm Ca(NO3).4H2O, 300ppm MgSO4.7H2O and 100ppm KNO3, completing a final volume of 10ml with sterile water. The solution was heated in a microwave and afterwards a drop of culture medium in previously identified slides was added. In a laboratory the petals were removed and the anthers exposed to incandescent light; after the pollen was released it was then distributed in slides containing culture medium and a cover slip was placed on top of the medium, which were maintained in a controlled environment (25 °C) for 24 hours and then analyzed in an optical microscope. Four slides per population were evaluated by counting the number of randomly selected germinated and non-germinated grains in 250 pollen grains per slide, totaling 1,000 grains per accessions.

To count the number of anthers, ten flowers from each accession were selected, the petals and sepals were removed with the aid of needles and a scalpel under a magnifying glass and then separated and later counted. The results of the cytological analyses were subjected to multivariate analysis of variance (MANOVA) using SAS (SAS Institute).

RESULTS AND DISCUSSIONS

Of the 20 RAPD primers tested, 15 were selected and used for the analysis of genetic diversity of the 35 cherry tree accessions, since they were polymorphic and had good amplification profile. The fragments ranged from 200 to 1600 base pairs, with a total of 136 bands being obtained. The primers OPA-04, OPA-10 and OPN-02 produced the highest number of bands (12), whereas the OPA-03 and OPN-09 primers produced six bands. The average number of bands per primer was 8.93 and the average of polymorphic bands was 7.87 (Table 1).

The results obtained with these markers were excellent and resulted in 136 bands and 88.08% of polymorphism (Table 1). Similar values were obtained with RAPD markers by Gomes Filho et al. (2010) in guava trees (87.97%) and by Aguiar et al. (2013), with 70 to 79% of polymorphism in Brazilian cherry tree populations. These results may be associated to the different populations evaluated as well as to the different collecting sites, because plants were sampled with considerable distance between them. Oliveira et al. (2007) reported that the expected genetic variability is distributed among populations with greater uniformity among individuals within a same population and greater variability between populations. In this evaluation, the use of 15 RAPD primers produced a total of 136 bands with an average of 8.93 fragments by primer (Table 1). Gomes Filho et al. (2010) evaluated the genetic divergence among six cultivars and

Table 1. Number of bands and percentage of polymorphism generated by 15 selected RAPD primers and used in 35 cherry accessions collected in the state of Rio Grande do Sul

Primer	Sequence (5'- 3')	Size pb	Total of amplified bands	Total of polymorphic bands	Polymorphism (%)
OPA-02	TGCCGAGCTG	200 – 1600	8	7	87.50
OPA-03	AGTCAGCCAC	200 – 1300	6	6	100.00
OPA-04	AATCGGGCTG	220 – 1200	12	11	91.67
OPA-07	GAAACGGGTG	200 – 1500	8	7	87.50
OPA-08	GTGACGTAGG	280 – 1600	10	8	80.00
OPA-09	GGGTAACGCC	200 – 1600	8	7	87.50
OPA-10	GTGATCGCAG	200 – 1400	12	11	91.67
OPA-11	ACGCGTCTGG	250 – 1000	7	6	85.71
OPA-13	AGTCAGCCAC	220 – 1600	10	9	90.00
OPN-02	ACCAGGGGCA	300 – 1500	12	10	83.33
OPN-04	GACCGACCCA	250 – 1600	7	6	85.71
OPN-05	ACTGAACGCC	200 – 1500	9	8	88.89
OPN-06	GAGACGCACA	280 – 1300	11	10	90.91
OPN-08	ACCTCAGCTC	220 – 1600	8	7	87.50
OPN-09	TGCCGGCTTG	220 – 1300	6	5	83.33
Mean			8.93	7.87	88.08

19 guava trees accessions with RAPD markers and obtained 117 polymorphic brands, using 28 primers. Oliveira et al. (2014), however, evaluated 37 *Psidium* accessions with 17 ISSR (Inter Simple Sequence Repeats) markers and obtained 216 polymorphic bands with an average of 13 bands per primer.

In plant breeding programs success depends on genetic variability. Therefore, it is necessary to identify it through analyses with morphological or molecular markers (Ferreira and Grattapaglia 1998). Traditionally, morphological and agronomic descriptors are used; however, these can be influenced by the environment and some can only be evaluated during the adult stage of the plants (Vieira et al. 2009). In this study, the genetic diversity analyses were performed using molecular markers because, according to Carvalho (2008), the cherry tree is a perennial species and only reaches reproductive age between six and seven years old, which would require too much time for analyses by morphological markers, particularly to evaluate the fruit. Thus, RAPD markers were used because, according to Aguiar et al. (2013), they have a low cost and can be used on any species regardless of prior knowledge of the genome, being simpler and more suitable for rapid analysis of structure and diversity in natural populations.

The similarity between accessions allowed the construction of a dendrogram, which showed the formation of four groups, with group I being comprised by two accessions (1 and 2), group II by three (28, 29 and 30), group III by one accession (34) and group IV by 30 accessions. The groups did not have a direct relationship with the source material, since Group I consisted of a plant collected in Venâncio Aires and another collected in Porto Alegre; group II consisted of three accessions, one collected in Porto Alegre, one in General Câmara and another in Eldorado do Sul; Group III consisted of an accession collected in São Borja; the remaining accessions are in group IV (Figure 1).

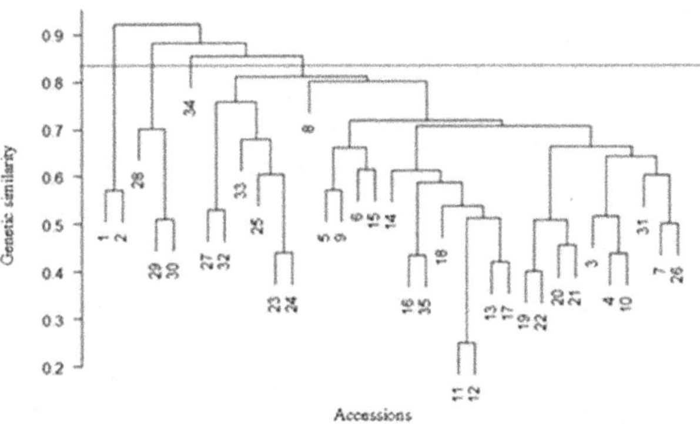

Figure 1. Dendrogram of genetic similarity between the 35 cherry accessions, obtained from RAPD markers. The dashed line indicates the 84% cut-off point based on the average similarity between populations.

In this study, the genetic similarity of the accessions evaluated with RAPD markers was 84% (Figure 1). Similar values were obtained by Franzon et al. (2008), with 60.4% genetic similarity in Brazilian cherry tree populations with AFLP (Amplified Fragment Length Polymorphisms) markers. Salgueiro et al. (2004) also obtained 78.9% of genetic variability with AFLP markers in Brazilian cherry populations that were at least 99 km from each other and had no known interaction between them. Still, Margis et al. (2002) found 88% of variability within populations of this same species in the state of Rio de Janeiro. The high genetic similarity of accessions of this study (84%) may be associated with evolutionary and natural selection factors, as well as the opening of new agricultural areas which led to the loss of many plant populations and their reproductive systems. According to Saavedra and Spoor (2002), most species had their diversity reduced as a result of plant domestication, selection and breeding. According to Borém and Miranda (2009), with the opening of new agricultural areas many genotypes were lost, often resulting in the disappearance of local varieties, causing current populations to present narrow genetic bases which can lead to genetic erosion. According to Odum (1988), these factors may compromise the continuity of the species in face of climate variations, diseases and insects or plagues. Moreover, genetic diversity is considered lower in modern cultivars and self pollinated

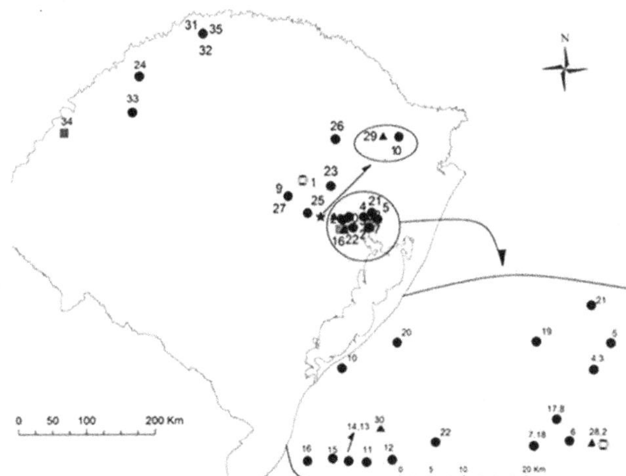

Figure 2. Collection sites of the 35 cherry accessions in the state of Rio Grande do Sul with the cluster according to genetic similarity obtained with RAPD markers. Group I = ▫ ; Group II = ▲ ; Group III = ▪ ; Group IV ●).

species, which would be in line with the results obtained in this study when correlated to the reproductive mode of the species since, as observed by Carvalho (2008) and Sarmento et al. (2012), the cherry tree is a self pollinated (autogamous) species and therefore this factor might explain the high similarity observed.

The distribution in groups (I, II, III and IV) according to the genetic similarity obtained with the RAPD molecular markers allowed us to observe a good distribution in the collection sites (Figure 2). These results are very significant because they will allow the selection of genotypes with greater diversity to be used as parents in directed crosses. According to Amaral Junior and Thiébaut (1999), knowledge of variability amongst materials of interest is an advantage when identifying new gene sources. Also, according to Freitas and Bered (2003), knowledge of genetic variability and maintenance of genotypes in germplasm banks is very important, as they can provide genes that confer adaptation to environmental stress and resistance to diseases and plagues. In this study, the accessions from groups I (1, 2), II (28, 29 and 30) and III (34) can be used as parents in crosses with plants from group IV, since they showed higher genetic diversity (Figure 1). The data corroborate with Amaral Junior and Thiébaut (1999) and Freitas and Bered (2003), whom consider that, for plant breeding, which seeks variability in the progeny with the purpose of selection of superior cultivars, the most dissimilar accessions are the most appropriate for crosses since they broaden the chances of obtaining the heterotic effect in the hybrid generation and increase the probability of recovering superior segregants in advanced generations.

In the analysis of meiotic behavior, the average of normal cells, i.e. cells with chromosomal associations in bivalents was 82.12%, varying from 78.24% to 86.75%. In the analysis of pollen viability, the average of normal grains was 92.44%, ranging from 84.24% to 98.46%. In the analysis of in vitro pollen germination, the average was 40.26%, ranging from 30.45% to 44.11%. In the counting of anther numbers we observed an average value of 161.85 anthers/flower, ranging from 130.54 to 185.01 anthers/flowers (Table 2). In cytological analyses, the average of cells considered normal was high in all accessions but did not differ statistically, allowing us to infer that all accessions are meiotically stable, with high pollen viability and considerable in vitro germination.

The cytological analyses in gametic cells allowed the characterization of the accessions as diploid with n=11, since in most cells analyzed the presence of 11 bivalents at the diakinesis stage of meiotic prophase I of was observed. By analogy, we conlude that the chromosome number of accessions is 2n=22 (Figure 3). The data collected corroborated Pedrosa et al. (1999) and Costa and Forni-Martins (2007b). In the analysis of all phases of meiosis in this study we

Table 2. Cytological analysis of 35 cherry trees accessions collected in the state of Rio Grande do Sul

No.	Cells in meiosis considered normal (%)	Pollen Viability (%)	Pollen Germination (%)	Average No. Anthers	No.	Cells in meiosis considered normal (%)	Pollen Viability (%)	Pollen Germination (%)	Average No. Anthers
1	80.40	91.54	40.20	153.82	19	85.52	92.58	41.45	160.09
2	82.13	92.98	38.22	179.74	20	79.40	91.39	43.18	185.01
3	82.10	90.67	36.34	150.78	21	83.25	87.05	42.22	173.72
4	86.75	92.26	42.56	165.82	22	81.18	98.46	40.36	142.19
5	81.45	87.86	34.18	169.31	23	82.48	94.81	41.11	171.58
6	79.02	90.34	35.44	145.03	24	84.26	96.78	39.43	171.85
7	84.36	93.98	32.08	160.18	25	85.42	89.45	42.77	163.54
8	86.48	90.45	30.45	130.54	26	82.50	91.94	37.32	169.93
9	82.23	94.06	42.54	152.92	27	80.44	96.63	42.38	152.81
10	84.35	95.62	42.28	168.86	28	80.56	88.47	43.56	160.69
11	84.12	90.88	42.26	136.78	29	79.12	97.02	40.44	163.09
12	78.24	90.54	41.48	173.21	30	81.22	86.13	43.68	170.71
13	80.18	84.24	43.17	165.54	31	83.40	94.17	43.78	158.16
14	82.46	93.58	44.11	177.12	32	80.78	96.28	41.34	167.38
15	80.78	93.82	42.03	177.15	33	81.58	95.29	40.62	162.98
16	80.46	92.16	40.12	147.66	34	81.45	96.94	40.46	158.34
17	81.89	93.25	38.23	159.87	35	80.02	92.06	40.98	161.92
18	84.18	91.62	38.17	156.69					
					Mean	82.12^{ns}	92.44^{ns}	40.26^{ns}	161.85^{ns}

ns= no significant differences in average (5% probability) by Tukey test.

observed few cells containing abnormalities such as laggard chromosomes, early disjunction and micronuclei (Table 2). According to Souza et al. (2000) the meiotic behavior of a plant has a direct reflection on its degree of fertility, since the occurrence of failures during the process, with lagging chromosomes or disorganized spindles represent difficulties in producing hybrids, since they produce chromosomal variation due to the loss or gain of chromosomes in new generations. Moreover, according to Muñoz et al. (2006), meiosis becomes a source of genetic variability used by organisms for environmental adaptation. The meiotic normality and therefore the high viability of pollen grains are in accordance to Muñoz et al. (2006), who claim that the normal course of meiosis ensures gamete viability. The stable meiotic behavior of accessions in this study corroborates Loguercio and Battistin (2004), who evaluated nine jambolan accessions (*Syzygium cumini* L.) and observed regular meiotic behavior, with the formation of 11 bivalents and pollen viability higher than 93.19 %. Costa and Forni-Martins (2006a) evaluated the meiosis and pollen viability of four species of the *Campomanesia* and *Psidum* genus and did not observe any abnormalities and also identified a pollen viability higher than 86%. Franzon and Raseira (2004) observed cytological stability higher than 96% in *Acca sellowiana*, while in *Eugenia pyriformis* and *Campomanesia xanthocarpa* the stability was 89.5% and 89.3%, respectively.

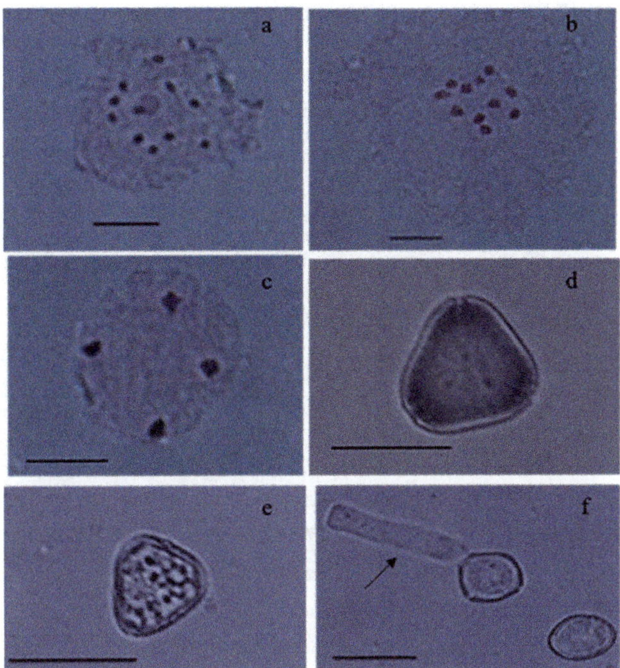

Figure 3. Cytological analysis of cherry accessions. a) Prophase I (diakinesis) with 11 II; b) Prophase I (diakinesis) with 11 II; c) Telophase I; d) viable pollen grain; e) unviable pollen grain; f) germinated (arrow) and non-germinated pollen grain. Scale 10 µm.

The in vitro pollen germination was 40.26% (Table 2). Franzon and Raseira (2004) and Franzon et al. (2007) evaluated the germination capacity of cherry tree pollen and found values higher than 60%. These results may be associated with the methodology since, according to Marcellán and Camadro (1996), many factors can influence germination, such as: culture medium, incubation temperature and length. Franzon et al. (2007) include others, such as flower development stage when collecting and storing the pollen.

When counting anther numbers we observed an average value of 161.85 anthers/flowers, but with a variation of 130.54 to 179.74 anthers/flower (Table 2). There was a considerable variation although the factor or factors responsible for this variation cannot be identified.

CONCLUSIONS

The accessions evaluated show high polymorphism, genetic similarity and cytological stability and can be used directly in commercial orchards, and also as male parents in directed crosses in breeding programs.

REFERENCES

Aguiar RV, Cansian RL, Kubiak GB, Slaviero LB, Tomazoni TA, Budke JC and Mossi AJ (2013) Variabilidade genética de *Eugenia uniflora* L. em remanescentes florestais em diferentes estádios sucessionais. **Revista Ceres 60**: 226-233.

Amaral Júnior AT and Thiébaut JTL (1999) **Análise multivariada na avaliação da diversidade em recursos genéticos vegetais**. Editora Universidade Estadual do Norte Fluminense - UENF, Rio de Janeiro, 55p.

Borém A and Miranda G (2009) **Melhoramento de Plantas.** Editora UFV, Viçosa, 523p.

Carvalho PE (2008) **Espécies arbóreas brasileiras**. Editora Colombo, Brasília, 593p.

Costa IR and Forni-Martins ER (2006a) Chromosome studies in Brazilian species of *Campomanesia* Ruiz & Pavon and *Psidium* (*Myrtaceae* Juss.). **Caryologia 59**: 7-13.

Costa IR and Forni-Martins ER (2006b) Chromosome studies in species of *Eugenia*, *Myrciaria* and *Plinia* (*Myrtaceae*) from south-eastern Brazil. **Australian Journal of Botany 54**: 409-415.

Costa IR and Forni-Martins ER (2007) Karyotype analysis in South American species of *Myrtaceae*. **Botanical Journal of the Linnean Society 155**: 571-580.

Donadio LC, Môro FV and Servidone AA (2004) **Frutas brasileiras**. Editora Novos Talentos, Jaboticabal, 288p.

Ferreira ME and Grattapaglia D (1998) **Introdução ao uso de marcadores moleculares em análise genética**. Editora Embrapa, Brasília, 220p.

Franzon RC, Gonçalves RS, Antunes LEC, Raseira MCB and Trevisan R (2008) Propagação da pitangueira através da enxertia de garfagem. **Revista Brasileira de Fruticultura 30**: 488-491.

Franzon RC and Raseira MCB (2004) Meiotic index in *Myrtaceae* native fruits trees from southern Brazil. **Crop Breeding and Applied Biotechnology 4**: 344-349.

Franzon RC, Raseira MCB and Wagner JA (2007) Testes de germinação *in vitro* e armazenamento de pólen de pitangueira (*Eugenia uniflora* L.). **Acta Scientiarum-Agronomy 29**: 251-255.

Freitas LB and Bered F (2003) **Genética e evolução vegetal**. Editora UFRGS, Porto Alegre, 463p.

Gomes Filho A, Oliveira JG, Viana AP, Siqueira APO, Oliveira MG and Pereira MG (2010) Marcadores moleculares RAPD e descritores morfológicos na avaliação da diversidade genética de goiabeiras (*Psidium guajava* L.). **Acta Scientiarum-Agronomy 32**: 627-633.

Guerra D, Schifino-Wittmann MT, Schwarz SF, Souza PVD and Campos SS (2013) Influence of greenhouse versus field conditions on reproductive characteristics of citrus rootstocks. **Crop Breeding and Applied Biotechnology 13**: 186-193.

Guerra D, Schifino-Wittmann MT, Schwarz SF, Weiler RL, Dahmer N and Souza PVD (2016) Tetrapolidization in citrus rootstocks: effect of genetic constitution and environment in chromosome duplication. **Crop Breeding and Applied Biotechnology 16**: 35-41.

Landrum LR and Kawasaki ML (1997) The Genera of *Myrtaceae* in Brazil: an Illustrated Synoptic Treatment and Identification Keys. **Brittonia 49**: 508-536.

Loguercio AP and Battistin A (2004) Microsporogênese de nove acessos de *Syzygium cumini* (L.) *Myrtaceae* oriundos do Rio Grande do Sul – Brasil. **Revista da FZVA 11:** p.95-106.

Lorenzi H (2002) **Árvores Brasileiras: Manual de identificação e cultivo de plantas arbóreas nativas do Brasil**. Editora Nova Odessa, São Paulo, 368p.

Marcellán ON and Camadro EL (1996) The viability of asparagus pollen after storage at low temperatures. **Scientia Horticulturae 67**: 101-104.

Margis R, Felix D and Caldas JF (2002) Genetic differentiation among three neighboring Brazil-cherry (*Eugenia uniflora* L.) populations within the Brazilian Atlantic rain forest. **Biodiversity and Conservation 11**: 149-163.

Muñoz AM, Caetano CM, Vallejo FA and Sanchez MS (2006) Comportamiento meiótico y descripción morfológica del pólen de pronto alivio. **Acta Agronomica 1**: 1-9.

Odum EP (1988) **Ecologia**. Editora Guanabara, Rio de Janeiro, 434p.

Oliveira F (2007) Aspectos da vegetação arbórea encontrada na orla da Praia da Alegria no município de Guaíba, RS, Brasil. **Caderno de Pesquisa 19**: 6-17.

Oliveira MSP, Amorim EP, Santos JB and Ferreira DF (2007) Diversidade genética entre acessos de açaizeiro baseada em marcadores RAPD. **Ciência e Agrotecnologia 31**: 1645-1653.

Oliveira NNS, Viana AP, Quintal SSR, Paiva CL and Marinho CS (2014) Análise de distância genética entre acessos do gênero *Psidium* via marcadores ISSR. **Revista Brasileira de Fruticultura 36**: 917-923.

Pedrosa A, Gataí J, Barros AE, Felix L and Guerra M (1999) Citogenética de angiospermas coletadas em Pernambuco. **Acta Botânica Brasílica 13**: 49-51.

Pillar VD (1999) Howsharp are classifications? **Ecology 80**: 2508-2516.

Rodrigues VEG and Carvalho DA (2001) Levantamento etnobotânico de plantas medicinais no domínio do cerrado na região do Alto Rio Grande, Minas Gerais. **Ciência e Agrotecnologia 25**: 102-123.

Romagnolo MB and Souza MC (2004) Os gêneros *Calycorectes* O. Berg, *Hexachlamys* O. Berg, *Myrcianthes* O. Berg, *Myrciaria* O. Berg e *Plinia* L. (*Myrtaceae*) na planície alagável do alto rio Paraná, Brasil. **Acta Botânica Brasilica 18**: 613-627.

Saavedra G and Spoor W (2002) Genetic base broadening in autogamouscrops: Lycopersion esculentum Mill. As a model. **Managing Plant Genetic Diversity 443**: 292-299.

Sahar N and Spiegel-Roy P (1984) In vitro germination of avocado pollen. **Scientia Horticulturae 19**: 886-888.

Salgueiro F, Felix D, Caldas JF, Margis-Pinheiro M and Margis R (2004) Even populations differentiations for maternal and biparental gene markers in Eugenia uniflora, a widely distributed species from the Braziliam coastal Atlantic rain forest. **Diversity and Distributions 10**: 201-210.

Sarmento MB, Silva ACS and Silva CS (2012) Recursos genéticos de frutas nativas da família Myrtaceae no Sul do Brasil. **Magistra 24**: 250-262.

Simioni C and Valle CB (2011) Meiotic analysis in induced tetraploids of *Brachiaria decumbens* Stapf. **Crop Breeding and Applied Biotechnology 11**: 43-49.

Souza MM, Pereira TNS, Rodrigues R, Dutras GA and Sudré CP (2000) Irregularidades meióticas em pimenta. **Horticultura Brasileira 18**: 748-749.

Vieira ESN, Pinho EVRV, Carvalho MGG and Da Silva PA (2009) Caracterização de cultivares de soja por descritores morfológicos e marcadores bioquímicos de proteínas e isoenzimas. **Revista Brasileira de Sementes 1**: 086-094.

Wunch A and Hormaza JI (2007) Characterization of variability and genetic similarity of European pear using microsatellite loci developed in apple. **Scientia Horticulturae 113**: 37-43.

24

Evaluation of SRAP markers for mapping of *Pisum sativum* L.

María Fernanda Guindon[1], Eugenia Martin[1*], Aldana Zayas[1], Enrique Cointry[1] and Vanina Cravero[1]

***Corresponding author:**
E-mail: eamartin@unr.edu.ar

[1] IICAR-CONICET, Instituto de Investigaciones en Ciencias Agrarias de Rosario, Zavalla, Argentina

Abstract: *Linkage maps have become important tools for genetic studies. With the aim of evaluating the SRAP (sequence-related amplified polymorphism) technique for linkage mapping in Pisum sativum L., a F_2 mapping population derived from an initial cross between cvs. DDR11 and Zav25 was generated. A total of 25 SRAP primer combinations were evaluated in 45 F_2 plants and both parental lines, generating 208 polymorphic bands/markers. The markers were analyzed by the chi-square goodness-of-fit test to check the expected Mendelian segregation ratio. The resulting linkage map consists of 112 genetic markers distributed in 7 linkage groups (LGs), covering a total of 528.8 cM. The length of the LGs ranged from 47.6 to 144.3 cM (mean 75.54 cM), with 9 to 34 markers. The linkage map developed in this study indicates that the SRAP marker system could be applied to mapping studies of pea.*

Key words: *Pea, plant breeding, linkage map, F_2 population, molecular markers.*

INTRODUCTION

Pea (*Pisum sativum* L.) is an autogamous, annual cool-season legume originated from areas in the Middle East, in the East of the Caucasus, Iran and Afghanistan, and West of the Mediterranean basin (Smýkal et al. 2011). Its genome is organized in seven chromosome pairs (2n = 2x = 14), and the haploid size estimated at 4.45 Gb (Smýkal et al. 2012). Peas were an important source of animal and human food for many centuries. The species is rich in protein, slowly digestible starch, soluble sugars, fiber, minerals, and vitamins (Dahl et al. 2012). The global dry pea production averages 10 million tonnes a year. Argentina is one of the top exporting countries, eighth in the global ranking between 2008 and 2011 (FAO 2012). The rising world population will require increased crop production. Moreover, some researchers suggest that the current rate of increase in crop yields will not be enough to meet this demand (Tester and Langridge 2010). Therefore, plant breeding programs are needed to further raise crop yields. In this context, linkage mapping will be useful to maximize the success probability. Genetic linkage maps are powerful tools for genetic research and breeding of plants. The linkage maps are the first step in: 1) the analysis of qualitative and quantitative traits; 2) the introgression of desirable genes and quantitative trait loci (QTLs); 3) positional or map-based cloning of genes responsible for economically important traits (Semagn et al. 2006). Different kinds of markers, such as simple sequence repeats (SSR; Loridon et al. 2005), single nucleotide polymorphisms (SNP; Deulvot et al. 2010), inter simple sequence repeats (ISSR; Mishra et al. 2009), and sequence tagged sites (STS; Barilli et al.

2010) have been used to develop moderate density linkage maps in pea. Several markers were common in different maps corresponding to different crosses; this allowed an integration of these maps (Loridon et al. 2005, Aubert et al. 2006), as well as the development of consensus linkage maps for the species (Weeden et al. 1998, Bordat et al. 2011).

In this study, we proposed the use of the sequence-related amplified polymorphism (SRAP) technique (Li and Quiros 2001) to generate a number of markers distributed across all pea chromosomes. Since its development, SRAP has been employed in a wide range of plant species for genetic diversity estimation (Cravero et al. 2007, Espósito et al. 2007, Aneja et al. 2012), gene tagging (Martin et al. 2008, Zhang et al. 2010), and map construction (Lin et al. 2005, Sun et al. 2007, Wang et al. 2008, Martin et al. 2013). The aim of the current study was to evaluate the usefulness of SRAP markers in the development of a genetic linkage map of *Pisum sativum* L.

MATERIAL AND METHODS

Plant material

The F_2 mapping population was derived from an initial cross between the cvs. DDR11 and Zav25. The latter is an experimental line obtained from the IICAR-CONICET breeding program (Espósito et al. 2007). For most yield-related traits, such as number of pods and seeds per plot, the values of 'DDR11' are lower than those of 'Zav25'.

Both parents and 45 plants of the F_2 population were sown in an experimental field of Universidad Nacional de Rosario (lat 33° 1′ S, long 60° 53′ W) in the winter of 2012, in a completely randomized design (inter-row spacing 70 cm, plant spacing 10 cm).

DNA extraction

The genomic DNA of each F_2 plant and both parents was isolated from fresh leaves by the CTAB method described by Doyle and Doyle (1990), with the following modifications: after DNA precipitation, the samples were stored at -20 °C for 30 min, the two washes of the final step were performed with ethanol 70%, and the resulting pellet was resuspended in distilled water.

After RNAase-treatment, each DNA sample was quantified using agarose gel electrophoresis (1% w/v) and comparison of band intensity with the standard λ DNA (76 ng μL^{-1}). The hybrid origin of each plant was checked using two microsatellite markers SSR: PSMPSAA135 and PSMPSAA205 (Tar'an et al. 2005), which were contrasting in both parental lines.

SRAP genotypic analysis

The F_2 population was genotyped using 25 SRAP primer combinations generated from five forward and five reverse primers developed by Li and Quiros (2001) (Table 1). The primers are 17-18 bases long and have a core sequence, which includes 10-11 non-specific bases at the 5′end and sequence CCGG in the forward and AATT in the reverse primer. The core sequence is followed by three selective nucleotides at the 3′ end of each primer. The primers were selected based on the results of Espósito et al. (2007) in a characterization of pea accessions with this type of molecular markers.

Polymerase chain reactions were carried out in a final volume of 20 µL containing 15 ng genomic DNA, 0.2 mM dNTPs, 1.5 mM $MgCl_2$, 0.5 µM of each primer, 1X Taq buffer (Invitrogen, California, USA), and 1 U of Taq recombinant polymerase (Invitrogen). Samples were subjected to the following thermal profile: 5 min denaturing at 94 °C and five cycles of three steps: 1 min denaturing at 94 °C, 1 min annealing at 35 °C, and 1 min elongation at 72 °C; for the following 35 cycles, annealing temperature was elevated to 50 °C with a final elongation step of 10 min at 72 °C.

Table 1. Primer sequences used for SRAP (Sequence-related amplified polymorphism) analysis

Forward	Reverse
me1 5′-TGAGTCCAAACCGGATA-3′	em1 5′-GACTGCGTACGAATTAAT-3′
me2 5′-TGAGTCCAAACCGGAGC-3′	em2 5′-GACTGCGTACGAATTTGC-3′
me3 5′-TGAGTCCAAACCGGAAT-3′	em3 5′-GACTGCGTACGAATTGAC-3′
me4 5′-TGAGTCCAAACCGGACC-3′	em4 5′-GACTGCGTACGAATTTGA-3′
me5 5′-TGAGTCCAAACCGGAAG-3′	em5 5′-GACTGCGTACGAATTAAC-3′

The resulting amplicons were separated on 6% (w/v) denaturing polyacrylamide gels and then visualized by silver staining (Bassam et al. 1991). The SRAP fragments were treated as dominant markers. Each marker was labeled according to the primer combination used for its generation plus the estimated amplicon size.

Linkage analysis and linkage map construction

Linkage analyses were performed using JoinMap v4 (van Ooijen 2006). Each segregating marker was tested for deviations from the expected 3:1 segregation ratio using Chi-square tests. Markers with Mendelian segregation ($x^2 \leq x^2_{\alpha=0.1}$) or with minor distortion ($x^2_{\alpha=0.1} < x^2 \leq x^2_{\alpha=0.01}$) were used for the construction of the linkage map. Markers with highly distorted segregation ($x^2 > x^2_{\alpha=0.01}$) were included in a second step of mapping only when their presence did not affect the local marker order. Linkage groups (LGs) were established at a minimum LOD (logarithm of odds) value of 3.0. The marker order for each LG was determined at LOD = 1.0, REC = 0.40 and Jump = 5. Recombination values were converted to genetic distances using the Kosambi (1994) mapping function. Linkage groups were numbered sequentially according to their length in cM. Linkage maps were drawn using MapChart 2.2 software (Voorrips 2002).

RESULTS AND DISCUSSION

Since the development of molecular marker techniques, the number of marker loci identified on genetic maps is increasing at a high rate. The SRAP marker system designed by Li and Quiros (2001) is a simple and efficient technique. It has several advantages over other molecular markers, namely its simplicity and reasonable throughput rate. It also allows easy isolation of bands for sequencing and, most importantly, it targets ORFs (open reading frames). Elsewhere, SRAPs were established as a powerful tool for construction of genetic linkage maps, e.g., of *Brassica* (Li and Quiros 2001, Sun et al. 2007), *Gossypium* (Lin et al. 2005, Yu et al. 2007), *Cucumis melo* L. (Wang et al. 2008), and more recently of *Cynara cardunculus* (Martin et al. 2013). Our study presents the first application of SRAP markers for the construction of a linkage map of *P. sativum* L.

The parental lines to develop the mapping population ('DDR11' and 'Zav25') were selected based on observations of Espósito et al. (2007), who reported that these lines are divergent at the morphological (Euclidean distance = 0.47) and molecular levels (Dice distance = 0.66) and that they were grouped separately by hierarchical cluster analysis.

A total of 25 SRAP primer combinations were used of which 23 pairs were amplified. Most combinations produced clear bands without overlapping, but in some cases the scoring of the markers was somewhat cumbersome, because of the high number of bands and their different intensities, and the presence of minor bands in some plants. The number of fragments amplified by each primer combination ranged from 3 (Me1-Em1; Me1-Em4) to 24 (Me3-Em2), with an average of 9.96. A total of 208 polymorphic bands (PB) were generated. The most polymorphic primer combination was Me2-Em3, with 18 PB (Table 2).

The markers proved efficient for genetic studies in pea, producing an average of 8.32 PB/primer combination. This value is similar to those of the other species, where 3 to 14 bands per primer combination were reported (Li and Quiros 2001, Lin et al. 2003, Sun et al. 2007, Yu et al. 2007, Wang et al. 2008, Martin et al. 2013). Since one primer combination may detect a high number of polymorphic loci,

Table 2. Number of bands and polymorphic bands generated by SRAP primer combinations

Combination	Bands	Polymorphic bands
Me1-Em1	3	3
Me1-Em2	9	9
Me1-Em3	5	5
Me1-Em4	3	2
Me1-Em5	15	14
Me2-Em1	8	8
Me2-Em2	9	9
Me2-Em3	18	18
Me2-Em4	16	14
Me2-Em5	13	9
Me3-Em1	4	1
Me3-Em2	24	16
Me3-Em3	8	8
Me3-Em4	10	8
Me3-Em5	8	7
Me4-Em1	0	0
Me4-Em2	10	7
Me4-Em3	6	6
Me4-Em4	17	12
Me4-Em5	5	5
Me5-Em1	0	0
Me5-Em2	6	3
Me5-Em3	16	16
Me5-Em4	18	14
Me5-Em5	18	14
Total	249	208

Table 3. Characteristics of the linkage groups (LG) generated by SRAP markers

LG	Number of markers	Number of highly distorted markers	Size	Smallest distance between markers	Highest distance between markers	Average distance between markers
				cM (CentiMorgan)		
1	34	6	144.309	0.031	25.392	4.244
2	21	6	87.134	0.778	15.588	4.149
3	18	1	79.119	0.204	16.06	4.395
4	10	1	61.401	1.608	17.801	6.140
5	11	4	54.555	1.973	15.329	4.959
6	9	2	54.674	1.653	14.435	6.074
7	9	2	47.601	2.207	15.1	5.289

this technique can be used to construct ultra-dense genetic maps. Furthermore, SRAP markers can be combined with next-generation techniques to enhance their capacity and effectiveness. Li et al. (2011) combined SRAP with Illumina/Solexa sequencing to directly integrate genetic loci in the *B. rapa* genetic map based on paired-end Solexa sequencing. Results of the SRAP technique obtained in this way may prove invaluable for QTL analysis and map-based cloning.

Chi-square analysis of the 208 loci revealed that 116 loci (~ 55.8%) were consistent with the expected Mendelian 3:1 segregation ratio ($x^2 \leq x^2_{\alpha=0.1}$), the distortion of 37 loci (~ 17.8%) was minor ($x^2_{\alpha=0.1} < x^2 \leq x^2_{\alpha=0.01}$) and 55 loci (~ 26.4%) were highly distorted ($x^2 > x^2_{\alpha=0.01}$).

Initially, the 153 markers with Mendelian or slightly distorted segregation were used for the construction of linkage groups (LG), using a minimum LOD value of 3.0 and the mapping parameters Rec = 0.40, LOD = 1.0, and Jump = 5. Under

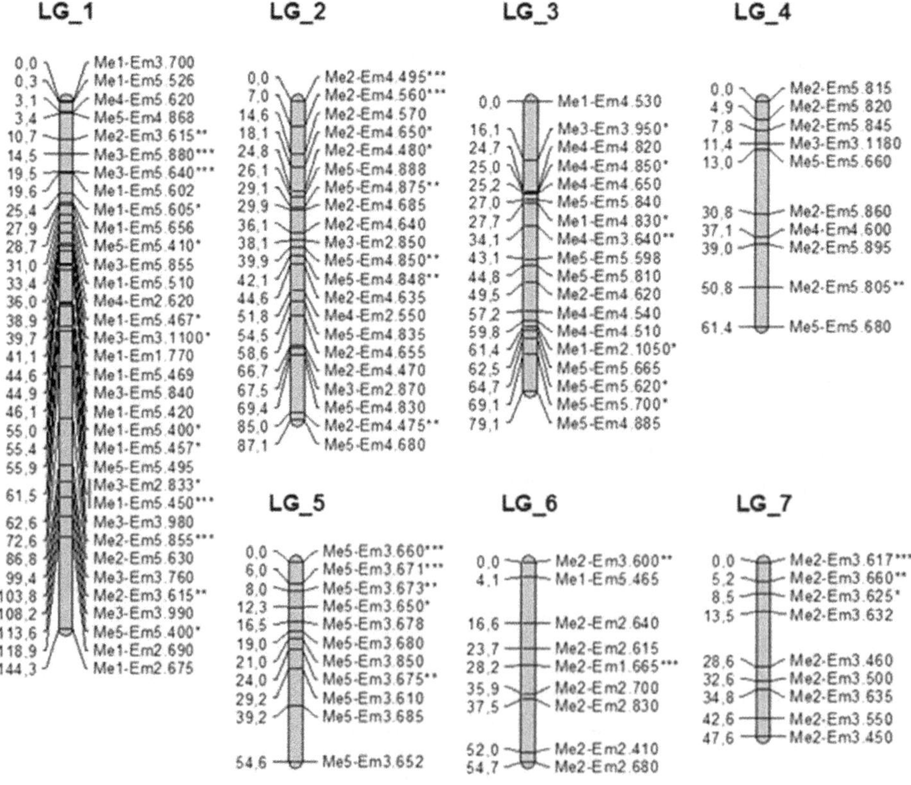

Figure 1. Genetic linkage map of pea. Marker names are shown on the right of each LG (linkage group) and map distances (in cM) on the left. Markers with significant levels of segregation distortion are indicated by asterisks (minor distortion * 0.05 ≥ P > 0.01, highly distorted segregation: **0.01 ≥ P > 0.001, ***P ≤ 0.001).

these conditions, an initial framework map with 62 loci was constructed. Then the framework order was fixed and a second round of mapping performed, including markers with distorted segregation. These markers were only included in the final map if their presence did not alter the surrounding marker order in a given linkage group. By this step, we incorporated 29 Mendelian markers and 22 distorted markers in the previously established linkage groups. This strategy ensures the accuracy and a high coverage of the final map. Similar strategies were successfully used for linkage mapping in different species, e g., of olive (Khadari et al. 2010), globe artichoke and cardoon (Martin et al. 2013), lentil (Verma et al. 2015), poplar (Zhou et al. 2015), and wheat (Li et al. 2015).

The resulting map comprises 112 loci distributed over seven linkage groups (LGs), which is equal to the haploid number of chromosomes in the pea genome (Figure 1). The overall map length was 528.8 cM and the mean inter-marker distance 4.72 cM. The LG length varied from 47.6 cM to 144.3 cM. The number of markers included in each LG ranged from 9 to 34 (Table 3). A total of 22 distorted markers (19%) were included in the genetic map.

Although several linkage maps for pea have been developed using different kinds of markers, this is the first linkage map of this species constructed with SRAP markers. The length of our map (528.8 cM) is shorter than that of previous ones generated with other molecular markers, which covered 1430, 1458, and 1283 cM, respectively (Loridon et al. 2005, Aubert et al. 2006, Barilli et al. 2010). All these genetic maps were constructed using mapping populations with different sizes, derived from different crosses. Both, the size of the mapping population and their origin affect the marker coverage and map length because an increasing divergence between parents generates a greater number of possible recombinations and the possibility of finding recombinant plants is higher when populations are large. Then, the small size of our F_2 population (45 plants) could be the cause of the smaller size of the map obtained in this study (Ferreira et al. 2006). To enhance accuracy and reduce the statistical error, a great number of plants should be evaluated. On the other hand, the large number of unlinked markers with Mendelian or slightly distorted segregation (62) reflects the need to enrich this map with additional markers to cover the entire genome.

CONCLUSIONS

The linkage map generated in this study provided basic information for assistance of future molecular marker application in the local breeding program of *Pisum sativum* L. Since pea has no reference genome, molecular markers that do not require sequence information must be evaluated. In this context, sequence-related amplified polymorphism (SRAP) represents an efficient tool for genetic analysis of pea even though the proposed linkage map was only partly saturated. For the first time, SRAP markers were applied in this study to develop a linkage map of pea. Moreover, since these markers target coding regions of the genome, they can potentially identify markers with inherent biological significance. Additional markers are required to expand the coverage of this map for QTL analysis. The segregating population used to develop this linkage map is currently being phenotyped for yield-related traits to detect QTLs associated to this character.

ACKNOWLEDGEMENTS

This research was supported by Consejo Nacional de Investigaciones Científicas y Técnicas (CONICET, Argentina) and Fondo para la Investigación Científica y Tecnológica (FONCyT, Argentina).

REFERENCES

Aneja B, Yadav NR, Chawla V and Yadav RC (2012) Sequence-related amplified polymorphism (SRAP) molecular marker system and its applications in crop improvement. **Molecular Breeding 30**: 1635-1648.

Aubert G, Morin J, Jacquin F, Loridon K, Quillet MC, Petit A, Rameau C, Lejeune-Hénaut I, Huguet T and Burstin J (2006) Functional mapping in pea, as an aid to the candidate gene selection and for investigating synteny with the model legume *Medicago truncatula*. **Theoretical and Applied Genetics 112**: 1024-1041.

Barilli E, Satovic Z, Rubiales D and Torres AM (2010) Mapping of quantitative trait loci controlling partial resistance against rust incited by *Uromyces pisi* (Pers.) Wint. in a *Pisum fulvum* L. intraspecific cross. **Euphytica 175**: 151-159.

Bassam BJ, Caetanoanolles G and Gresshoff PM (1991) Fast and sensitive silver staining of DNA in polyacrylamide gels. **Analytical Biochemistry 196**: 80-83.

Bordat A, Savois V, Nicolas M, Salse J, Chauveau A, Burgeois M, Potier J, Houtin H, Rond C, Murat F, Marget P, Aubert G and Burstin J (2011) Translational genomics in legumes allowed Placing in silico 5460 unigenes on the pea functional map and identified candidate genes in *Pisum sativum* L. **G3: Genes, Genomes, Genetics 1**: 93-103.

Cravero VP, Martin EA and Cointry EL (2007) Genetic diversity in *Cynara Cardunculus* determined by sequence-related amplified polymorphism markers. **Journal of the American Society for Horticultural Science 132**: 208-212.

Dahl W, Foster L and Tyler R (2012) Review of the health benefits of peas (*Pisum sativum* L.) **British Journal of Nutrition 108**: 3-10.

Deulvot C, Charrel H, Marty A, Jacquin F, Donnadieu C, Lejeune-Hénaut I, Burstin J and Aubert G (2010) Highly-multiplexed SNP genotyping for genetic mapping and germplasm diversity studies in pea. **BMC Genomics 11**: 468-478.

Doyle JJ and Doyle JL (1990) Isolation of plant DNA from fresh tissue. **Focus 12:** 149-151.

Espósito MA, Martin EA, Cravero VP and Cointry EL (2007) Characterization of pea accessions by SRAP's markers. **Scientia Horticulturae 113**: 329-335.

FAO (2012) **The Statistics Division of the FAO. Food and Agriculture Organization of the United Nations**. Available at <http://faostat.fao.org/site/567/default.aspx#ancor.> Accessed on June 6, 2014.

Ferreira A, Flores da Silva M, Silva LC and Cruz CD (2006) Estimating the effects of population size and type on the accuracy of genetic maps. **Genetics and Molecular Biology 29**: 187-192.

Khadari B, El Aabidine AZ, Grout C, Sadok IB, Doligez A, Moutier N, Santoni S and Costes E (2010) A genetic linkage map of olive based on amplified fragment length polymorphism, intersimple sequence repeat and simple sequence repeat markers. **Journal of the American Society for Horticultural Science 135**: 548-555.

Li G and Quiros CF (2001) Sequence-related amplified polymorphism (SRAP), a new marker system based on a simple PCR reaction: its application to mapping and gene tagging in *Brassica*. **Theoretical and Applied Genetics 103**: 455-461.

Li W, Zhang, J, Mou Y, Geng J, McVetty P, Hu S and Li G (2011) Integration of Solexa sequences on an ultradense genetic map in *Brassica rapa* L. **BMC Genomics 12**: 249-263

Li C, Bai G, Chao S and Wang Z (2015) A high-density SNP and SSR consensus map reveals segregation distortion regions in wheat. . **BioMed Research International 2015**: Article ID 830618, 10 pages.

Lin Z, Zhang X, Nie Y, He D and Wu M (2003) Construction of a genetic linkage map for cotton based on SRAP. **Chinese Science Bulletin 48**: 2064-2068.

Lin Z, He D, Zhang X, Nie Y, Guo X and Feng C (2005) Linkage map construction and mapping QTL for cotton fibre quality using SRAP, SSR and RAPD. **Plant Breeding 124**: 180-187.

Loridon K, McPhee K, Morin J, Dubreuil P, Pilet-Nayel ML, Aubert G, Rameau C, Baranger A, Coyne C, Lejeune-He`naut I and Burstin J (2005) Microsatellite marker polymorphism and mapping in pea (*Pisum sativum* L.). **Theoretical and Applied Genetics 111**: 1022-1031.

Martin E, Cravero V, Espósito M, López Anido F, Milanesi L and Cointry E (2008) Identification of markers linked to agronomic traits in globe artichoke. **Australian Journal of Crop Science 1**: 43-46.

Martin E, Cravero V, Portis E, Scaglione D, Acquaviva E and Cointry E (2013) New genetic maps for globe artichoke and wild cardoon and their alignment with an SSR-based consensus map. **Molecular Breeding 32**: 177-187

Mishra RK, Kumar A, Chaudhary S and Kumar S (2009) Mapping of the *multifoliate pinna (mfp)* leaf-blade morphology mutation in grain pea *Pisum sativum*. **Journal of Genetics 88**: 227-232.

Semagn K, Bjornstad A and Ndjiondjop MN (2006) Principles, requirements and prospects of genetic mapping in plants. **Africal Journal of Biotechnology 5**: 2569-2587.

Smýkal P, Kenicer G, Flavell AJ, Corander J, Kosterin O, Redden RJ, Ford R, Coyne CJ, Maxted N, Ambrose MJ and Ellis NTH (2011) Phylogeny, phylogeography and genetic diversity of the *Pisum* genus. **Plant Genetic Resources: Characterization and Utilization 9**: 4-18.

Smýkal P, Aubert G, Burstin J, Coyne CJ, Ellis NTH, Flavell AJ, Ford R, Hýbl M, Macas J, Neumann P, McPhee KE, Redden RJ, Rubiales D, Weller JL and Warkentin TD (2012) Pea (*Pisum sativum* L.) in the genomic era. **Agronomy 2**: 74-115.

Sun Z, Wang Z, Tu J, Zhang J, Yu F, McVetty P and Li G (2007) An ultradense genetic recombination map for *Brassica napus*, consisting of 13551 SRAP markers. **Theoretical and Applied Genetics 114**: 1305-1317.

Tar'an B, Zhang C, Warkentin T, Tullu A and Vandenberg A (2005) Genetic diversity among varieties and wild species accessions of pea (*Pisum sativum* L.) based on molecular markers, and morphological and physiological characters. **Genome 48**: 257-272.

Tester M and Langridge P (2010) Breeding technologies to increase crop production in a changing world. **Science 327**: 818-822.

van Ooijen JW (2006) Software for the calculation of genetic linkage maps in experimental populations. Demo Version. Kyazma B.V., Wageningen. Available at <https://www.kyazma.nl/index.php/JoinMap/Evaluate/> Accessed in June 2015.

Verma P, Goyal R, Chahota RK, Sharma TR, Abdin MZ and Bhatia S (2015) Construction of a genetic linkage map and identification of QTLs for seed weight and seed size traits in lentil (*Lens culinaris* Medik.). **PLoS ONE 10.** doi:10.1371/journal.pone.0139666.

Voorrips RE (2002) MapChart: Software for the graphical presentation of linkage maps and QTLs. **The Journal of Heredity 93**: 77-78.

Wang J, Yao J and Li W (2008) Construction of a molecular map for melon (*Cucumis melo* L.) based on SRAP. **Frontiers of Agriculture in China 2**: 451-455.

Weeden NF, Ellis THN, Timmerman-Vaughan GM, Swiecicki WK, Rozov SM and Berdnikov VA (1998) A consensus linkage map for *Pisum sativum*. **Journal of Heredity 83**: 123-129.

Yu J, Yu S, Lu C, Wang W, Fan S, Song M, Lin Z, Zhang J and Zhang X (2007) High-density linkage map of cultivated allotetraploid cotton based on SSR, TRAP, SRAP and AFLP markers. **Journal of Integrative Plant Biology 49**: 716-724.

Zhang W, He H, Guan Y, Du H, Yuan L, Li Z, Yao D, Pan J and Cai R (2010) Identification and mapping of molecular markers linked to the tuberculate fruit gene in the cucumber (*Cucumis sativus* L.) **Theoretical and Applied Genetics 120**: 645-654.

Zhou W, Tang Z, Hou J, Hu N and Yin T (2015) Genetic map construction and detection of genetic loci underlying segregation distortion in an intraspecific cross of *Populus deltoides*. **PLoS ONE 10.** doi:10.1371/journal.pone.0126077.

25

Genetic engineering of cotton with a novel cry2AX1 gene to impart insect resistance against Helicoverpa armigera

Karunamurthy Dhivya[1], Sundararajan Sathish[1], Natarajan Balakrishnan[1], Varatharajalu Udayasuriyan[1] and Duraialagaraja Sudhakar[1*]

***Corresponding author:**
E-mail: dsudhakar@hotmail.com

[1] Tamil Nadu Agricultural University, Department of Plant Biotechnology, Centre for Plant Molecular Biology and Biotechnology, Coimbatore, India 641 003

Abstract: *Embryogenic calli of cotton (Coker310) were cocultivated with the Agrobacterium tumefaciens strain LBA4404 harbouring the codon-optimised, chimeric cry2AX1 gene consisting of sequences from cry2Aa and cry2Ac genes isolated from Indian strains of Bacillus thuringiensis. Forty-eight putative transgenic plants were regenerated, and PCR analysis of these plants revealed the presence of the cry2AX1 gene in 40 plants. Southern blot hybridisation analysis of selected transgenic plants confirmed stable T-DNA integration in the genome of transformed plants. The level of Cry2AX1 protein expression in PCR positive plants ranged from 4.9 to 187.5 ng g^{-1} of fresh tissue. A transgenic cotton event, TP31, expressing the cry2AX1 gene showed insecticidal activity of 56.66 per cent mortality against Helicoverpa armigera in detached leaf disc bioassay. These results indicate that the chimeric cry2AX1 gene expressed in transgenic cotton has insecticidal activity against H. armigera.*

Keywords: *cry2AX1, Agrobacterium tumefaciens, somatic embryogenesis.*

INTRODUCTION

Cotton is one of the most important crop species, valued around the world by the textile industry. Besides being the backbone of the textile industry, cotton and its by-products are also part of livestock feed, seed oil, fertilizers, paper, and other consumer products (Wilkins et al. 2000). Since cotton is highly susceptible to biotic and abiotic stresses, it requires intensive crop management. In cotton, bollworms, namely *Helicoverpa armigera* (Hubner), the American bollworm, *Earias vittella* (Fabricius), the spotted bollworm, and *Pectinophora gossypiella* (Saunders), the pink bollworm, pose serious threats to cotton production (Agarwal et al. 1984), causing yield loss of more than 50%. Conventional plant breeding methods have been extensively applied to improve these traits. However, these approaches have been limited by the lack of sufficient genetic variability in the existing germplasm pool (Wu et al. 2004). The control of insect pests has been accomplished primarily through the application of chemical pesticides, which lead to severe environmental problems. Moreover, many insects have developed resistance to different chemical pesticides, resulting in inefficient insect control programs. Considering the problems related to the action of these insecticides on non-target/beneficial organisms, environment and human health, it has become necessary to find alternative methods of control, which can be a part of Integrated Pest Management (IPM).

In this context, genetic engineering provides an enormous scope for widening the genetic diversity of crop plants through stable expression of foreign genes from divergent sources, including bacteria. *Bacillus thuringiensis* (Bt) is perhaps the most important source of insect resistance genes. Resistance to insects through deployment of Bt genes is one of the most successful strategies in modern agriculture. Genetically engineered (GE) crops were grown on 175.2 million hectares globally in 2013 (James 2013). Among them, transgenic cotton expressing insecticidal proteins from Bt has been one of the most rapidly adopted GE crops in the world (James 2012, Lu et al. 2012). Cultivation of Bt transgenic cotton has allowed a significant decrease in the use of chemical insecticides and, consequently, in environmental pollution and human exposure to toxins (Bennett et al. 2004). Bt toxins are also highly specific against insect pests, without affecting predators and other beneficial insects (Christou 2005).Due to these advantages, Bt cotton varieties or hybrids are recognized as a valuable component of an integrated pest management system.

Integration of genes encoding proteins from Bt has made it possible to obtain cotton lines that are resistant to several polyphagous insects (Perlak et al. 1990, Perlak et al. 1991). The ultimate goal is to obtain durable protection, which specifically requires stability of gene expression during the course of selfing or backcrossing, and also requires reducing the probability of development of resistant insects. Transgenic Bt cotton expressing *cry1Ac* has been registered for commercial cultivation in India since 2002, and it primarily targets *Helicoverpa armigera*.

A major concern of using transgenics with a single Bt toxin is the possibility of breakdown of insect resistance in plants. However, Tabashnik et al. (2003) have shown that insects that developed resistance against Cry1A are still susceptible to the Cry2A protein. Therefore, pyramiding two or more *Bt* genes with different modes of action is one of the strategies to delay development of resistance in insects. The combination of *cry1Ac* and *cry2Ab* in the second version of Bt cotton (BGII) exhibited superior control of lepidopteran pests and delayed development of resistance in insects (Perlak et al. 2001). Commercial Bt crops expressing Cry1Ab, Cry1Ac, Cry1F, Cry2Ab, and Cry3Bb proteins, with different modes of action, either individually or in combination, are now being grown worldwide with protection against a variety of insect pests.

With a view toward developing an alternative gene belonging to the *cry2* group, a novel chimeric *cry2AX1* gene was constructed, consisting of sequences from *cry2Aa* and *cry2Ac* (Udayasuriyan et al. 2010). Due to differences in structural and insecticidal mechanisms, *cry2A* genes are potential candidates for management of resistance in insects when deployed in combination with *cry1*. In this study, the codon-optimised synthetic *cry2AX1* gene was used in cotton transformation to develop cotton plants exhibiting insecticidal activity against *H. armigera*.

MATERIAL AND METHODS

Plasmid constructs and binary vector

The codon-optimised 1902 bp synthetic *cry2AX1* gene (Acc. No. GQ332539.1) fused downstream of the cotton transit peptide sequence (186 bp) from the Ribulose bisphosphate carboxylase small subunit (*rbcS1b*) gene family (Acc. No. JN608790.1) was cloned under the control of a double enhancer version of *CaMV35S* promoter and *nos* termination signal in pCAMBIA 2300 backbone (Ruturaj et al. 2014). The *Agrobacterium* strain LBA4404 harbouring the above construct (p2300-tp*2AX1*) (Figure 1) was used for cotton transformation.

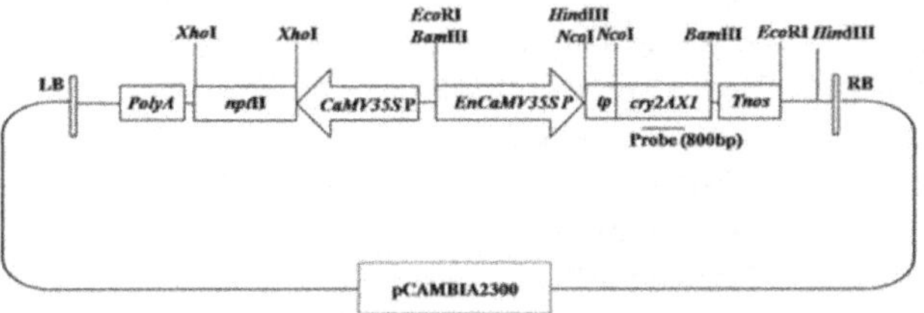

Figure 1. T-DNA region of plant transformation construct p2300-tp*2AX1*. Cotton chloroplast transit peptide (tp) was fused to the *cry2AX1* gene. The tp-*cry2AX1* gene is driven by a double enhancer *CaMV35S* promoter and terminated by the nopaline synthase (*nos*) terminator. The plant selectable marker gene *npt*II is under the control of the duplicated *CaMV35S* promoter and tailed by the *CaMV35S* polyA. LB: left border of T-DNA region; RB: right border of T-DNA region.

Plant material

Seeds of *Gossypium hirsutum cv.* Coker310 were surface sterilized and germinated in half-strength MS medium. Cotyledons and hypocotyls derived from one week old seedlings were used as explants for callus induction.

Callus induction and somatic embryogenesis

Initiation and maintenance of embryogenic calli

Cotyledonary (10-16 mm^2) and hypocotyl (3-5 mm) segments were used as explants for callus induction. Calli were cultured on MS medium supplemented with 0.5 mgL^{-1} kinetin and 0.1 mgL^{-1} 2,4-D (2,4-Dichlorophenoxyacetic acid) (Trolinder and Goodin 1988a) and 3% maltose. The proliferated calli were cultured on MS medium containing 1.9 gL^{-1} KNO_3 for initiation of somatic embryogenesis (Trolinder and Goodin 1988b) (Figure 2).

Genetic transformation and plant regeneration

Friable embryogenic calli were cultured with *Agrobacterium* strain LBA4404 (p2300-tp*2AX1*) (Leelavathi et al. 2004). After cocultivation, the calli were subcultured on MS medium supplemented with 1.9 gL^{-1} KNO_3, 25 mgL^{-1} kanamycin, and 250 mgL^{-1} cefotaxime until initiation and maturation of somatic embryos. The mature somatic embryos with true cotyledons were germinated on MS medium containing 0.1 mgL^{-1} GA3 (Giberellic acid), 1.0 mgL^{-1} IAA(Indole-3-acetic acid), and 3% sucrose. The regenerated plantlets were transferred to soil (soil:sand:peat mixture 1:1:1) for hardening (Figure 2).

Figure 2. *Agrobacterium*-mediated transformation of cotton. a. Cotyledon explant inoculation (Explant from 7 day old seedlings). b. Callus induction on cotyledon explants (30 days). c. Pre-embryogenic calli for cocultivation (180 day old calli). d. Proliferation of embryogenic calli (30-45 days). e. Proliferated somatic embryos (45-60 days). f. Friable embryogenic cultures with cotyledonary embryos (30 days). g. Regeneration of complete plant (30-45 days). h. Plant with well-developed roots after transfer to soil (15-20 days).

Molecular analysis of putative transgenic plants

Genomic DNA was isolated from the leaves of putative transgenic and non-transgenic control plants (Stewart and Via 1993). PCR analysis was performed to analyse the presence of *cry2AX1* and *npt*II (*Neomycin phosphotransferase*) genes in the putative transgenic lines of Coker310 using gene specific primers, *cry2AX1*: FP 5'-AACGTTCTTAACTCTGGAAGGA-3'; RP 5'-GCAGAAATTCCCCACTCATCAG-3' and *npt*II:FP 5'-CTGATGCTCTTCGTCCAGAT-3'; RP 5'-AGAGGCTATTCGGCTATGACT-3'. The presence of the actin gene was checked as an internal control.

Southern blot hybridisation analysis was done to confirm the integration of the transgene in putative T_0 transgenic plants. For Southern blot hybridisation analysis, 10 µg of genomic DNA was digested with *Eco*RI, which releases the *cry2AX1* gene (~3.1 kb size). The digested products were gel electrophoresed on 0.8% agarose gel and blotted onto a positively charged nylon membrane. For hybridisation, the 800 bp internal region of the *cry2AX1* gene was used as a probe. The probe DNA was radio-labelled with α^{32}-P dCTP (Deoxycytidine triphosphate) by random priming using the Decalabel DNA labelling kit (Thermo Scientific Inc.). The blot was washed with 3X SSC (Saline-sodium citrate) + 0.1% SDS (*Sodium dodecyl sulphate*) and 2X SSC + 0.1% SDS for 15 min each, followed by 10 min in 0.5X SSC + 0.1% SDS at 60°C after hybridisation. The membrane was exposed to X-ray film for a week.

Estimation of Cry2AX1 protein by ELISA

The Cry2AX1 protein levels in the putative transgenic plants were analysed using the Enzyme Linked Immunosorbent Assay quantitative (ELISA) kit (Envirologix, USA) as per manufacturer's instructions. Protein extracts were made by grinding 30 mg of leaf tissues (fully expanded terminal leaf) in 500 mL of extraction buffer (provided in the kit) and centrifuged at 10,000 rpm for 10 min at 4 °C. An aliquot of 100 µL of leaf extracts was loaded into the ELISA plate. Colour intensity was observed at 450 nm. Proper negative and positive controls (standards provided in the kit) were included in the experiment. The Cry2AX1 protoxin was quantified based on standards provided with the kit.

Detached leaf disc insect bioassay

Putative T_0 transgenic plants were subjected to insect bioassay to assess insecticidal activity of the Cry2AX1 protein against cotton bollworm, *H. armigera*. Leaf discs (3 cm diameter) of putative transgenic (T_0) and non-transformed (control) plants were placed on a wet filter paper placed inside a sterile petriplate. Ten first instar larvae of *H. armigera* were released per replication. Three replications were maintained in each line, and the bioassay was carried out at 26-28°C with 60% relative humidity. Larval mortality and larval growth were recorded for five days.

Statistical analysis

The experimental data values of Cry2AX1 protein concentration and mortality of *H. armigera* were mean values from three replicates, and the results were presented as mean ± SD. All mortality data were subjected to arcsine transformations before analysis. Data analysis was done by analysis of variance (ANOVA) following the AGRES statistical package. Mean values were separated by Duncan's multiple range test (DMRT) at a 5 per cent probability level (Duncan 1955).

RESULTS AND DISCUSSION

Generation and evaluation of transgenic plants

The cultivation of transgenic cotton (*G. hirsutum* L.) rapidly gained a great deal of ground in the late 1990s and now accounts for most cotton production in the US and many other countries, including India. The transformation and regeneration of cotton *via* somatic embryogenesis is a long process, and cotton remains one of the recalcitrant species to be manipulated in culture (Wilkins et al. 2000). Somatic embryogenesis in cotton is hampered by its extended culture period, low frequency of embryos, and high incidence of abnormal embryos (Kumria et al. 2003). Low conversion rates of somatic embryos into complete plantlets in cotton

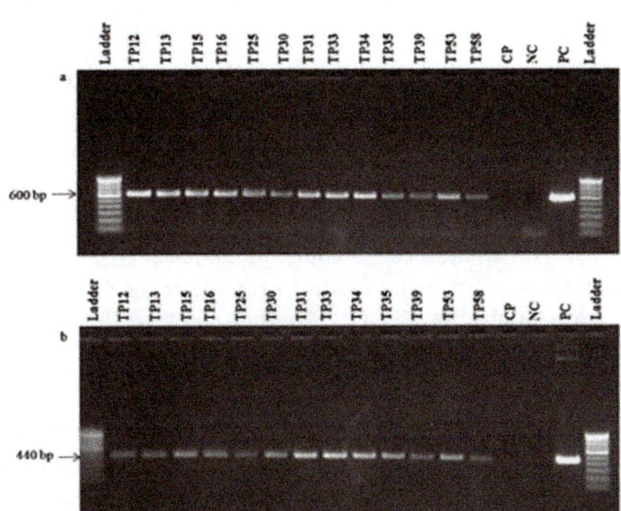

Figure 3. PCR analysis of *cry2AX1* transgenic cotton plants. a. A 600 bp internal sequence of the *cry2AX1* gene amplified by PCR from the DNA isolated from putative transgenic plants. b. A 440 bp internal sequence of the *npt*II gene amplified by PCR from the DNA isolated from putative transgenic plants. Ladder: 100 bp ladder, Lane 2-14: Putative cotton transgenic plants, CP: non-transformed plant, NC: negative control, PC: positive control (p2300-tp*2AX1* plasmid).

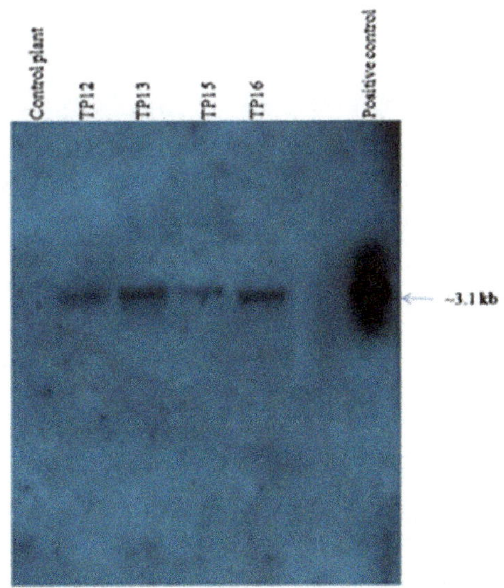

Figure 4. Southern blot analysis of putative T_0 transgenic plants. DNA digested with *Eco*RI and probed with a radioactively labelled 800 bp internal sequence of the *cry2AX1* gene. Lane 1: Control plant, Lane 2: TP12, Lane 3: TP13, Lane 4: TP15, Lane 5: TP16, Lane 6: blank, Lane 7: positive control (p2300-tp*2AX1*plasmid).

tissue culture have been reported (Voo et al. 1991, Zhang et al. 1993). In germinating somatic embryos, rooting is an inefficient process and one that has received scant attention. In the Coker genotypes, only ~5-6% of somatic embryos root sufficiently to allow recovery of complete plants (Wilkins et al. 2000).

We introduced a synthetic *cry2AX1* gene into cotton plants (Coker310) through *Agrobacterium* mediated transformation to evaluate its efficacy against cotton bollworm, *H. armigera*. Six month old friable embryogenic calli that were cocultivated with the *Agrobacterium* strain LBA4404 (p2300-tp*2AX1*) construct proliferated into somatic embryos, whereas no proliferation was observed in non-transformed calli. Transformation of cotton with the synthetic *cry2AX1* gene (consisting of sequences of *cry2Aa* and *cry2Ac* genes) resulted in generation of 48 putative transgenic plants.

In PCR analysis carried out with gene specific primers to confirm the presence of transgenes in putative transgenic plants, 40 plants were found to possess *cry2AX1* and *npt*IIgenes (Figure 3a, b). No amplification was observed in non-transformed control plants. Southern blot hybridisation analysis of *cry2AX1* transformants indicated stable integration of the *cry2AX1* gene in the genome of transgenic cotton plants, whereas the untransformed control plant did not show any sign of hybridisation (Figure 4).

The selected PCR positive transgenic plants were subjected to ELISA for quantification of the insecticidal Cry2AX1 protein. Thirteen out of 40 PCR positive plants were found to be positive for expression of the Cry2AX1 protein. However, we observed a wide variation of expression in the transgenic lines developed, ranging from 4.9 to 187.5 ng g^{-1} of fresh leaf tissue (Table 1). A wide range of Bt protein expression in transgenic plants with the same genetic background and gene construct were reported by several earlier studies (Maqbool et al. 2001, Breitler et al. 2000, Ramesh et al. 2004,

Table 1. Bioassay on putative T_0 transgenic cotton lines against *Helicoverpa armigera*

S. No	Lines	**Concentration of Cry2AX1 protein** (ng $g^{-1}$of fresh leaf tissue)* (Mean ± SD)	**Larval mortality** (%)# (Mean ± SD)
1	TP12	6.60 ± 0.0	NT
2	TP13	9.90 ± 0.0	NT
3	TP15	49.0 ± 0.0	33.33±4.71 (35.22)[ef]
4	TP16	16.6 ± 0.0	23.33±4.71 (28.78)[g]
5	TP25	94.9 ± 0.0	46.66±4.71 (43.08)[c]
6	TP30	23.0 ± 0.0	26.66±4.71 (30.99)[fg]
7	TP31	187.5 ± 0.0	56.66±4.71 (48.85)[b]
8	TP33	54.0 ± 0.0	43.33±4.71 (41.15)[cd]
9	TP34	4.90 ± 0.0	NT
10	TP35	17.0 ± 1.41	26.66± 4.71 (30.99)[fg]
11	TP39	61.0 ± 11.31	43.33±4.71 (41.15)[cd]
12	TP53	49.0 ± 0.0	36.66±4.71 (37.22)[de]
13	TP58	32.0 ± 1.41	30.0± 0.0 (33.21)[efg]
14	Control (Non-transformed plant)	0.0 ± 0.0	0.0 ± 0.0 (0.91)[h]
15	BGII (Positive control)	27000.0 ± 0.0	100.0 ± 0.0 (89.09)[a]

* Mean of two replicates; # Mean of three replicates; NT – Not tested; Figures in parentheses are arcsine transformed values; Standard Error of Deviation SEd = 2.50; Critical Difference CD (0.05) = 5.16; Coefficient of Variation CV% = 7.98; Means followed by the same small letters within a column are not significantly different at the 5% level in Duncan's multiple range test (DMRT)

Meiyalaghan et al. 2006). We reason that the variation in the level of transgene expression or lack of expression may be due to mutations in the transgene, truncation of T-DNA during integration, post transcriptional gene silencing, or transcriptional gene silencing (integration of T-DNA into genomic regions such as the heterochromatin that repress transgene expression) (Francis and Spiker 2005). Inactivation of the transgene is often shown to be accompanied by an increase in DNA methylation (Amasino et al. 1984). Inactivation also very frequently correlates with the number of copies of integrated transgenes (Jones et al. 1987).

Insect bioassay of *cry2AX1* transgenic cotton plants

Laboratory biotoxicity assays with the first instar *H. armigera* larvae were conducted to analyse the efficacy of Cry2AX1 protein in cotton transformants. The results of leaf disc bioassay studies showed larval mortality ranging from 23.33 to 56.66 % in the selected ELISA positive plants (Table 1). The surviving larvae on transgenic plants showed severe growth inhibition and significant reduction in leaf feeding, whereas larvae released on control plants were alive with normal growth (Figure 5). Earlier reports suggest that the protein concentration is directly related to the level of insect resistance in Bt transgenic plants (Chen et al. 2005, Rashid et al. 2008, Mehrotra et al. 2011). Differences in the level of toxicity (mortality) observed among the different transgenic lines could be attributed to differences in the level of Bt gene expression. Variation of a single amino acid can also significantly influence the level of toxicity in Cry proteins (Udayasuriyan et al. 1994, Rajamohan et al. 1996).

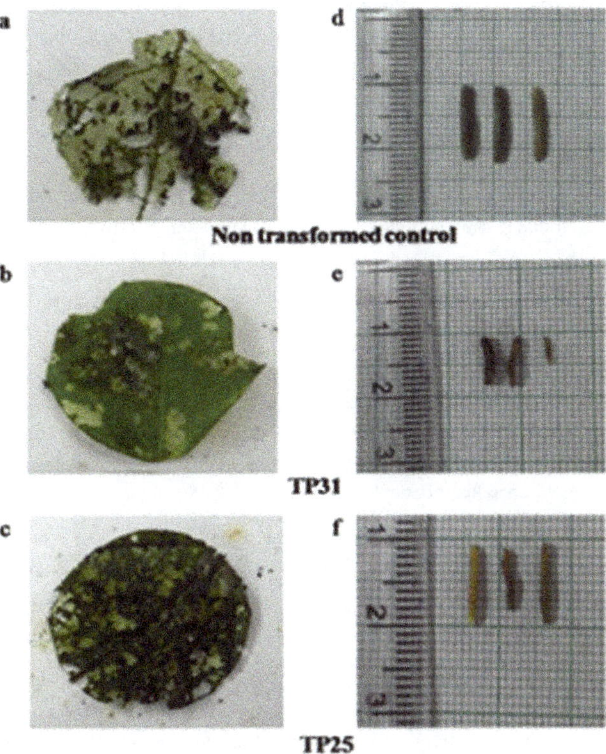

Figure 5. Detached leaf disc bioassay against cotton boll worm (*Helicoverpa armigera*) in transgenic cotton plants expressing the Cry2AX1 protein. a. Non-transformed control (Coker310), b. Transformed cotton plant TP31, c. Transformed cotton plant TP25; d, e, f. Surviving larvae on control and transformed plants.

The Cry2Aa protein is known to be toxic to prominent lepidopteran insect pests like rice stem borers and brinjal fruit and shoot borer (Maqbool et al. 1998, Rao et al. 1999). Pesticidal activity of *B. thuringiensis* δ-endotoxins, Cry1Aa, Cry1Ab, Cry1Ac, and Cry2A against *Helicoverpa zea* was determined by Karim et al. (2000). *H. zea* was susceptible to Bt toxins in the order Cry1Ac, Cry1Ab, Cry1Aa, and Cry2A with 63.60, 89.04, 159.65, and 375.78 ng/larvae, respectively. Indian populations of *H. armigera* were 5- to 30-fold less susceptible to *cry2Aa* than *cry1Ac* (Chakrabarti et al. 1998, Babu et al. 2002, Misra et al. 2002). The expression level of Cry2Ab in commercialized Bt cotton ranges from 16.8 to 22.7 $\mu g\ g^{-1}$ in fresh leaf tissue (Li et al. 2011). These findings indicated that a greater amount of Cry2AX1 protein expression may be needed to achieve the desirable level of insecticidal activity in plants.

In the current investigation, an attempt was made to study the insecticidal activity of the *cry2AX1* gene against *H. armigera* in cotton. The *H. armigera* neonates showed mortality up to 56.66% in the transgenic plants, even with a relatively lower level of Cry2AX1 protein in the plant (about 187.5 ng g^{-1} of leaf tissue). The expression of a higher level of Cry2AX1 protein and high insecticidal activity can possibly be achieved by generating or screening more transgenic plants. A transgenic plant carrying the *cry2AX1* gene with higher insecticidal activity could be an alternative management strategy for *H. armigera* in cotton.

ACKNOWLEDGEMENTS

The authors express their gratitude to the Council of Scientific and Industrial Research (CSIR), Government of India, New Delhi, for financial assistance in the form of a network project (CSIR Order No. 5/258/53/2006- NMITLI dt. 18.3.2008).

We thank Prof. D. Pental and Prof. P.K. Burma, University of Delhi South Campus, Delhi for providing the *rbcs1b* cotton transit peptide sequence. We also thank Dr. P. Nandeesha (currently working in Indian Institute of Horticultural Research, Bengaluru) for making the construct.

REFERENCES

Agarwal RA, Gupta GP and Grag DO (1984) Cotton pest management in India. Research Publication, Azad Nagar, Delhi, p. 1-191.

Amasino RM, Powell ALT and Gordon MP (1984) Changes in T-DNA methylation and expression are associated with phenotypic variation and plant regeneration in a crown gall tumor line. **Molecular and General Genetics 197**: 437-446.

Babu BG, Udayasuriyan V, Mariam MA, Sivakumar NC, Bharathi M and Balasubramanian G (2002) Comparative toxicity of Cry1Ac and Cry2Aa δ-endotoxins of *Bacillus thuringiensis* against *Helicoverpa armigera* (H.) **Crop Protection 21**: 817-822.

Bennett RM, Ismael Y, Kambhampati V and Morse S (2004) Economic impact of genetically modified cotton in India. **AgBioForum 7**: 96-100.

Breitler JC, Marfa V, Royer M, Meynard D, Vassal JM, Vercambre B, Frutos R, Messeguer J, Gabarra R and Guiderdoni E (2000) Expression of a *Bacillus thuringiensiscry1B* synthetic gene protects Mediterranean rice against the striped stem borer. **Plant Cell Reports 19**: 1195-1202.

Chakrabarti SK, Mandaokar AJ, Kumar PA and Sharma RP (1998) Toxicity of lepidopteran specific delta endotoxins of *Bacillus thuringiensis* towards neonate larvae of *Helicoverpa armigera*. **Journal of Invertebrate Pathology 72**: 336-337.

Chen H, Tang W, Xu C, Li X, Lin Y and Zhang Q (2005) Transgenic *indica* rice plants harbouring a synthetic *cry2A* gene of *Bacillus thuringiensis* exhibit enhanced resistance against lepidopteran rice pests. **Theoretical and Applied Genetics 111**: 1330-1337.

Christou P (2005) Sustainable and durable insect pest resistance in transgenic crops. **ISB News Report.** 3p.

Duncan DB (1955) Multiple range and multiple *F* test. **Biometrics 11**: 1-42.

Francis KE and Spiker S (2005) Identification of *Arabidopsis thaliana* transformants without selection reveals a high occurrence of silenced T-DNA integrations. **The Plant Journal 41**: 464-477.

James C (2012) Global review of commercialized transgenic crops: 2012 feature: Bt cotton. ISAAA Briefs No. 44. ISAAA, Ithaca, 329p.

James C (2013) Global review of commercialized transgenic crops: 2013 feature: Bt cotton. ISAAA Briefs No. 46. ISAAA, Ithaca, 332p.

Jones JDG, Gilbert DE, Grady KL and Jorgensen RA (1987) T-DNA structure and gene expression in petunia plants by *Agrobacterium tumefaciens* C58 derivatives. **Molecular and General Genetics 207**: 478-485.

Karim S, Gould F and Dean DH (2000) *Bacillus thuringiensis* endotoxin proteins show a correlation in toxicity and short circuit current inhibition against *Helicoverpa zea*. **Current Microbiology 41**: 214-219.

Kumria R, Sunnichan VG, Das DK, Gupta SK, Reddy VS, Bhatnagar RK and Leelavathi S (2003) High-frequency somatic embryo production and maturation into normal plants in cotton (*Gossypium hirsutum*) through metabolic stress. **Plant Cell Reports 21**: 635-639.

Leelavathi S, Sunnichan VG, Kumria R, Vijaykanth GP, Bhatnagar RK and Reddy VS (2004) A simple and rapid *Agrobacterium*-mediated transformation protocol for cotton (*Gossypium hirsutum* L.): embryogenic calli as a source to generate large numbers of transgenic plants. **Plant Cell Reports 22**: 465-470.

Li Y, Romeis J, Wang P, Peng Y and Shelton AM (2011) A comprehensive assessment of the effects of Bt cotton on *Coleomegilla maculata* demonstrates no detrimental effects by Cry1Ac and Cry2Ab. **PLoS One 6(7)**: e22185.

Lu Y, Wu K, Jiang Y, Guo Y and Desneux N (2012) Widespread adoption of Bt cotton and insecticide decrease promotes biocontrol services. **Nature 487**: 362-365.

Maqbool SB, Husnain T, Raizuddin S and Christou P (1998) Effective control of yellow rice stem borer and rice leaf folder in transgenic rice Indica varieties Basmati 370 and M7 using novel δ-endotoxin *cry2A Bacillus thuringiensis* gene. **Molecular Breeding 4**: 501-507.

Maqbool SB, Raizuddin S, Loc TN, Gatehouse AMR, Gatehouse JA and Christou P (2001) Expression of multiple insecticidal genes confers broad resistance against a range of different rice pests. **Molecular Breeding 7**: 85-93.

Mehrotra M, Sanyal I and Amla DV (2011) High efficiency *Agrobacterium*-mediated transformation of chickpea (*Cicer arietinum* L.) and regeneration of insect- resistant transgenic plants. **Plant Cell Reports 30**: 1603-1616

Meiyalaghan S, Jacobs JME, Butler RC, Wratten SD and Conner AJ (2006) Transgenic potato lines expressing *cry1ba1*or *cry1ca5* genes are resistant to potato tuber moth. **Potato Research 49**: 203-216.

Misra HS, Khairnar NJP, Mathur M, Vijiyalakshmi N, Hire RS, Dongre TK and Mhanan SK (2002) Cloning and characterization of an insecticidal crystal protein gene from *Bacillus thuringiensis* subspecies *kenyae*. **Journal of Genetics 81**: 5-11.

Perlak FJ, Deaton RW, Armstrong TA, Fuchs RL, Sims SR, Greenplate JT and Fischoff DA (1990) Insect resistant cotton plants. **Nature Biotechnology 8**: 939-943.

Perlak FJ, Fuchs RL, Dean DA, McPherson SL and Fischhnff DA (1991) Modification of the coding sequence enhances plant expression of insect control protein genes. **Proceedings of the National Academy of Sciences 88**: 3324-3328.

Perlak FJ, Oppenhuizen M, Gustafson K, Voth R, Sivasubramanian S, Heering D, Carey B, Ihrig RA and Roberts JK (2001) Development and commercial use of Bollgard® cotton in the USA early promises versus today's reality. **The Plant Journal 27**: 489-501.

Rajamohan F, Alzate O, Cotrill JA, Curtiss A and Dean DH (1996) Protein engineering of *Bacillus thuringiensis* δ endotoxin: Mutations at domain II of Cry1Ab enhance receptor affinity and toxicity towards gypsy moth larvae. **Proceedings of the National Academy of Sciences 93**: 14338-14343.

Ramesh S, Nagadhara D, Pasalul C, Kumari AP, Sarma NP, Reddy VD and Rao KV (2004) Development of stemborer resistant transgenic parental lines involved in the production of hybrid rice. **Journal of Biotechnology 111**: 131-141

Rao NGV, Majumdar A, Mandaokar AD, Nimbalkar SA and Kumar PA (1999) Susceptibility of brinjal shoot and fruit borer to the d-endotoxins of *Bacillus thuringiensis*. **Current Science 77**: 336-337.

Rashid B, Zafar S, Husnain T and Riazuddin S (2008) Transformation and inheritance of *Bt* genes in *Gossypium hirsutum*. **Journal of Plant Biology 51**: 248-254.

Ruturaj RB, Naveenkumar A, Nandeesha P, Manikandan R, Balakrishnan N, Balasubramani V, Sudhakar D and Udayasuriyan V (2014) Transformation of tomato with *cry2AX1* gene of *Bacillus thuringiensis*. **International Journal of Tropical Agriculture 32**: 577-585.

Stewart CN and Via LE (1993) A Rapid CTAB DNA isolation technique useful for RAPD fingerprinting and other PCR Applications.**BioTechniques 14**: 748-749.

Tabashnik BE, Carriere Y, Dennehy TJ, Morin S, Sisterson MS, Roush RT, Shelton AM and Zhao JZ (2003) Insect resistance to transgenic Bt crops: lessons from the laboratory and field. **Journal of Economic Entomology 96**: 1031-1038.

Trolinder NL and Goodin JR (1988a) Somatic embryogenesis in cotton (*Gossypium*). I. Effects of source of explant and hormone regime. **Plant Cell, Tissue and Organ Culture 12**: 178-181.

Trolinder NL and Goodin JR (1988b) Somatic embryogenesis in cotton (*Gossypium*). II. Requirements for embryo development and plant regeneration. **Plant Cell, Tissue and Organ Culture 12**: 43-53.

Udayasuriyan V, Indra Arulselvi P, Balasubramani V, Sudha DR, Balasubramanian P and Sangeetha P (2010) Construction of new chimeric *cry2AX1* gene of *Bacillus thuringiensis* encoding protein with enhanced insecticidal activity. (Indian patent number 244427).

Udayasuriyan V, Nakamura A, Mori A, Masaki H and Uozumi T (1994) Cloning of a new *crylA(a)* gene from *B. thuringiensis* strain FU-2-7 and analysis of chimeric Cry1A(a) proteins of toxicity. **Bioscience, Biotechnology, and Biochemistry 58**: 830-835.

Voo KS, Rugh CL and Kamalay JC (1991) Indirect somatic embryogenesis and plant recovery from cotton (*Gossypium hirsutum* L.). **In Vitro Cellular & Developmental Biology - Plant 27**: 117-124.

Wilkins T, Rajasekaran K and Anderson DM (2000) Cotton biotechnology. **Critical Reviews in Plant Sciences 19**: 511-550.

Wu JH, Zhang XL, Nie YC, Jin SX and Liang SG (2004) Factors affecting somatic embryogenesis and plant regeneration from a range of recalcitrant genotypes of Chinese cottons (*Gossypium hirsutum* L.) **In Vitro Cellular & Developmental Biology - Plant 40**: 371-375.

Zhang BH, Li XL, Li FL and Li FG (1993) Plant recovery from cotton somatic embryos. **Acta Agricultural Boreali-occidentalis Sinica 2**: 24-28.

26

Characterization and selection of interspecific hybrids of Brachiaria decumbens for seed production in Campo Grande – MS

Lenise Castilho Monteiro[1], Jaqueline Rosemeire Verzignassi[2], Sanzio Carvalho Lima Barrios[2], Cacilda Borges do Valle[2], Celso Dornelas Fernandes[2], Gleiciane de Lima Benteo[1] and Cláudia Barrios de Libório[1]

***Corresponding author:**
E-mail:

[1] Instituto Federal Goiano-Campus Rio Verde, Av. Sul Goiânia, km 1, Zona Rural, 75.901-970, Rio Verde, GO, Brazil
[2] Embrapa Gado de Corte, Av. Rádio Maia, 830, Zona Rural, 79.106-550, Campo Grande, MS, Brazil

Abstract: *The breeding program Brachiaria developed by Embrapa Beef Cattle provides studies to obtain forage with agronomic characters desired by farmers. In this regard, in 2013 and 2014, a study was carried out in order to select the best Brachiaria decumbens intraspecific hybrids, which are superior in relation to production, forage nutritional value, and resistance to spittlebugs. Estimates of genetic parameters and gains with selection were carried out. It was found that there was significant variability between genotypes for all characters. Gain with selection (GS%) ranged from 12 to 324%, and the highest percentage was found for weight of pure seeds (PS) of seed collectors of the second harvest. For reproductive tillers (RT), C001 and R091 hybrids had better performance than the control, and this character may be considered as a parameter to estimate production of pure seeds before flowering starts.*

Key words: *Seed production potential, genetic variability, genotypes, forage improvement.*

INTRODUCTION

At first, selection of forage plants was based only on mass production potential and on forage quality under pasture conditions. However, in recent years, some changes have been observed in strategies of breeding programs of these plants, aiming at obtaining superior cultivars in all aspects. Cultivars development process is long, comprising several stages, and the evaluated characters should be correlated in order to result in crops that have good performance at all stages and variables of study and research. In order to reach the release of a new cultivar, the development process involves several research lines, such as improvement; cytogenetics of the reproductive system; plant nutrition; microbiology; plant health; pasture management and nutritional quality of plants; and seed production and technology, among others (Karia et al. 2006, Valle et al. 2009).

Improvement of tropical forage plants is relatively new compared to other cultures (Karia et al. 2006, Araújo et al. 2008, Valle et al. 2009), and aims to release more productive plants in production and forage quality, and in various agronomic aspects, such as production of good quality seeds and in satisfactory quantity, adaptation to different soil and climate conditions, and especially resistance to spittlebug.

Brachiaria plants have been widely used as forage in tropical America, and most species found in Brazil are considered exotic. Recently, there has been a reclassification of the *Brachiaria* genus to *Urochloa*. However, discussions remain among researchers regarding its characteristics and new taxonomic classification. *Brachiaria decumbens*, for its high forage production potential and high adaptability to acid soils and of low fertility, has great importance for national beef cattle. Even being economically very relevant, there is only one cultivar in the market, cv. Basilisk, which was released in the 1960s. One of the limitations of this cultivar is its susceptibility to spittlebugs.

It is extremely important that cultivars from breeding programs solve problems such as susceptibility to spittlebugs. Cv. Marandu is an example, which has been available since the 1980s. A few options of cultivars available to producers have good seed production, so that the ratio production cost x price of seed production is satisfactory. Seed production may vary depending on the location of production, management, and on the form of harvest, and it may reach up to 300 kg ha^{-1} (Monteiro et al. 2016).

This work aims to characterize and evaluate the potential of seed production by intraspecific hybrids of *Brachiaria decumbens*, developed and pre-selected by the breeding program of Embrapa Beef Cattle, aiming at selecting superior genotypes, which could be candidates for new cultivars or potential parents of sexual propagation to be used in new crosses.

MATERIAL AND METHODS

Genetic materials used in the development of this work were selected based on previous experimental results obtained by the Forage Cultivars Breeding and Production Program of Embrapa Beef Cattle. Promising intraspecific hybrids of *Brachiaria decumbens* (sexual and apomictic), parents and candidates for new cultivars were selected among 324 hybrids, which were obtained by crossing between three plant of sexual reproduction of *B. decumbens* artificial tetraploids (D24/2, D24/27 e D24/45) with cv. *Basilisk* (apomictic). Selection was based on agronomic characteristics of production, forage nutritional value, and resistance to spittlebugs. *B. brizantha* cv. Marandu was used as control. Table 1 shows the genotypes evaluated during the first (2013) and the second (2014) harvest.

Trials were carried out for two consecutive years in Campo Grande - MS, at Embrapa Beef Cattle (lat 20º 25' 03" S, long 54° 42' 20" W, alt 530 m asl). Local climate is classified as Aw tropical rainy savanna, characterized by rainy summer and dry winter. Trials were carried out under normal environmental conditions, in rainy condition. Soil was classified as clayey dystroferric Red Latosol (Oxisol) (53% clay, 38% sand and silt 9%).

Plants were transferred to the field in early 2012 in the

Table 1. Genotypes evaluated during the first and the second years of production

Genotypes	1st year	0	2nd year
A001	A036		B006
A002	A038		C001
A003	A041		R025
A004	A042		R033
A005	A043		R041
A007	A044		R044
A008	B005		R071
A009	B009		R078
A011	B010		R087
A012	B026		R091
A013	C001		R101
A015	R184		R107
A017	S044		R110
A018	T038		R120
A019	X121		R124
A020	-		R126
A021	-		R144
A023	-		R181
A024	-		S018
A025	-		S031
A026	-		S036
A027	-		T005
A028	-		T012
A029	-		T026
A030	-		T054
A031	-		X030
A032	-		X072
A033	-		X117
A035	-		Y021

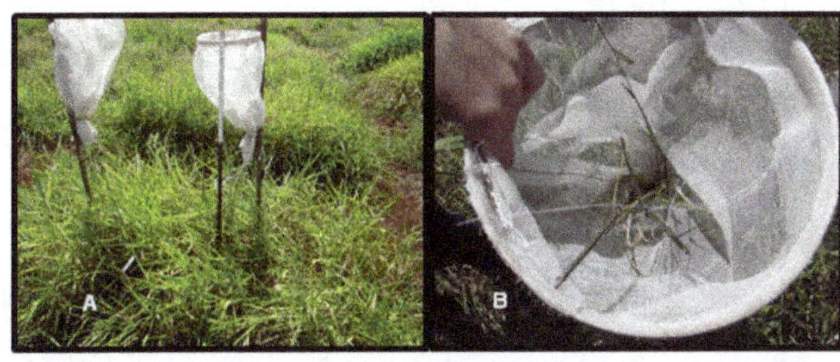

Figure 1. Seed collectors installed in the plot (A). Inflorescence inside the collector (B).

form of seedlings. Each plot contained two seedlings, with useful area of 2 m^2, spaced 1m between plots. Evaluations were carried out in the first quarter of 2013, with first-year plants, and between December 2013 and the first quarter of 2014, with second-year plants. For the first year of evaluations (2012/2013 crops), it was carried out uniform cutting of plants for seed production in October 2012. For the second year of evaluation, standardization was carried out on October 15, 2013. The experimental design was a complete block with two replications, and two observations per plot. After uniform cuttings (2012 and 2103), fertilization of plants was carried out based on soil chemical analysis. Evaluations were carried out in plots and in seeds collectors. Two collectors per plot were installed and, inside them, it was placed five inflorescences in complete anthesis without threshing (Figure 1).

To determine the typical inflorescences of genotype, ten inflorescences selected at random were collected. When at least six of them were of the same type, it was considered typical inflorescence. Eight typical inflorescences were collected in each plot, and from them, the following characteristics were determined: number of seeds per raceme (NSR), number of racemes (NR), number of seeds per inflorescence (NSI), length of racemes (LR), and length of inflorescences (LI). To determine these characteristics, it was carried counting in increasing order from the inflorescence base to the apex. LR was determined with the aid of a graduated scale. To this end, each raceme was measured from its insertion point in the rachis to the apex. LI was obtained with the aid of the graduated scale, by measuring the distance between the insertion point of the first and last raceme in the inflorescence.

It was also carried out evaluations in the plots at the beginning of the formation of the first reproductive tillers, and before the formation of the first panicle (inflorescence), in 0.25 m^2 in relation to the number of reproductive tillers (RT). After threshing of inflorescences' seeds, collectors were removed from the plots with their respective seeds and flowers. The collected material was taken to the Seed Laboratory of Embrapa Beef Cattle and evaluated for weight of pure seeds (PS) and weight of empty seeds (ES).

To determine the quality of seeds, it was carried out harvest of the total area of the plot (2 m^2). To this end, in the maturation point of each plot, it was carried out plant cutting of the entire plot (except the inflorescences that were inserted in the collectors). Cutting was carried out manually at 20 cm from the soil, with a rice cutter, and was carried out at the beginning of threshing, of 15 to 20% of inflorescences' seeds. The material collected was packed in paper bags, which were closed and placed to dry in the shade. After drying, samples were submitted to manual separation of inflorescences' seeds. Seeds were then subjected to processing by pre-cleaning and cleaning using mesh and air column blower. After seeds processing, which were harvested in 2013, it was carried out the germination standard test (G%), according to RAS (Seed Analysis Rules) (Brasil 2009).

The collected data were submitted to analysis of deviance (ANADEV) using the Selegen-Reml/Blup software (Resende 2006) in order to simultaneously estimate genetic parameters and predict breeding values (BLUP). For the evaluations carried out in 2013, it was used the model of the 20 Selegen-Reml/Blup software, as follows:

$$y = Xr + Zg + e$$

In which y is the data vector; r is the vector of replication effects (fixed) added to the mean; g is the vector of genotypic effects (random); e is the error vector or residues (random); X and Z are the incidence matrices for these effects. For the other characters, evaluated in 2014, it was used the model 2 of the software, as follows:

$$y = Xr + Zg + Wp + e$$

In which y is the data vector; r is the vector of replication effects (fixed), added to the mean; g is the vector of genotypic effects (random); p is the vector of plot effects; e is the vector error or residue (random); X, Z and W are the incidence matrices for these effects.

RESULTS AND DISCUSSION

For germination percentage (G%) of the first harvest, there was genotypic difference. Block effect, which was considered as fixed effect, was tested by the Snedecor F test. When genotype effects are significant, respective components of variance are significantly different from zero, and so are their coefficients of determination (Resende and Duarte 2007). Heritability between genotypes means (h^2mc) exceeded in 50%, and it can be considered of medium magnitude, showing that more than half of the observed phenotypic variation, on average, was due to genetic causes (Table 2).

Of the seeds trashed in the collectors, it was observed accuracy (Acc) of 84 and 97% (Table 2). Thus, it was within the stablished standards, and was classified as high (Resende and Duarte 2007), which means good reliability for the prediction of genotypic values of treatments. Both for weight of pure seeds (PS) and for weight of empty seeds (ES), it was found significant differences by the Chi-square test (LRT) for the genotype effect, and their heritability estimates (h^2mc) presented percentage above 0.50 (Table 2).

Table 2. Deviance Analysis (ANADEV), variation components of genetic parameters for germination (G%) in the first harvest, weight of pure seeds (PS), weight empty seeds (ES), number of reproductive tillers (RT), number of seeds per inflorescence (NSi), number of seeds per racemes (NSR), number of racemes (NR), length of raceme (LR) and length of inflorescences (LI) in the second harvest

	ANADEV			Variation Components			Genetic Parameters				
	Genotypes	**Plot**	**Blocks**	**Genotypes**	**Plot**	**Blocks**	h^2g	h^2mc	**Acc**	**General Mean**	***CVe*** **(%)**
[1] G%	4.52*	-	1.63	11.93	-	-	0.42 +- 0.24	0.59	0.77	10.32	39.06
[2] PS ($g\ m^{-2}$)	26.42**	0.01	0.00	0.30	8.10^{-4}	0.06	0.83 +- 0.30	0.95	0.97	0.25	70.98
[2] ES ($g\ m^{-2}$)	7.14**	0.00	0.00	0.05	7.10^{-4}	0.08	0.37 +- 0.20	0.7	0.84	0.83	24.08
[3] RT	11.86**	0.00	157.55**	871.77	0.50	115.13	0.88 +- 0.31	0.97	0.98	38.61	19.73
[3] NSI	33.24**	22.98**	0.05	1070.14	388.28	973.40	0.44 +- 0.11	0.77	0.88	30.77	18.61
[3] NSR	43.43**	0.21	0.10	33.18	1.34	51.74	0.38 +- 0.10	0.82	0.91	30.77	12.27
[3] NR	235.46**	2667.90**	250000**	0.54	0.48	10^{-6}	0.53 +- 0.12	0.69	0.83	4.42	15.68
[3] LR	64.85**	5.08*	224.64**	5.32	0.47	3.17	0.59 +- 0.13	0.89	0.95	6.23	18.05
[3] LI	32.02**	132.55**	366.99**	8.16	4.44	2.33	0.55 +- 0.12	0.76	0.87	11.34	19.77

[1] Evaluations carried out in beds during the first harvest.[2] Evaluations carried out in seed collectors during the second harvest.[3] Evaluations carried out in beds during the second harvest. * Significant at the likelihood ratio test, considering 5% probability by the X^2 test. ** Significant at the likelihood ratio test, considering 1% probability by the X^2 test.

Table 3. Predicted genotypic values (BLUP) of 44 hybrids (Hib) and gain with selection (GS% at 30, 20 and 10% selection intensity) for the germination (G%) of *B. decumbens* hybrids in the first harvest

G (%)					
	Treatments	BLUP (LL-UL)		Treatments	BLUP (LL-UL)
1	A044	19.95 (15.55 - 24.35)	23	A015	9.95 (4.76 - 15.15)
2	A029	13.70 (9.30 - 18.10	24	A032	9.95 (4.76 - 15.15)
3	A030	13.70 (9.30 - 18.10)	25	T038	9.95 (4.76 - 15.15)
4	A004	13.34 (8.15 - 18.53)	26	C001	9.83 (5.43 - 14.23)
5	A025	13.10 (8.70 - 17.50)	27	A009	9.57 (4.35 - 14.79)
6	A023	12.49 (7.30 - 17.69)	28	B026	9.57 (4.35 - 14.79)
7	A031	12.49 (7.30 - 17.69)	29	R184	9.57 (4.35 - 14.79)
8	A042	12.49 (7.30 - 17.69)	30	A033	9.53 (4.33 - 4.72)
9	A036	11.69 (6.47 - 16.90)	31	B009	9.53 (4.33 - 14.72)
10	X121	11.69 (6.47 - 16.90)	32	A035	9.24 (4.84 - 13.64)
11	A020	11.65 (6.45 - 16.84)	33	A012	9.15 (3.93 - 14.36)
12	A008	11.62 (7.22 - 16.02)	34	B005	9.10 (3.91 - 14.30)
13	A026	11.32 (6.92 - 15.72)	35	A017	8.34 (3.94 - 12.74)
14	S044	10.84 (5.62 - 16.06)	36	A024	8.30 (3.08 - 13.52)
15	A013	10.80 (5.60 - 15.99)	37	A041	8.30 (3.08 - 13.52)
16	A019	10.80 (5.60 - 15.99)	38	A007	8.26 (3.06 - 13.45)
17	A027	10.43 (6.03 - 14.83)	39	A002	7.75 (3.35 - 12.15)
18	A028	10.37 (5.18 - 15.57)	40	A005	7.41 (2.22 - 12.60)
19	B010	10.37 (5.18 - 15.57)	41	A011	7.41 (2.22 - 12.60)
20	A021	10.13 (5.73 - 14.53)	42	A018	7.41 (2.22 - 12.60)
21	A038	09.99 (4.78 - 15.21)	43	A043	6.86 (2.46 - 11.26)
22	A003	09.95 (4.76 - 15.15)	44	A001	6.14 (0.95 - 11.33)
GS%	30%	20%	10%		
	26	32	47		

LL: Lower limit of the withdrawal period; UL: Upper limit of the withdrawal period.

Heritability of individual plots (h^2g) takes into account the existence of only one genotype replication, while among genotypes means (h^2mc), the presence of replications assists in reducing the environmental effect (Resende and Duarte 2007), contributing to higher h^2mc estimates. Coefficient of variation (CVe) was considered as low for the characters, except for weight of pure seeds (PS), which was 70.98%

For evaluations carried out in beds, accuracy of the morphological characterization of the inflorescence ranged from 83 to 91%. It is considered satisfactory and is classified of high magnitude (Resende and Duarte 2007). Heritability of individual plots (h^2g) for number of seeds in the inflorescences (NSI) and number of seeds per racemes (NSR) did not exceed 0.50. Broad sense heritabilities (h^2mc) were 0.77 and 0.82, respectively.

For the number of reproductive tillers (RT), number of racemes (NR), length of raceme (LR) and length of inflorescences (LI), heritability of individual plots (h^2g) and broad sense heritability (h^2mc) were greater than 0, 50. Genotypic and plot effects were significant for all characters, except for number of seeds per raceme (NSR), which showed no difference in the plot effect. Estimates of heritability between genotypes means (h^2mc) showed high magnitude, exceeding 0.50 (Table 2). From the predicted breeding values (BLUP), it is possible to obtain the ranking of hybrids for each character, and to reliably identify hybrids with superior genotypic values (Resende 2006).

For evaluations during the first year, hybrid A044 was ranked in the first position for G%, differing from the other hybrids, when the confidence interval is observed. The range of BLUP values was high, from 19, 95 for the first ranked, to 6.14, for the last ranked; thus, variation was of 70%. Gain with selection (GS%), at 30%, 20% and 10% of selection intensity, in relation to the population mean, ranged from 26 to 47%. It should be noted that the higher the gain with selection in relation to the population mean, the greater the progress in the selection of superior genotypes (Table 3).

For evaluations carried out in seeds collectors during the second harvest, both for weight of pure seeds (PS) and for weight of empty seeds (ES), control cv. Marandu presented the highest genotypic values (BLUP). Low production of pure seed of the genotypes may be inherited by the next generation, since both the narrow-sense heritability (h^2g)

Table 4. Predicted genotypic values (BLUP), (GS% at 30, 20 and 10% selection intensity) regarding weight of pure seeds (PS) and weight of empty seeds (ES) in relation to cv. Marandu (Mar) for *B. decumbens* hybrids evaluated in seed collectors in the second harvest

			PS (g m^{-2})				**ES** (g m^{-2})	
	Treatments		**BLUP** (LL-UL)		**Treatments**		**BLUP** (LL-UL)	
1	Mar		2.38 (2.04 - 2.72)		Mar		1.17 (0.93 - 1.41)	
2	R087		0.76 (0.35 - 1.16)		R181		1.15 (0.90 - 1.39)	
3	R120		0.41 (0.08 - 0.75)		R091		1.02 (0.77 - 1.260	
4	C001		0.30 (-0.03 - 0.64)		S036		1.00 (0.75 - 1.24)	
5	R126		0.30 (-0.04 - 0.64)		R120		0.94 (0.70 - 1.19)	
6	X030		0.20 (-0.13 - 0.54)		S031		0.93 (0.68 - 1.17)	
7	R124		0.09 (-0.25 - 0.43)		X030		0.92 (0.68 - 1.17)	
8	Y021		0.07 (-0.27 - 0.40)		X072		0.82 (0.58 - 1.07)	
9	R091		0.06 (-0.27 - 0.40)		X117		0.82 (0.58 - 1.07)	
10	B006		0.05 (-0.31 - 0.41)		R110		0.81 (0.56 - 1.05)	
11	R041		0.05 (-0.29 - 0.39)		B006		0.80 (0.53 - 1.06)	
12	T012		0.05 (-0.29 - 0.38)		R041		0.79 (0.55 - 1.04)	
13	X117		0.05 (-0.29 - 0.38)		R126		0.77 (0.53 - 1.02)	
14	R181		0.04 (-0.36 - 0.44)		T012		0.77 (0.52 - 1.01)	
15	S031		0.04 (-0.30 - 0.38)		T026		0.77 (0.52 - 1.01)	
16	S036		0.04 (-0.36 - 0.44)		C001		0.70 (0.46 - 0.94)	
17	R110		0.03 (-0.30 - 0.37)		R124		0.69 (0.45 - 0.93)	
18	T054		0.03 (-0.31 - 0.37)		Y021		0.58 (0.34 - 0.83)	
19	X072		0.03 (-0.31 - 0.36)		R087		0.56 (0.27 - 0.85)	
20	T026		0.02 (-0.31 - 0.36)		T054		0.50 (0.26 - 0.75)	
	GS (%)	30%	20%	10%	GS (%)	30%	20%	10%
		149	221	324		23	27	34

LL: Lower limit of the withdrawal period; UL: Upper limit of the withdrawal period.

Table 5. Predicted genotypic values (BLUP) and gain with selection (GS% in 30, 20 and 10% selection intensity) in relation to the number of reproductive tillers (RT), number of seeds per inflorescence (NSI), number of seeds per raceme (NSR) and number of racemes (NR) compared to the treatments (Treat) cv. Marandu (Mar) and *B. decumbens* hybrids evaluated in the beds in the second harvest

	RT		NSI		NSR		NR	
	Treat	**BLUP** (LL-UL)	**Treat**	**BLUP** (LL-UL)	**Treat**	**BLUP** (LL-UL)	**Treat**	**BLUP** (LL-UL)
1	C001	140.84 (124.15 - 157.54)	R120	201.50 (174.53 - 228.48)	Mar	47.61 (43.28 - 51.94)	R120	6.12 (5.75 - 6.50)
2	R091	71.41 (54.71 - 88.11)	Mar	189.63 (162.66 - 216.61)	T012	36.83 (32.50 - 41.17)	R126	5.30 (4.93 - 5.68)
3	Y021	68.99 (52.29 - 85.69)	R181	160.27 (133.30 - 187.25)	S031	35.57 (31.24 - 39.90)	R181	5.30 (4.93 - 5.68)
4	R124	49.63 (32.93 - 66.33)	R091	151.07 (124.10 - 178.05)	S036	33.50 (29.17 - 37.84)	R091	4.89 (4.52 - 5.27)
5	Mar	36.57 (19.87 - 53.27)	T012	150.36 (123.39 - 177.34)	C001	32.75 (28.41 - 37.08)	T026	4.89 (4.52 - 5.27)
6	B006	36.08 (19.39 - 52.78)	S036	148.95 (121.97 - 175.92)	R120	31.78 (27.45 - 36.11)	T054	4.89 (4.52 - 5.27)
7	S031	33.91 (17.21 - 50.60)	R126	147.69 (120.72 - 174.67)	R091	30.70 (26.36 - 35.03)	R110	4.49 (4.11 - 4.86)
8	S036	30.52 (13.82 - 47.22)	T026	140.34 (113.37 - 167.32)	R041	30.21 (25.88 - 34.54)	R041	4.49 (4.11 - 4.86)
9	R041	29.79 (13.10 - 46.49)	R110	134.30 (107.32 - 161.27)	R181	29.93 (25.60 - 34.26)	S036	4.49 (4.11 - 4.86)
10	T012	29.31 (12.61 - 46.01)	R041	134.08 (107.10 - 161.05)	R110	29.78 (25.45 - 34.12)	T012	4.49 (4.11 - 4.86)
11	T054	28.58 (11.89 - 45.28)	S031	125.96 (98.99 - 152.94)	X072	29.69 (25.36 - 34.03)	R124	4.08 (3.70 - 4.45)
12	X030	25.68 (8.98 - 42.38)	X072	120.46 (93.49 - 147.44)	B006	28.67 (24.34 - 33.00)	Mar	4.08 (3.70 - 4.45)
13	X072	25.68 (8.98 - 42.38)	T054	119.48 (92.51 - 146.46)	T026	28.47 (24.14 - 32.81)	X072	4.08 (3.70 - 4.45)
14	R126	23.99 (7.29 - 40.68)	C001	118.56 (91.58 - 145.53)	R126	27.77 (23.43 - 32.10)	B006	3.67 (3.29 - 4.04)
15	T026	23.50 (6.81 - 40.20)	B006	104.99 (78.02 - 131.97)	X030	27.15 (22.82 - 31.48)	C001	3.67 (3.29 - 4.04)
16	R120	22.78 (6.08 - 39.48)	X030	101.35 (74.37 - 128.32)	Y021	25.11 (20.78 - 29.44)	S031	3.67 (3.29 - 4.04)
17	R110	21.81 (5.11 - 38.51)	R124	100.09 (73.12 - 127.07)	R124	24.24 (19.91 - 28.58)	X030	3.67 (3.29 - 4.04)
18	R087	17.71 (-1.64 - 37.06)	Y021	081.41 (54.44 - 108.39)	T054	24.15 (19.82 - 28.48)	Y021	3.26 (2.88 - 3.63)
19	R181	16.73 (0.03 - 16.00)						
	GS (%)	30% 20% 10%	GS (%)	30% 20% 10%	GS (%)	30% 20% 10%	GS (%)	30% 20% 10%
		73 114 174		23 26 37		14 16 22		19 22 29

LL: Lower limit of the withdrawal period; UL: Upper limit of the withdrawal period.

and the broad-sense heritability (h^2mc) were of high magnitude (Table 2). For PS, gains with selection (GS%) at 30, 20 and 10% of selection intensity were respectively 149, 221 and 324%; and for ES, it was 23, 27 and 34%, compared to population means (Table 4).

Weight of pure seeds (PS) of the hybrids was, in general, of low magnitude than that of control (Table 3). Three factors may have contributed to this: the high abortion rate in function of hybridization, as suggested by Lutts et al. (1991); the interference by spittlebugs infestation; and the possibility of being a characteristic of the genotype, producing only vegetative tillers, and only under extreme stress conditions it would produce reproductive tillers.

Non-viability of hybrid seeds may be caused by environmental interference, affecting the viability of pollen. Also, it could happen for the hybrid's genotype present gametophytic incompatibility allele, and thus blocking endosperm formation, and consequently seed filling.

Mateus et al. (2015) evaluated *B. decumbens* intraspecific hybrids, including R033, R126, R181, S036, T026 and X117, and also found high nymphal survival of spittlebugs, exceeding 60%. It is noteworthy that all hybrids tested in this study were considered susceptible or mid-resistant in previous works, with spittlebug nymphal survival ranging from 44 to 90%. For values above 40% nymphal survival, genotypes are not subjected to screening for genetic resistance.

For RT, cv. *B. brizantha* was ranked in the fifth position, differing from genotypes C001, R091 and Y021, taking into account the confidence interval. Gains with selection (GS%) at 30, 20 and 10% intensity were, respectively, 73, 114 and 174% (Table 5). The high potential for production of reproductive tillers has heritability of high magnitude of 0.88 and 0.97 in individual plots (h^2g) and in the broad-sense heritability (h^2mc), respectively. Comparing the BLUP of the morphological characters, cv. Marandu was ranked in the second, first and twelfth positions in number of seeds per inflorescence (NSI), number of seeds per raceme (NSR), and number of racemes (NN), respectively. However, the cultivar did not present better performance than R120 genotype for NSI. R120. R120, R126, R181, R091, T026, T054, R110, R041, S036, T012, and R124 genotypes showed higher BLUP in relation to the control, presenting gain in relation to the

Table 6 Predicted genotypic values (BLUP) and gain with (GS% at 30, 20 and 10% selection intensity) and length of racemes (LR) and length of inflorescence (CLI) compared to cv. Marandu (Mar) *B. decumbens* for hybrids (Hib) evaluated in the beds in the second harvest

	LR				LI			
	Treatments	**BLUP** (LL-UL)			**Treatments**	**BLUP** (LL-UL)		
1	Mar	14.87 (13.38 - 16.36)			Mar	17.63 (15.24 - 20.02)		
2	R181	6.88 (5.39 - 8.37)			R120	14.63 (12.24 - 17.02)		
3	X030	6.57 (5.08 - 8.06)			T026	13.93 (11.55 - 16.32)		
4	S036	6.50 (5.01 - 7.99)			T054	13.77 (11.39 - 16.16)		
5	S031	6.18 (4.69 - 7.67)			R126	13.03 (10.65 - 15.42)		
6	C001	5.98 (4.49 - 7.47)			S036	11.77 (9.38 - 14.15)		
7	T026	5.89 (4.40 - 7.38)			X072	11.58 (9.19 - 13.96)		
8	T012	5.83 (4.34 - 7.32)			R041	11.57 (9.18 - 13.95)		
9	X072	5.83 (4.34 - 7.32)			R110	11.33 (8.95 - 13.72)		
10	R120	5.79 (4.30 - 7.28)			S031	11.14 (8.76 - 13.53)		
11	R041	5.67 (4.18 - 7.16)			R091	10.99 (8.60 - 13.38)		
12	T054	5.55 (4.06 - 7.04)			R124	10.42 (8.04 - 12.81)		
13	R110	5.48 (3.99 - 6.97)			T012	10.25 (7.86 - 12.64)		
14	B006	5.45 (3.96 - 6.94)			C001	10.18 (7.80 - 12.57)		
15	R124	5.45 (3.96 - 6.94)			B006	9.52 (7.13 - 11.90)		
16	R126	5.19 (3.70 - 6.68)			X030	8.09 (5.70 - 10.47)		
17	Y021	4.66 (3.17 - 6.15)			R181	7.88 (5.50 - 10.27)		
18	R091	4.42 (2.93 - 5.91)			Y021	6.38 (3.99 - 8.76)		
	GS (%)	30%	20%	10%	GS (%)	30%	20%	10%
		12	14	17		22	26	30

LL: Lower limit of the withdrawal period; UL: Upper limit of the withdrawal period.

number of racemes (NR). R120 genotype stood out both for number of seeds per inflorescence (NSI) and number of racemes (NR), taking the first position in the two variables of seed production. This fact may be a promising indicative for its production potential. Although the genotypes present genotypic values of high magnitude when compared with the control in NSI and NR, they had heritability in individual plots (h^2g) inferior to 0.50.

Cultivar Marandu, the control, was ranked in the first position for both length of racemes (LR) and length of inflorescences (LI). However, it appears that a character does not influence the other, since R120 genotype takes the second position in length of inflorescences, and its racemes have the tenth shorter length (Table 6).

CONCLUSIONS

The number of reproductive tillers presents high heritability both in the broad sense (h^2mc) and in individual plots (h^2g), and can be a parameter to predict seed production potential. However, it is necessary further studies on which components really influence seed production of forage plants.

The genotypes that had better performance for production of reproductive tillers (RT) and weight of pure seeds (PS) were C001 and R091.

ACKNOWLEDGEMENTS

The authors thank the Instituto Federal Goiano - Campus Rio Verde, Embrapa Beef Cattle, Fapeg, CNPq, Fundect, Unipasto, Fundapam, and the members of the Technology and Tropical Forage Seed Production Team of Embrapa Beef Cattle, especially Mr. Luiz de Jesus, Hugo Corado and Vagner Martins.

REFERENCES

Araújo SAC, Deminics BB and Campos PRSS (2008) Melhoramento genético de plantas forrageiras tropicais no Brasil. **Archivos de Zootecnia 57**: 61-76.

Brasil (2009) **Ministério da Agricultura, Pecuária e Abastecimento: Regras para análise de sementes**. Ministério da Agricultura, Pecuária e Abastecimento. Secretaria de Defesa Agropecuária. MAPA/ACS, Brasília, 399p.

Karia CT, Duarte JB and Araújo ACG (2006) **Desenvolvimento de cultivares do gênero *Brachiaria* (trin.) Griseb. no Brasil**. Embrapa Cerrados, Planaltina, 163p.

Lutts S, Ndikumana J and Louant BP (1991) Fertility of *Brachiaria ruziziensis* in interspecific crosses with *Brachiaria decumbens* and *Brachiaria brizantha*: meiotic behavior, pollen viability and seed set. **Euphytica 57**: 267-274.

Mateus RG, Barrios SCL, Valle CB, Valério JR, Torres FZV, Martins LB and Amaral PN (2015) Genetic parameters and selection of Brachiaria decumbens hybrids for agronomic traits and resistance to spittlebugs. **Crop Breeding and Applied Biotechnology 15**: 227-234.

Monteiro LC, Verzignassi JR, Barrios SCL, Valle CB, Fernandes CD, Benteo GL and Libório CB (2016) *Brachiaria decumbens* intraspecific hybrids: characterization and selection for seed production. **Journal of Seed Science 38**: 62-67.

Resende MDV (2006) **O software Selegen-Reml/Blup**. Embrapa Gado de Corte, Campo Grande, 299p.

Resende MDV and Duarte JB (2007) Precisão e controle de qualidade em experimentos de avaliação de cultivares. **Pesquisa Agropecuária Tropical 27**: 182-194.

Valle CB, Jank L and Resende RMS (2009) O melhoramento de forrageiras tropicais no Brasil. **Revista Ceres 56**: 460-472.

27

IPR Aimoré - Triticale cultivar of early maturity and wide adaptation

Carlos Roberto Riede[1], Luiz Alberto Cogrossi Campos[1], Klever Márcio Antunes Arruda[1*], Deoclécio Domingos Garbuglio[1] and Avahy Carlos da Silva[2]

Abstract: *IPR Aimoré is a new triticale cultivar that combines earliness and high yield. In Value for Cultivation and Use trials carried out in 33 environments, involving the main triticale producing states, IPR Aimoré surpassed 4% the controls means, with mean yield of 4.955 kg ha^{-1}.*

Key words: *x Triticosecale Wittmack, earliness, grain yield.*

***Corresponding author:**
E-mail: klever@iapar.br

[1] 1 Instituto Agronômico do Paraná (IAPAR), Rod. Celso Garcia Cid, km 375, CP 10.030, 86.047-902, Londrina, PR, Brazil
[2] IAPAR, Av. Euzébio de Queirós, s/no, Uvaranas, CP 129, 84.001-970, Ponta Grossa, PR, Brazil

INTRODUCTION

Characteristics such as high potential of grain and biomass yield, resistance to diseases, such as leaf rust and powdery mildew, adaptation to acid and low fertile soils, and tolerance to drought and cold (Mergoum et al. 2009, Arseniuk 2015), make triticale (x *Triticosecale* Wittmack) a great alternative for succession and rotation in agricultural production systems, mainly for no-tillage system on straw, since it provides good vegetation cover for the successor culture, even under stress conditions.

Triticale grains can be used for several purposes. In human feeding, its flour has been used to compose mixtures with wheat flour for the manufacture of low fermentation products, such as cookies, pizza dough, meatballs, waffles, and cakes. However, although this cereal presents potential use in human feeding, it is mainly used in Brazil and in the world as animal feeding, in full or partial substitution of other cereals in feed composition (Nascimento Junior et al. 2004). Recently, triticale has also aroused interest for biofuel production, and it has presented competitiveness when compared with other winter cereals (McGoverin et al. 2011).

Triticale production estimates in Brazil, for the year of 2014, was of 95,800 tons, with mean grain yield of about 2,450 kg ha^{-1} (CONAB 2015). The national production is mainly concentrated in the states of São Paulo, Paraná and Rio Grande do Sul. In these states, winter crops have been used to diversify agricultural activity, to distribute the cultivation costs throughout the year, and even to increase the yield potential of summer crops.

However, farmers require cultivars that are suitable to the production system, in order to not only allow the installation of summer crops at the most appropriate time, but also to allow the adoption of procedures that reduce risks associated with winter crops. In this context, the availability of cultivars with different cycles gives the farmer an increase in the seeding period, and consequently in the harvest period, reducing risks of losses by frost and rain.

BREEDING METHOD

The triticale cultivar IPR Aimoré was obtained by the Agronomic Institute of Paraná (IAPAR), in partnership with the International Center for Maize and Wheat Research (CIMMYT) based in Mexico. The material originated from the cross 804/BAT/3/MUSX/LYNX//STIER_12-3/4/VARSA_3-1/5/FAHAD_8-1*2//HARE_263/CIVET, whose genealogy or obtainment process is coded as CTSS98Y00236S-0M-1Y-0M-0Y-8B-1Y-0B. This code means that the cross n. 236 originated the material, in chronological order of triticale crosses at CIMMYT, carried out in the year of 1998. The letters refer to Mexican locations where crosses and selections were carried out, i.e., Yaqui Valley (Y), located to the South of the state of Sonora, and Toluca (M) and El Batan (B), located in the state of Mexico. The modified pedigree method was used, in which from the F_3 generation onwards, individual plants were selected, whose progenies were alternately evaluated in the form of bulk (0M, 0Y, 0B), or in the form of individual plants (1Y, 8B), re-selected, or discarded in successive inbreeding generations until the F_8 generation. Afterwards, the advanced breeding line was mass selected and included in a collection named *38th International Triticale Yield Nursery* (38 ITYN), received by IAPAR in 2006. The accession CTSS98Y00236S-0M-1Y-0M-0Y-8B-1Y-0B was initially evaluated by IAPAR in Ponta Grossa - Paraná, in a collection, in the year of 2007, and in Grain Yield Preliminary Trials, from 2008 to 2009, together with the other accessions previously selected, and which had already been internally coded as TPOLO 0608. Since the line TPOLO 0608 had stood out in previous reviews, it was promoted to Regional VCU trials (Triticale Meridional Trial), carried out by the network partnership between IAPAR, Fundação Meridional and Embrapa Soybean, in different homogeneous regions of cultivars adaptation (Brasil 2008).

The main agronomic characteristics of TPOLO 0608 line were presented in internal meeting of IAPAR, held in 2013, in order to decide regarding its possible release as a new cultivar. Given the presented yield potential and early maturity, which had already been a demand for cereal producers, it was decided for the release as a new Triticale cultivar, IPR Aimoré, and for the registration at the Ministry of Agriculture, Livestock and Supply (MAPA).

CULTIVAR CHARACTERISTICS

The characteristics of the cultivar IPR Aimoré were evaluated in 33 Value for Cultivation and Use (VCU) trials in homogeneous regions of adaptation 1 (cold, wet and high altitude), 2 (moderately warm, wet and low altitude), and 3 (hot, moderately dry, and low altitude) in Paraná (PR), São Paulo (SP), Santa Catarina (SC), and Mato Grosso do Sul (MS), from 2010 to 2012. Table 1 shows the Brazilian municipalities where the VCU trials were carried out.

The mean grain yield superiority of cultivar IPR Aimoré was 3, 6 and 3% in relation to the two best controls, cultivars BRS 203 and IPR 111, in regions 1, 2 and 3, respectively (Table 2). Considering all the environments, the mean overall

Table 1. Locations where the Value for Cultivation and Use Trials of cultivar IPR Aimoré (line TPOLO 0608) were carried out

State	Locations	Geographic coordinates	Altitude (m)	Homogeneous region
PR	Cambará	23° 00′ S, 50° 01′ W	460	3
	Cascavel	24° 57′ S, 53° 28′ W	613	2
	Cruzmaltina	24° 00′ S, 51° 24′ W	752	3
	Guarapuava	25° 23′ S, 51° 30′ W	1098	1
	Irati	25° 28′ S, 50° 38′ W	850	1
	Londrina	23° 22′ S, 51° 10′ W	543	3
	Londrina (Warta)	23° 11′ S, 51° 10′ W	567	3
	Mauá da Serra	23° 53′ S, 51° 15′ W	948	2
	Pato Branco	26° 08′ S, 52° 39′ W	736	2
	Ponta Grossa	25° 05′ S, 50° 01′ W	969	1
	Tibagi	24° 31′ S, 50° 32′ W	832	2
SP	Itaberá	23° 49′ S, 49° 03′ W	675	2
SC	Abelardo Luz	26° 34′ S, 52° 21′ W	850	2
	Campos Novos	27° 24′ S, 51° 12′ W	962	1
MS	Antônio João	22° 08′ S, 55° 41′ W	600	3
	Maracaju	21° 38′ S, 55° 08′ W	390	3

Table 2. Mean grain yield (kg ha^{-1}) of cultivar IPR Aimoré, evaluated in three regions of VCU trials in the states of PR, SP, SC and MS, in the years of 2010 to 2012

Region	Year	N. of Trials	IPR Aimoré	IPR 111	BRS 203	CM[1]	CM[2]
1	2010	2	4.680	2.379	5.436	3.907	120
	2011	4	5.841	6.168	5.781	5.975	98
	2012	3	4.323	4.607	3.938	4.273	101
	Mean		5.077	4.806	5.090	4.948	103
2	2010	4	6.946	5.775	6.688	6.231	111
	2011	3	5.363	6.105	4.677	5.391	99
	2012	5	5.160	5.602	4.265	4.933	105
	Mean		5.806	5.785	5.175	5.480	106
3	2010	4	4.608	3.495	3.361	3.428	134
	2011	5	4.094	4.255	3.621	3.938	104
	2012	3	2.957	4.634	3.969	4.302	69
	Mean		3.981	4.096	3.621	3.859	103

[1] Controls means (IPR 111 and BRS 203)
[2] Performance in relation to the controls means, in percentage

gain was 4%, and the higher yields were obtained in the regions of VCU 1 and 2. Table 2 presents the mean grain yield per homogeneous region of adaptation and cultivation year.

In the mean of both experiments, IPR Aimoré proved to present very early maturity, since heading occurs about 53 days after emergence, and maturation occurs at 112 days after emergence. When compared with the other seven triticale cultivars registered at MAPA, which were recommended for cultivation in the year of 2014 (Reunião 2014), IPR Aimoré presented higher earliness in all environments evaluated, especially the homogeneous region of adaptation 3, where this cultivar had an mean cycle of only 102 days. For comparison, BRS Harmonia, which has the second shortest cycle among the evaluated cultivars, had 11-days longer cycle in this region (Table 3).

The ears of the cultivar IPR Aimoré are awned, of clear color and fusiform, with long, light red colored soft grains. In relation to other characteristics of interest, the cultivar presented medium height (99 cm), moderate resistance to lodging and natural threshing. It also presented moderate tolerance to aluminum toxicity, and susceptibility to preharvest sprouting. In relation to physical parameters of the grains, IPR Aimoré presented Hectoliter weight variation between 61 and 81, with a mean of 75 kg hL^{-1}. Regarding the weight of 1000 grains, there was variation between 36 and 55, with mean of 46 grams, which is a very good value for triticale. Table 3 shows the mean data per homogeneous region of adaptation, of some characteristics of interest of cultivar IPR Aimoré, as well as of other triticale cultivars.

In relation to the major diseases that occur in the region, IPR Aimoré is moderate resistant to yellow spot (*Drechslera tritici-repentis*) and moderately susceptible to leaf rust (*Puccinia triticina*), brown spot (*Bipolaris sorokiniana*), and Septoria (*Septoria nodorum*). Similarly to all other triticale cultivars, IPR Aimoré proved to be susceptible to head blast (*Magnaporthe grisea)* and scab (*Gibberella zeae*).

The cultivation of IPR Aimoré is recommended for the regions of VCU 1 (Paraná and Santa Catarina), 2 (Paraná, Santa Catarina and São Paulo) and 3 (Paraná and Mato Grosso do Sul). In function of its earliness, sowing should be carried out from the middle of the time indicated to the region, consequently reducing the risk of loss by frost in the critical stage of the culture, in cold regions, and in order to escape from head blast, in warmer regions.

With the increasing demand for food resources, either for human or animal feeding, and for renewable energy sources, it has been intensified the search for species which are able to diversify the production system and to increase profitability and sustainability of agriculture. Most triticale cultivars available in Brazil, e.g., BRS Harmonia (Nascimento Junior et al. 2015) and IPR 111 (Silva et al. 2006), are of intermediate cycle. Additionally, the number of triticale cultivars developed by breeding programs is still limited, and consequently the number of varieties available to farmers is also limited. In this context, the new triticale cultivar IPR Aimoré may represent an excellent contribution to the Cereal Supply Chain.

Table 3. Mean of different agronomic characteristics of cultivar IPR Aimoré and of other triticale cultivars, per homogeneous region of adaptation

Region	Cultivar	PH (cm)	DEH (days)	DEM (days)	HW (kg hL^{-1})	WTG (g)
1	IPR Aimoré	96	64	117	74	42
1	IPR 111	99	68	122	71	44
1	BRS 203	97	68	124	71	34
1	BRS Harmonia	93	65	119	69	44
1	BRS Minotauro	104	70	124	72	36
1	BRS Saturno	101	69	123	74	39
1	BRS Ulisses	90	67	121	72	36
1	Embrapa 53	98	69	123	69	37
2	IPR Aimoré	103	55	117	72	50
2	IPR 111	108	66	123	71	48
2	BRS 203	106	67	125	72	38
2	BRS Harmonia	101	65	123	72	49
2	BRS Minotauro	110	70	125	71	38
2	BRS Saturno	112	70	126	72	44
2	BRS Ulisses	98	67	121	72	43
2	Embrapa 53	111	71	124	72	41
3	IPR Aimoré	96	42	102	78	43
3	IPR 111	101	57	114	77	47
3	BRS 203	103	62	123	75	30
3	BRS Harmonia	89	56	113	78	44
3	BRS Minotauro	107	63	120	76	33
3	BRS Saturno	107	71	123	76	37
3	BRS Ulisses	94	56	114	78	37
3	Embrapa 53	106	59	116	76	35

PH: Plant Height; DEH: Days from Emergency to Heading; DEM: Days from Emergency to Maturation; HW: Hectoliter Weight; WTG: Weight of 1000 grains.

BASIC SEEDS PRODUCTION

The triticale cultivar IPR Aimoré is registered at the National Register of Plant Varieties of the Ministry of Agriculture, Livestock and Supply (RNC/MAPA), under the registration number 31412. IAPAR is responsible for the production and marketing of basic seeds. IPR Aimoré is the first triticale cultivar released by the partnership between IAPAR and Fundação Meridional, which supports the evaluation network of VCU trials and the logistics for the distribution and dissemination of new cultivars.

REFERENCES

Arseniuk E (2015) Triticale abiotic stresses–An overview. In Eudes F (ed) **Triticale**. Springer, New York, p. 69-81.

Brasil - Ministério da Agricultura, Pecuária e Abastecimento (2008) Instrução normativa nº 3, de 14 de outubro de 2008. **Diário Oficial [da] República Federativa do Brasil**, 15 outubro 2008. Seção 1, p. 31.

CONAB - Companhia Nacional de Abastecimento (2015) **Acompanhamento da safra brasileira de grãos 2014/2015 - quinto levantamento**, fevereiro 2015. Available at < http://www.conab.br >. Accessed on February 21, 2015.

McGoverin CM, Snyders F, Muller N, Botes W, Fox G and Manley M (2011) A review of triticale uses and the effect of growth environment on grain quality. **Journal of the Science of Food and Agriculture 91**: 1155-1165.

Mergoum M, Singh PK, Peña RJ, Lozano-del Río AJ, Cooper KV, Salmon DF and Gómez Macpherson H (2009) Triticale: a "new" crop with old challenges. In Carena MJ (ed) **Cereals**. Springer, New York, p. 267-287.

Nascimento Junior A, Baier AC, Teixeira MCC and Wiethölter S (2004) Triticale in Brazil. In Mergoum M and MacPherson HG (Org) **Triticale improvement and production**. 1st edn, FAO, Roma, p. 93-98.

Nascimento Junior A, Bassoi MC, Silva MS, Caierão E and Miranda MZ (2015) BRS Harmonia - triticale cultivar. **Crop Breeding and Applied Biotechnology 15**: 40-42.

Reunião da Comissão Brasileira de Pesquisa de Trigo e Triticale (2014) **Informações técnicas para trigo e triticale - safra 2014**. Fundação Meridional, Londrina, 235p.

Silva AC, Riede CR, Campos LAC, Pola JN and Shioga PS (2006) "IPR 111" - Triticale cultivar. **Crop Breeding and Applied Biotechnology 6**: 250-252.

28

RB975242 and RB975201 - Late maturation sugarcane varieties

Monalisa Sampaio Carneiro[1,2*], Roberto Giacomini Chapola[2], Antonio Ribeiro Fernandes Junior[3], Danilo Eduardo Cursi[2], Fernanda Zatti Barreto[1,2], Thiago Willian Almeida Balsalobre[2] and Hermann Paulo Hoffmann[2]

Abstract: *The sugarcane varieties RB975201 and RB975242 were developed and released for harvest at the end of the season (late maturation) in the Central-South region of Brazil. In specific environments, these varieties were compared with commercial standards in sugar yield per area. They are resistant to major sugarcane diseases and present the Bru1 gene of resistance to brown rust.*

Key words: *Saccharum spp., high production, disease resistance.*

***Corresponding author:**
E-mail: monalisa@cca.ufscar.br

[1] Universidade Federal de São Carlos (UFSCar), 13.600-970, Araras, SP, Brazil
[2] UFSCar, Departamento de Biotecnologia e Produção Vegetal e Animal
[3] UFSCar, Estação Experimental de Valparaíso, 16.880-000, Valparaíso, SP, Brazil

INTRODUCTION

Sugarcane (*Saccharum* spp.) originated from crosses between *Saccharum officinarum*, *S. barberi*, *S. sinense*, *S. robustum*, *S. edule* and wild species of *S. spontaneum*, forming the *Saccharum* complex (Sreenivasan et al. 1987). The cultivated sugarcane is predominantly allogamous, highly heterozygous, and maintained by vegetative propagation.

Currently, Brazil has three sugarcane breeding programs: Agronomic Institute of Campinas (variety abbreviation - IAC), Sugarcane Technology Center (variety abbreviation - CTC), and Inter-University Network for the Development of Sugarcane Industry, RIDESA (variety abbreviation - RB, that is Brazilian Republic).

RIDESA is a cooperation between IFES (Federal Institutions of Higher Education) to develop improved sugarcane varieties by taking advantage of researchers training and regional bases of the former IAA/PLANALSUCAR. RIDESA's creation was an important step in order to promote nationwide coordinated actions of technological support to one of the most important segments of the Brazilian economy. RIDESA consists of 10 federal universities (UFSCar, UFRPE, UFAL UFRRJ, UFV, UFG, UFPR, UFS, UFPI and UFMT) which share the experimental station of Serra do Ouro, based in Murici/AL, and experimental units installed in major producing regions, forming a national network of trials aimed at sugarcane breeding. RIDESA has produced, since 1990, 75 varieties, which summed to the varieties released by PLANALSUCAR constitute 94 RB varieties, produced in 45 years of research on sugarcane. Currently, RB varieties account for 68% of the total area of sugarcane cultivation in Brazil (Barbosa et al. 2015, Carneiro et al. 2015, Iaia et al. 2015;,Oliveira et al. 2015).

The Federal University of São Carlos (UFSCar) develops breeding studies in the states of São Paulo and Mato Grosso do Sul, both located in Central-South Brazil. This region has the largest cultivated area (58.9%) and the highest sugarcane production in Brazil. RB975201 and RB975242 varieties were developed and studied for 18 years, and were released in 2015 by the breeding program of UFSCar.

BREEDING PROGRAM

Figure 1 shows the pedigrees of RB975242 and RB975201 varieties. In 1997, crosses were carried out in the Flowering and Crossing Station Serra do Ouro, located in the municipality of Murici, Alagoas (lat 09° 18'S, long 35° 56' W, alt 450m asl). The seeds obtained were germinated and then seedlings were planted in the field, establishing the first selection stage (T1). At this stage, the genotypes in a single clump were selected based on mass selection method in ratoon cane cycle (Breaux et al. 1963). The general morphological criteria were agroindustrial traits such as Brix and number of stalks (Hogarth 1987, Berding et al. 2004), flowering, pith and reaction to major diseases (Matsuoka et al. 1999, Morais et al. 2015).

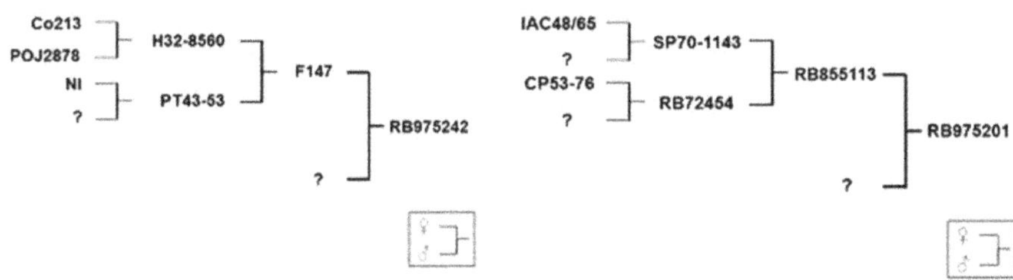

Figure 1. Pedigree of sugarcane varieties RB 975242 and RB975201.

Based on these criteria, clones were selected and, together with standard commercial varieties, they constituted the second selection stage (T2). At this stage, clones were established in Araras (lat 22° 18' S, long 47° 23' W, alt 690m asl) and Valparaiso (lat 21° 13' S, long 50° 52' W, alt 390m asl), in São Paulo. The experiment consisted of augmented blocks design (Federer 1956), and plots consisted of two 2.5 m furrows with one replication. Clones were evaluated in plant cane, and ratoon, using the same criteria of T1stage, adding the variable weight per plot (WP) and kilos of brix per plot (KBP) (Kang et al. 1983).

Clones selected during the T2 stage advanced to the third selection stage (T3), and were also established in augmented blocks (Federer 1956). Plots consisted of two 5m furrows, spaced 1.40m apart, with two replications. Genotypes at T3 were tested in 10 locations with different climate and soil conditions. Selection was carried out based on the mean performance of both plant and ratoons in several environments. The selection criteria were similar to those used in stage T2, sucrose content (PC, in %) in sugarcane and kilos of pol per plot (KPP) were considered additional variables (Matsuoka et al. 1999).

After the T3 stage, the clones selected followed for experimentation stage (ES), in which they were evaluated in 10 trials distributed in regions with different soil and climatic conditions, considering the data of four cycles (plant cane and three ratoons). The experiments were established in a randomized block design with four replications, with standard commercial varieties of intermediate/late maturation distributed in the blocks as controls. The evaluated variables were ton of stalks per hectare (TSH), sucrose content (PC in %) in sugarcane, ton of pol per hectare (TPH), and % of fiber content. The coefficient of environmental variation, the effects of the genotype x environment interaction, and the adaptability and stability of clones (Eberhart and Russell 1966) were estimated from individual (each site) and joint (all sites) analyses of variance (Steel and Torrie 1960). The promising clones of ES were subjected to maturation curve, in order to identify the best harvest time, according to the amount of sugar per ton of stalks (kg t^{-1}). The genotypes with better performance were multiplied and evaluated in units which are partner of the breeding program of UFSCar, aiming to verify the behavior of genotypes in production conditions (Barbosa et al. 2001, Barbosa et al. 2004).

PERFORMANCE

RB975201

RB975201 has fast development and upright to semidecumbent growth habit. Its leaves (trash) are easily removed; its stalk diameter is medium, with yellowish green color under straw, and purple color when exposed to the sun. The sheaths are green with little wax. It has high tillering in plant and ratoons, with good canopy cover, great ratooning ability in mechanical harvesting, and sugarcane longevity (additional crops from one planting).

It presents high agricultural yield, with great production stability, medium fiber content, medium PIU, and late maturation. RB975201, under the conditions of the Central-South region, presents a very difficult flowering and little pith. These characteristics, in combination with its fast growth, enable the recommendation of RB975201 for harvest at the end of the season in the Central-South region, between the months of August and November (Figure 2).

RB975201 variety is responsive to improvement of soil and climatic conditions; in the experiments, it presented agricultural yield (TCH) higher than RB867515 in intermediate to favorable environments; in restrictive environments, yield of RB975201 was similar to that of RB867515 (Figure 3). This response of RB975201 to different production environments has been validated in pre-commercial areas; therefore, in production conditions of the Central-South region, RB975201's cultivation is recommended preferably in intermediate and favorable environments, according to the classification of Prado (2008).

With agricultural yield above 125t ha^{-1}, and sucrose content in stalks (PC in %) of about 15.0%, RB975201 variety

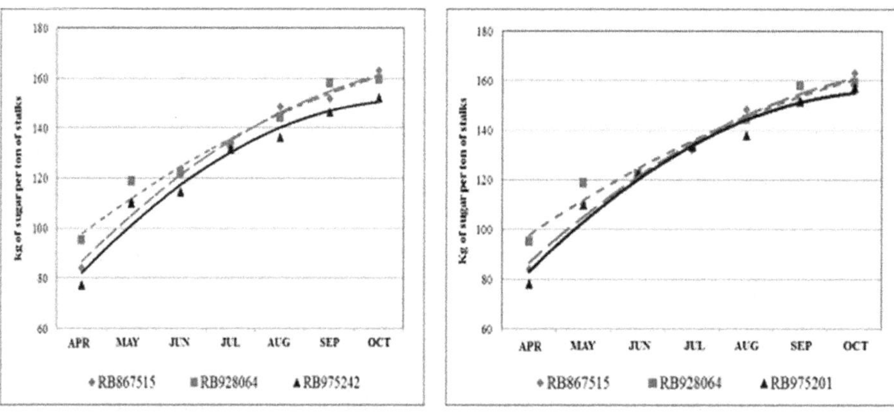

Figure 2. Maturation curves for RB975242 and RB975201 sugarcane varieties when compared with standard commercial varieties RB867515 and RB928064.

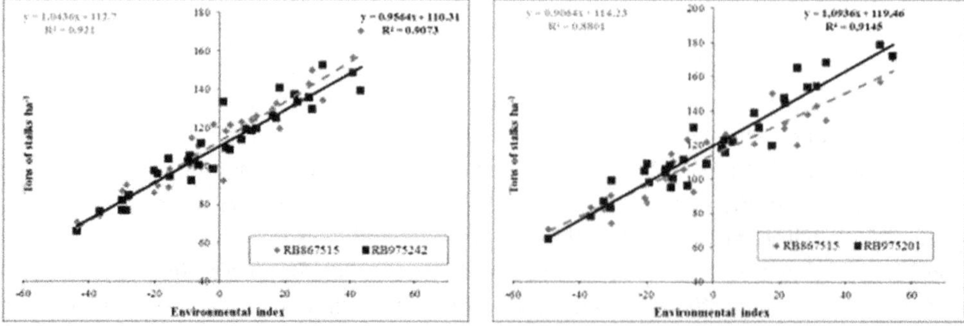

Figure 3. Adaptability and stability of RB975242 and RB975201 varieties when compared with the standard commercial variety RB867515. The mean data of tons of cane per hectare (TCH) were adjusted based on the regression method (Eberhart and Russell 1966). The points refer to the data of the 2nd, 3rd and 4th cuttings in a set of 10 experiments.

is placed in yield components agro-industrial productivity line, in tons of pol per hectare (TPH), superior to standard commercial varieties of intermediate/late maturation, considering the mean data of 10 experiments carried out by three to four cuttings (Figure 4).

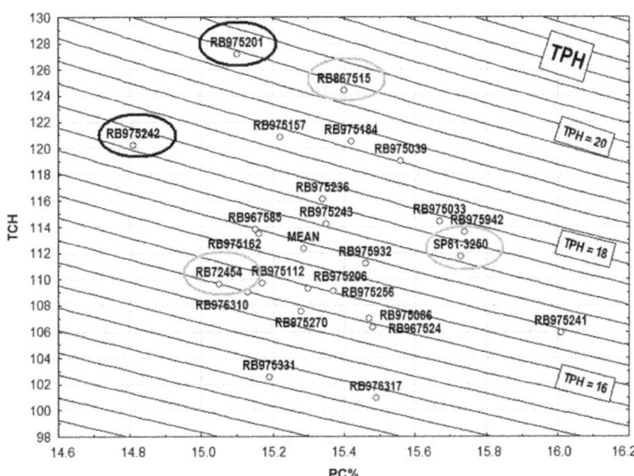

Figure 4. Isoquants of mean of tons of pol per hectare (TPH) in function of sucrose content (PC in %). in sugarcane (PC) and tons of cane per hectare (TCH) in different production environments. In the black circles, RB975242 and RB975201 varieties are compared with standard commercial varieties (gray circles) and clones.

RB975242

RB975242 has intermediate growth speed and upright growth habit. Its leaves (trash) are easily removed; its stalk diameter is medium, with green purplish color under straw, and purple color when exposed to the sun. The sheaths are light green and waxy. It has high tillering in plant and ratoons, with excellent closing between lines, great ratooning ability in mechanical harvesting, and sugarcane longevity.

It presents high agricultural yield, high production stability, medium fiber content, medium PIU, and late maturation. RB975242 does not flower nor does it pith, which are characteristics that allow its recommendation for harvest at the end of the season in the Central-South region, between the months of August and November (Figure 2).

RB975242 variety is rustic; in the experiments, it presented agricultural yield (TCH) similar to that of RB867515 in intermediate to restrictive environments and, in favorable environments, RB975242 yield was slightly lower than that of RB867515 (Figure 3). This behavior of RB975242 in different production environments has been validated in pre-commercial areas; therefore, in production conditions of the Central-South region, RB975242's cultivation in recommended preferably in the intermediate and restrictive environments, according to the classification of Prado (2008).

With agricultural yield above 120 t/ha and sucrose content in stalks (PC, in %) between 14.5% and 15.0%, RB975242 is placed in an agro-industrial productivity line, in tons of pol per hectare (TPH), superior to the standard commercial varieties of intermediate/late maturation, considering the mean data of 10 experiments carried out with three to four cuttings (Figure 4).

OTHER CHARACTERISTICS

Reaction to diseases

RB975201 and RB975242 varieties were subjected, together with other genotypes, to testing for artificial inoculation and natural infection of diseases. These tests were carried out in order to verify the reaction of clones and varieties against the major diseases of sugarcane in the Central-South region of Brazil.

Natural tests were carried out in areas favorable to infections by several diseases, both by weather conditions and by high inoculum pressure. The main diseases evaluated under natural conditions of infection are: Orange Rust (*Puccinia kuehnii*), Brown Rust (*Puccinia melanocephala*), Smut (*Sporisorium scitamineum*), Mosaic (Sugar cane mosaic virus) and Leaf Scald (*Xanthomonas albilineans*). RB975201 and RB975242 varieties, as well as others, were evaluated based on the number of infected clumps (% incidence) for Smut, Mosaic and Leaf Scald, and based on the percentage of leaf area with symptoms (% severity) of Orange and Brown Rusts (Amorim et al. 1987).

Table 1. Reaction to diseases and presence (+) or absence (-) of *Bru*1 gene in RB975242 and RB975201 sugarcane varieties in Central-South Brazil

Disease	Cultivar RB975242	Cultivar RB975201
Smut	R	R
Brown rust	R	R
*Bru*1	+	+
Orange rust	R	R
Mosaic	R	R
Leaf Scald	R	R

R = resistant
+ = Presence of *Bru*1 molecular markers (haplotype 1: presence of R12H16 and 9O20-F4- *RsaI* markers)

Artificial tests were carried out in a greenhouse with inoculation of the fungus spores of the Smut disease, and suspension contaminated with mosaic virus, according to methods described by Matsuoka (1979). RB975201 and RB975242 varieties were evaluated based on the grading scale established for each disease, which considers the amount of infected plants (% incidence) and classifies the genotypes into resistant, intermediate and susceptible. The results obtained in the tests of natural infection and of artificial inoculations indicated that RB975201 and RB975242 have adequate resistance levels for the diseases evaluated; therefore, in this regard, both are recommended without restriction for planting.

To determine the presence of *Bru*1 gene, which confers resistance to Brown Rust, the genomic DNA was extracted according to the CTAB method described by Aljanabi et al. (1999). The two molecular markers for diagnosis to *Bru1*, R12H16 and 9O20-F4-*RsaI*, strongly associated with the *Bru*1 gene (Costet et al. 2012), were used to evaluate the presence of the gene. All PCR reactions were carried out following the protocol: 20 µL final volume containing 50 ng DNA, 0.4 M of each primer, 0.4 mM of each dNTP, 2.5 mM MgCl2, 0.5 unit Taq DNA Polymerase (Invitrogen), with 1 X PCR buffer supplied with the enzyme.

Amplification conditions and restriction steps using the *RsaI* enzyme were carried out according to Costet et al. (2012). RB975201 and RB975242 were classified regarding the simultaneous presence (haplotype 1) or absence (haplotype 4) of both molecular markers. Results obtained using the two molecular markers associated the presence of the *Bru*1 gene in RB975201 and RB975242 varieties (Table 1).

MAINTENANCE AND DISTRIBUTION OF BASIC SEEDS

RB975201 and RB975242 varieties were produced and are available for research purposes in the Sugarcane Breeding Program of UFSCar, located in the Agricultural Sciences Center, in Araras, São Paulo state, where they will be kept for at least 5 years from the date of this publication.

REFERENCES

Aljanabi S, Forget L and Dookun A (1999) An improved and rapid protocol for the isolation of polysaccharide-and polyphenol-free sugarcane DNA. **Plant Molecular Biology Reporter 17:** 281-281.

Barbosa MHP, Silveira LCI, Oliveira MW, Souza VFM and Ribeiro SNN (2001) RB867515 Sugarcane cultivar. **Crop Breeding and Applied Biotechnology 1**: 437-438.

Barbosa MHP, Silveira LCI, Souza VFM and Ribeiro SNN (2004) RB928064 - Sugarcane cultivar. **Crop Breeding and Applied Biotechnology 4**: 356-359.

Barbosa GVS, Oliveira RA, Cruz MM, Santos JM, Silva PP, Viveiros AJA, Sousa AJR, Ribeiro CAG, Soares L, Teodoro I, Filho FS, Diniz CA and Torres VLD (2015) RB99395: Sugarcane cultivar with high sucrose content. **Crop Breeding and Applied Biotechnology 15**: 187-190.

Berding N, Hogarth M and Cox M (2004) Plant improvement of sugarcane. In James GL (ed) **Sugarcane**. Blackwell Science, Oxford, p. 1-19.

Breaux RD, Hebert LP and Fanguy HP (1963) Defects for which sugarcane seedlings are eliminated at the U.S. Sugar Cane Field Station, Houma, Louisiana. In **Proceedings of congress of international society of sugarcane technologists.** Elsevier, Amsterdam, p. 421-424.

Carneiro MS, Chapola RG, Júnior ARF, Cursi DE, Barreto FZ, Balsalobre TWA and Hoffmann HP (2015) RB975952 – Early maturing sugarcane cultivar. **Crop Breeding and Applied Biotechnology 15**: 193-196.

Costet L, Cunff L LE, Royaert S, Raboin LM, Hervouet C, Toubi L, Telismart H, Garsmeur O, Rouselle Y, Pauquet J, Nibouche S, Glaszmann JC, Hoarau JY and D'Hont A (2012) Haplotype structure around Bru1 reveals a narrow genetic basis for brown rust resistance in modern sugarcane cultivars. **Theoretical and Applied Genetics 125**: 825-836.

Dice LR (1945) Measures of the amount of ecological association between species. **Ecology 26**: 297-307.

Eberhart SA and Russell WA (1966) Stability parameters for comparing varieties. **Crop Science 6**: 36-40.

Federer WT (1956) Augmented (or Hoonuiaku) designs. **Hawaiian Planters' Record 55**: 191-208.

Iaia AM, Oliveira RA, Melo LJOT, Daros E, Neto DES, Bastos GQ, Oliveira FJ, Chaves A and Melo TTAT (2015) RB002504 - New early-maturing sugarcane cultivar. **Crop Breeding and Applied Biotechnology 14**: 45-47.

Oliveira RA, Daros E and Hoffmann HP (2015) **Liberação nacional de variedades RB de cana-de-açúcar**. Graciosa, Curitiba, 72p.

Hogarth DM (1987) Genetics of sugarcane. In Heinz DJ (ed) **Sugarcane improvement through breeding**. Elsevier, Amsterdam, p. 255-271.

Kang MS, Miller JD and Tai PYP (1983) Genetic and phenotypic path analysis and heritability in sugarcane. **Crop Science 23**: 643-647.

Matsuoka S (1979) Método para pré-testagem de clones de cana-de-açúcar ao carvão e ao mosaico conjuntamente. In: **I Congresso nacional da sociedade dos técnicos açucareiros e alcooleiros do Brasil**. STAB, Maceió, p. 231-233.

Matsuoka S, Garcia AAF and Arizono H (1999) Melhoramento da cana-de-açúcar. In Borém A (ed) **Melhoramento de espécies cultivadas**. Editora UFV, Viçosa, p. 205-252.

Morais LK, Aguiar MS, Silva PA, Câmara TMM, Cursi DE, Fernandes Júnior AR, Chapola RG, Carneiro MS and Bespalhok Filho JC (2015) Breeding of sugarcane. In Cruz VMV and Dierig DA (eds) **Industrial crops: breeding for bioenergy and bioproducts.** Springer, New York, p. 29-42.

Prado H (2008) **Pedologia fácil: aplicações na agricultura**. Hélio do Prado, Piracicaba, 45p.

Sreenivasan TV, Ahloowalia BS and Heinz DJ (1987) Cytogenetics. In Heinz DJ (ed) **Sugarcane improvement through breeding**. Elsevier, Amsterdam, p. 211-253.

Steel RGD and Torrie JH (1960) **Principles and procedures of statistics**. McGraw-Hill Book Company, New York, 481p.

29

Genetic parameters, adaptability and stability to selection of yellow passion fruit hybrids

Alírio José da Cruz Neto[1], Raul Castro Carriello Rosa[2], Eder Jorge de Oliveira[2], Sidnara Ribeiro Sampaio[3], Idália Souza dos Santos[3], Plácido Ulisses Souza[3], Adriana Rodrigues Passos[1] and Onildo Nunes de Jesus[2*]

Abstract: *The objective of this study was to evaluate the stability and adaptability, using the method of harmonic means of the relative performance of genetic values and to estimate the genetic components of variance and average via mixed models of 14 genotypes of passion fruit in three environments. Data were obtained in a random block design with three replicates and nine plants per plot. For the hybrids in the final validation phase, the estimates of heritability and genetic gains in the evaluated environments showed good prospects for selection of superior genotypes. There was a pronounced effect of genotype-environment interaction (GxE) for all traits investigated except fruit length, percentage of pulp, soluble solids, titratable acidity and SS/TA ratio. The most stable and adaptable hybrids in the evaluated environments were BRS Gigante Amarelo, HFOP-09, H09-09, GP09-02, GP09-03 and BRS Sol do Cerrado.*

Key words*: P. edulis* Sims*, GxE interaction, breeding, mixed models.*

***Corresponding author:**
E-mail: onildo.nunes@embrapa.br

[1] Universidade Estadual de Feira de Santana, 44.036-900, Feira de Santana, BA, Brazil
[2] Embrapa Mandioca e Fruticultura, 44.380-000, Cruz das Almas, BA, Brazil
[3] Universidade Federal do Recôncavo da Bahia, Centro de Ciências Agrárias, Ambientais e Biológicas, 44.380-000, Cruz das Almas, BA, Brazil

INTRODUCTION

Brazil is highlighted as a large producer of yellow passion fruit (*Passiflora edulis* Sims), and the demand for it has grown both in the markets for processed juice and natural fruits (Gonçalves et al. 2007). The latest official figures show Brazilian production of 694,539 tons in an area of 50,837 hectares (IBGE 2015), with the Northeast region standing out, accounting for 64.90% of the production. Considering the Northeast Region, Bahia State is responsible for approximately 65.96% and 42.81% of the national production (IBGE 2015). Despite such numbers, productivity in Bahia is considered low (12.21 t ha^{-1}) compared to the culture's potential, which is estimated to be 40 to 50 t ha^{-1} (Freitas et al. 2011).

Among the factors which limit passion fruit productivity is the use of local, low-yielding varieties of unknown genetic origin. The breeding programs aim at developing more productive varieties with stronger disease resistance. For the development of new varieties, it is essential to know the variance components to better predict genetic values and to maximize selection accuracy (Farias Neto and Rezende 2001). To succeed in the selection and identification of promising genotypes, it is essential to evaluate the agronomic performance of genotypes in multiple locations. In such evaluations, the genotypes are subject to the genotype–environment interaction (GxE), which reflects differentiated behavior of the individuals in the evaluated locales. When developing varieties, it is important for genotypes to be more stable and adapted to the adverse environmental conditions of the region for which they are being bred (Cruz et al. 2004, Silva et al. 2014).

Phenotypic stability is related to choosing the genotypes that are least affected by environmental variations, whereas adaptability is based on the identification of genotypes with predictable behaviors that can adjust to the environmental variations (Cruz et al. 2004). Currently there are several models to evaluate stability and adaptability of genotypes (Oliveira et al. 2014). The REML/BLUP method has been widely used in this type of study, as the genetic evaluation is conducted by predicting the genotypic values of selection candidates. This method provides better experimental accuracy, and it is more efficient than analysis of variance, especially in cases with unbalanced data (Resende 2004). The predicted genetic values can be used to estimate the adaptability and stability of genotypes using the harmonic mean of the relative performance of genetic values (HMRPGV), allowing estimating adaptability and stability simultaneously in a single parameter (Resende 2004). These methods have been successfully used on soybean, coffee, sugarcane and cashew crops, and their results have been found to be superior to conventional methods (Carvalho et al. 2008, Maia et al. 2009, Borges et al. 2010, Silva et al. 2015).

The objective of this study was to evaluate the stability and adaptability using the MHPRVG method and to estimate the genetic components of variance and average via mixed models (REML/BLUP) of 14 genotypes of passion fruit in three environments.

MATERIAL AND METHODS

Fourteen genotypes were evaluated - nine (GP09-02, GP09-03, H09-02, H09-07, H09-09, H09-14, and H09-30) of which from crosses between selected parents, through the Passion Fruit Plant Genetic Improvement Program of the Embrapa Cassava and Fruits (Embrapa Mandioca e Fruticultura), and five commercial hybrids (*BRS Sol do Cerrado*, *BRS Rubi do Cerrado*, *BRS Gigante Amarelo*, *FB200,* and *FB300*). The study was conducted in three producing centers in Bahia State: Dom Basílio (lat 13° 45′ S, long 41°46′ W and alt 200m asl), Rio de Contas (lat 3° 34′ S, long 41° 48′ W and alt 1300m asl), and Lençóis (lat 12° 36′ S, long 41° 20′ W and alt 402m asl).

Each plant was spaced at 2.0 m in the row. The rows were spaced 2.5 m apart. The training system consisted of vertical espaliers with 12 wire, 2.0 m above the ground. The agricultural traits, the number of fruits per plot, and the productivity (TCP) – expressed in t ha^{-1}, – were evaluated. Regarding the physical and chemical characteristics of fruits, five fruits were considered per plot, and the following traits measured: fruit length (FL) in cm; fruit diameter (FD) in cm; peel thickness (PT) in mm; fruit mass (FM) in g; peel mass (PM) in g; pulp mass (PUM) in g; soluble solids (SS) in ºBrix, as measured with a digital refractometer; total titratable acidity (TA), expressed in mg citric acid per 100 mL juice, as determined through titration with NaOH at 0.1mol L^{-1}; percentage of pulp (PP), measured through the PUM/FM ratio in %; and the SS/TA ratio. The measurements were performed at the peak of production, once in the areas studied, produced plants are kept in the field for a maximum of one year due to the incidence of foliar diseases.

A randomized block design was chosen, composed of 14 treatments distributed in three replicates with nine plants per plot. The following statistical model was adopted for the evaluation of genotypes with one observation per plot in three environments: $y = X_r + Z_g + W_i + e$, where: y: is the data vector, r: is the vector of block means (fixed), g: genotypic effects (random), i: effects of the genotype x environment interaction (random), and e: is the vector of error (random). X, Z and W: are the matrixes of incidence of *r*, *g* and *i*, respectively. The estimates for variance components were produced based on the restricted maximum likelihood method (REML) where the following components of variance and genetic parameters were estimated: σ^2_g : genotypic variance. σ^2_{int}: variance of the genotype x environment interaction. σ^2_r: residual variance. σ^2_p: individual phenotypic variance. h^2_g: broad-sense heritability of individual plots; that is, of the total genotypic effects. h^2_{mg}: genotype average heritability, assuming full survival. A_{cgen}: accuracy of the genotype selection, assuming full survival. c^2_{int}: coefficient of determination of the genotype x environment interaction effects. r_{gloc}: genotypic correlation between performances in the various environments. CV_{gi}: genotypic coefficient variation in %. CV_r: residual coefficient variation in %. μ: general experiment average. The genetic values were predicted through the best linear unbiased prediction (BLUP) method. The interaction-free (μ+g) genotypic values of each hybrid were obtained by adding each genotypic effect (g) to the general experiment average (μ). The genetic gain corresponded to the average of genetic effect vectors that were predicted for the selected hybrids. The sum of the general average (μ) and the genetic gain resulted in the new average for the improved population. The estimates of genetic parameters and the adaptability and stability (HMRPGV - harmonic mean of the relative performance of genetic values) were obtained by mixed models (REML/BLUP), through the use of the SELEGEN genetic and statistical software, model 54 (Resende 2007).

RESULTS AND DISCUSSION

The estimation of genetic parameters is important to guide genetic breeding programs, as these predict genetic values and maximize selection, thus helping the selective process for recommendation of new commercial materials (Farias Neto and Rezende 2001, Maia et al. 2009). The estimates for variance components are shown in Table 1. In general, the contribution of the genotypic variance (σ^2_g) for the phenotypic variance (σ^2_p) was 11.44 to 35.05% for productivity traits (TCP) and peel mass (PM), respectively. The variances of residual effects (σ^2_r) were the ones that most contributed to σ^2_p, with variations from 45.71 to 75.90% for fruit length and soluble solids, respectively. Since the evaluated traits are quantitative and therefore highly influenced by the environment, residual variance tends to be high (Atroch et al. 2013). Studies conducted with eucalyptus (Rosado et al. 2012) and clones of guarana plants (Atroch et al. 2013), were also found to have higher contributions from the residual variance than the phenotypic variance. Studying passion fruit, Santos et al. (2015) reported higher values of environmental variance for fruit mass. In yellow passion fruit plant populations, Viana et al. (2003) reported contributions of 64% and 48% (fruit length and diameter) for the environmental variance.

The variance from the GxE interaction (σ^2_{int}) had the smallest contribution to σ^2_p, ranging from 10.00 to 23.12% for traits titratable acidity and total cumulative productivity, respectively (Table 1). A low magnitude of GxE interaction (σ^2_{int}) indicates uniformity in the performance of genotypes according to environmental variations and thereby greater adaptability and genetic stability of individuals (Maia et al. 2009, Rosado et al. 2012). The σ^2_{int} is associated with the coefficient determination of the effects of GxE interaction (c^2_{int}), since it represents the percentage of σ^2_{int} present in σ^2_p and therefore exhibits the same value (Table 1). Due to the lower σ^2_{int}/ c^2_{int}, the genotypic correlation between performances in the various environments (r_{gloc}) showed values of 0.51 to 0.74 for most characteristics, except for total cumulative productivity and pulp mass with 0.33 and 0.49, respectively (Table 1). However, the values are not yet high enough for all the characteristics, which reinforces the need to further study the adaptability and stability.

The residual coefficients' variation (CV_r) ranged from 4.73% to 24.88% for FD and TCP, respectively (Table 1). In other fruit plants, such as the cashew tree, values of CV_r close to 34% were reported (Maia et al. 2009). Besides that, in passion fruit plants, Oliveira et al. (2008) reported CV_r ranging from 4.76% for fruit length to 20.48% for number of

Table 1. Estimates of variance components (REML) in 14 yellow passion fruit hybrids evaluated in three locations in the State of Bahia for traits: total cumulative productivity (TCP), fruit mass (FM), fruit length (FL), fruit diameter (FD), peel thickness (PT), peel mass (PM), pulp mass (PUM), percentage of pulp (PP), soluble solids (SS), titratable acidity (TA), and SS/TA ratio

Characters	**Parameters**											
	% of σ^2_p											
	σ^2_g	σ^2_{int}	σ^2_r	σ^2_p	h^2_g	h^2_{mg}	A_{cgen}	c^2_{int}	r_{gloc}	CV_{gi}%	CV_r%	**Average** (μ)
TCP	11.44	23.12	65.44	66.58	0.11±0.08	0.43	0.66	0.23	0.33	10.41	24.88	26.53
FM	28.10	21.60	50.29	1241.94	0.28±0.13	0.69	0.83	0.22	0.57	8.27	11.06	226.00
FL	31.91	12.77	55.32	0.47	0.31±0.14	0.75	0.86	0.13	0.70	3.96	5.31	9.63
FD	34.29	20.00	45.71	0.35	0.35±0.15	0.75	0.87	0.20	0.64	4.21	4.73	8.37
PT	30.77	15.38	53.85	1.04	0.30±0.14	0.74	0.86	0.15	0.67	8.23	10.83	6.89
PM	35.05	18.94	46.01	490.77	0.35±0.15	0.75	0.87	0.19	0.65	10.88	12.46	120.58
PUM	18.65	19.65	61.69	206.13	0.18±0.11	0.58	0.76	0.20	0.49	9.09	16.52	68.24
PP	13.28	11.97	74.75	15.96	0.13±0.09	0.52	0.72	0.12	0.53	4.96	11.78	29.32
SS	12.05	12.05	75.90	0.83	0.12±0.09	0.50	0.71	0.12	0.51	2.47	6.11	13.03
TA	30.00	10.00	60.00	0.20	0.31±0.14	0.75	0.87	0.11	0.74	6.62	9.03	3.78
SS/TA	25.62	11.90	62.48	0.22	0.25±0.12	0.70	0.84	0.12	0.68	6.59	10.30	3.62

σ^2_g: genotypic variance. σ^2_{int}: variance of the genotype x environment interaction. σ^2_r: residual variance. σ^2_p: individual phenotypic variance. h^2_g: broad-sense heritability of individual plots; that is, of the total genotypic effects. h^2_{mg}: genotype average heritability, assuming full survival. A_{cgen}: accuracy of the genotype selection, assuming full survival. c^2_{int}: coefficient of determination of the genotype x environment interaction effects. r_{gloc}: genotypic correlation between performances in the various environments. CV_{gi}: genotypic coefficient variation in % CV_r: residual coefficient variation in %. μ: General experiment average.

fruits. Although the adoption of a higher number of repetitions can contribute to environmental control and reduced residual coefficient variation (CV_r), the values are considered low for quantitative a characteristic, which indicates good experimental quality (Rosado et al. 2012).

The genotypic coefficient of variation (CV_{gi}), ranged from 2.47% for the variable soluble solids (SS) to 10.88% of peel mass (PM) (Table 1) that indicates there was genetic variability among the analyzed hybrids. In another study of passion fruit, Viana et al. (2003) found variations of 0.00 to 52.78% for the CV_{gi} for the characteristics percentage of pulp and number of fruits. The values of CV_r and CV_{gi} influence the accuracy statistics (Resende and Duarte 2007). The accuracy of the genotype selection (A_{cgen}), which is the square root of h^2_{mg}, reflects the correlation between the true genotypic value and the estimated values (Resende et al. 2002, Cargnelutti-Filho and Storck 2009) and shows a good experimental quality and security in the selection of superior genotypes (Carvalho et al. 2016). High accuracy values (0.71 to 0.87) were observed for most evaluated traits, except for total cumulative productivity (0.66), which presented moderate accuracy in accordance with the classification proposed by Resende and Duarte (2007) (Table 1). Furthermore, high values of this parameter indicate the existence of genetic variance among the genotypes (Maia et al. 2009).

Broad-sense heritability of individual plots (h^2_g) expresses the genetic variation between genotypes, and it also supports the definition of the most suitable improvement methods to be used in breeding programs (Resende 2002). The values of h^2_g ranged from 0.11 for total cumulative productivity to 0.35 for fruit diameter and peel mass, and were associated with high values of deviations (Table 1), which indicates individual selection may not be effective. Santos et al. (2015), evaluated interspecific progenies of passion fruit, and found similar results to this study, except for the traits PUM and SS, which were found to be 0.61 and 0.62, respectively, indicating the possibility of successful individual selection for those two traits. The estimated average heritability (h^2_{mg}) values were higher, varying from 0.43 to 0.75 (Table 1). The traits that were found to have the highest h^2_{mg} values (0.69 to 0.75) were fruit mass, peel thickness, fruit length, fruit diameter, titratable acidity, and peel mass. With the exception of fruit mass, which obtained a similar value of 0.68, the remaining traits were higher than those observed by Viana et al. (2003) evaluating yellow passion fruit plant populations in two distinct environments. Oliveira et al. (2008) evaluating in half-sibling progenies of yellow passion fruit observed values lower of h^2_{mg} of 0.28, 0.30, 0.51, and 0.57 for variables percentage of pulp, fruit diameter, fruit length, and fruit mass, respectively. Moraes et al. (2005) found a larger value of h^2_{mg} for fruit length and soluble solids, and a lower for fruit mass and fruit diameter, while evaluating an F_1 population of yellow passion fruit. High h^2_{mg} estimates demonstrate good genetic control of a trait, and high potential for selecting superior genotypes. Thus, most of the observed variation in the yellow passion fruit plants was genetic, allowing high selective accuracy rates for most agricultural traits. However, the traits total cumulative productivity, soluble solids, percentage of pulp, and pulp mass were found to have moderate values of 0.43, 0.50, 0.52, and 0.58, respectively (Table 1). These values reflect the low genetic variance (11.44 to 18.65% of σ^2_p) and the high residual variance of 61.69 to 75.90% of σ^2_p (Table 1). Thus, they tended to have low h^2_g, h^2_{mg} and A_{cgen}, and therefore smaller genetic gains for those traits (Table 1).

The differences observed in the heritability estimate values is acceptable, as that is an estimate that can fluctuate due to several factors, among them the genetic structure of the evaluated population (Santos et al. 2015), the changes in genetic and phenotypic parameters linked to the studied trait, the estimation method, the diversity of the population, the evaluated environment, the sample size, and the experimental accuracy (Hallauer and Miranda Filho 1988). It is important to point out that the inferences from the genetic parameters that are defined in this study originate from the evaluation of genotypes in three distinct environments. This makes the estimates more reliable (Resende and Dias 2000) and maximizes the genetic gains in the presence of the genotype x environment interaction (Costa et al. 2002). The studied population comprised genotypes in the final evaluation stage, from which genetic material was selected. However, even under these conditions, the genotypes still exhibited genetic variability that can be used for selection, through crosses aiming at increasing the frequency of favorable alleles.

Tables 2 and 3 show the genotypic effects (g), interaction-free genotypic values (μ+g), genetic gains, and new predicted average for the 14 hybrids evaluated. The genotypic values are the true values to be predicted and the new average values are predictions of the performance of selected hybrids (Borges et al. 2010). Some negative genotypic effects (- g) were identified in some hybrids and for most evaluated traits. Those values indicate the hybrids are below the general average (μ) and that they would be discarded in the selective process, because they may not carry desirable genetic complements in their genomes. It is possible to infer that such genotypes will interact significantly with their environment, which is not desirable for development of new varieties (Maia et al. 2009). For the total cumulative productivity trait (TCP), 57% of the hybrids were found to be superior to the general average (μ), which was 26.53 t ha^{-1} (Table 2). Genotypes BRS *Gigante Amarelo*, H09-09, GP09-03, BRS *Rubi do Cerrado*, and BRS *Sol do Cerrado* were, in that order, the five best for that trait. The new average for this trait varied from 28.92 to 28.17 t ha^{-1} and gains ranged from 2.39 to 1.65 t ha^{-1} (Table 2). Those gains are promising for the passion fruit productive sector, as the observed values for total cumulative productivity exceed the national average and the one for the State of Bahia, which are 13.66 t ha^{-1} and 12.21 t ha^{-1}, respectively (IBGE 2015).

Fruit mass (FM) is an important characteristic for the *in natura* consumption market, once larger fruits are preferred by consumers (Negreiros et al. 2007). Therefore, taking this information into consideration, the five best genotypes were BRS *Gigante Amarelo*, H09-30, HFOP-09, GP09-02, and H09-09, with gains from 38.97 to 14.60 g, respectively (Table

Table 2. Estimates of the genotypic effect; average predicted value in the environments, gain and new average in 14 yellow passion fruit hybrids evaluated in three locations in the State of Bahia for physical traits of fruits: total cumulative productivity (TCP), fruit mass (FM), fruit length (FL), fruit diameter (FD), peel thickness (PT), peel mass (PM), pulp mass (PUM), and percentage of pulp (PP)

Genotypes	Total cumulative productivity (TCP)				Fruit mass (FM)				Fruit length (FL)				Fruit diameter (FD)			
	g	μ+g	Gain	New Average	g	μ+g	Gain	New Average	g	μ+g	Gain	New Average	g	μ+g	Gain	New Average
GP09-03	1.66	28.19	2.06	28.58	-1.19	224.81	10.29	236.29	0.48	10.11	0.55	10.18	0.01	8.37	0.21	8.58
HFOP-08	-0.68	25.85	0.94	27.47	-2.97	223.04	7.44	233.44	-0.13	9.50	0.19	9.82	-0.01	8.35	0.16	8.53
HFOP-09	-1.31	25.22	0.74	27.26	13.81	239.82	23.06	249.06	0.15	9.78	0.36	9.99	0.02	8.39	0.29	8.66
H09-30	-3.78	22.74	0.00	26.53	16.40	242.40	27.69	253.69	-0.44	9.19	0.00	9.63	0.63	8.99	0.63	8.99
FB300	-1.88	24.65	0.52	27.04	-6.51	219.50	6.04	232.05	0.04	9.67	0.27	9.90	-0.24	8.13	0.09	8.46
H09-09	2.12	28.65	2.26	28.78	1.71	227.72	14.60	240.61	0.22	9.85	0.41	10.04	0.01	8.38	0.25	8.62
BRS-SC	0.98	27.51	1.65	28.17	-2.13	223.87	8.74	234.74	-0.32	9.31	0.10	9.73	0.13	8.50	0.36	8.73
H09-14	0.76	27.29	1.40	27.93	0.20	226.21	12.20	238.21	-0.23	9.40	0.15	9.78	0.14	8.51	0.44	8.81
FB200	-2.44	24.09	0.29	26.82	-24.65	201.36	0.00	226.00	-0.38	9.25	0.03	9.66	-0.50	7.87	0.00	8.37
BRS-GA	2.39	28.92	2.39	28.92	38.97	264.97	38.97	264.97	0.62	10.25	0.62	10.25	0.55	8.91	0.59	8.95
BRS- Rubi	1.08	27.60	1.81	28.34	-7.10	218.91	4.85	230.85	0.07	9.70	0.31	9.94	-0.20	8.17	0.13	8.49
H09-07	0.61	27.13	1.30	27.83	-15.24	210.76	1.90	227.90	-0.32	9.31	0.07	9.70	-0.28	8.09	0.04	8.41
H09-02	-0.33	26.19	1.12	27.65	-13.44	212.56	3.32	229.33	-0.07	9.56	0.23	9.86	-0.25	8.12	0.06	8.43
GP09-02	0.81	27.34	1.51	28.03	2.12	228.12	17.83	243.83	0.32	9.95	0.47	10.10	-0.01	8.36	0.18	8.55
Average (μ)				26.53				226.00				9.63				8.37

Genotypes	**Peel thickness** (PT)				**Peel mass** (PM)				**Pulp mass** (PUM)				**Percentage of pulp** (PP)			
	g	μ+g	Gain	New Average	g	μ+g	Gain	New Average	g	μ+g	Gain	New Average	g	μ+g	Gain	New Average
GP09-03	-0.18	6.71	0.34	7.22	1.06	121.64	8.81	129.40	-0.62	67.62	3.03	71.28	-0.22	29.10	0.55	29.87
HFOP-08	0.36	7.24	0.57	7.45	2.61	123.19	10.11	130.69	-4.27	63.97	1.04	69.28	-1.04	28.28	0.20	29.52
HFOP-09	0.44	7.32	0.64	7.52	7.38	127.97	15.43	136.02	4.48	72.72	7.93	76.17	0.41	29.73	0.74	30.06
H09-30	1.00	7.88	1.00	7.88	15.68	136.26	19.46	140.04	0.68	68.92	3.55	71.80	-0.81	28.51	0.43	29.75
FB300	0.09	6.98	0.41	7.30	-8.37	112.22	3.95	124.54	1.38	69.62	4.03	72.28	0.80	30.12	0.94	30.25
H09-09	-0.35	6.54	0.21	7.09	-2.09	118.50	7.45	128.04	2.26	70.51	5.29	73.53	0.76	30.08	0.87	30.19
BRS-SC	-0.30	6.58	0.27	7.15	-3.97	116.61	5.19	125.77	1.68	69.92	4.57	72.81	1.00	30.31	1.00	30.32
H09-14	0.48	7.37	0.74	7.62	5.68	126.27	11.61	132.19	-2.94	65.30	1.52	69.76	-1.02	28.30	0.31	29.63
FB200	-0.36	6.52	0.15	7.04	-19.08	101.50	0.00	120.59	-4.45	63.79	0.61	68.86	1.01	30.33	1.01	30.33
BRS-GA	0.24	7.12	0.47	7.35	23.24	143.83	23.24	143.83	11.37	79.61	11.37	79.61	0.17	29.49	0.64	29.96
BRS- Rubi	0.28	7.17	0.51	7.40	6.05	126.63	13.09	133.67	-7.97	60.27	0.00	68.24	-2.66	26.66	0.00	29.32
H09-07	-0.54	6.35	0.06	6.94	-12.96	107.62	1.47	122.05	-2.36	65.88	1.96	70.21	0.43	29.75	0.80	30.12
H09-02	-0.73	6.15	0.00	6.89	-11.44	109.15	2.67	123.26	-2.27	65.98	2.44	70.69	0.37	29.69	0.70	30.01
GP09-02	-0.42	6.46	0.11	6.99	-3.78	116.81	6.20	126.79	3.03	71.28	6.29	74.54	0.78	30.10	0.90	30.22
Average (μ)				6.89				120.58				68.24				29.32

g: genotypic effect; μ+g, average predicted value in the environments; μ, general experiment average. BRS *Sol do Cerrado* (BRS-SC); BRS *Gigante Amarelo* (BRS-GA). The five best hybrids for each evaluated trait are underlined.

2). In this case, the new average prediction for the above genotypes regarding fruit mass are 264.97 to 240.61 g and productivity of 28.92 to 28.78 t ha^{-1} (Table 2), therefore with estimates of higher values in comparison to the ones obtained for the UENF Rio Dourado cultivar (Pio Viana et al. 2016).

The characteristics for fruits length and diameter are also appreciated by consumers, because this means greater number of seeds and thus higher percentage of pulp (Negreiros et al. 2007). However, it is desirable for these traits to be accompanied by smaller peel thickness at the moment of selecting new varieties (Neves et al. 2013). Among the five best for fruit length and diameter, hybrids HFOP-09 and BRS *Gigante Amarelo* stood out, with respective gains of 0.36 and 0.62 for fruit length, and 0.29 and 0.59 for fruit diameter (Table 2). There were respective gains of 0.51 to 1.0, and 11.61 to 23.24 for the traits peel thickness and peel mass, with special mention to genotypes H09-30, HFOP-09, H09-14, and BRS *Rubi do Cerrado*, which were best for those traits. BRS *Gigante Amarelo* stood out with the highest gain for peel mass, with 23.24 (Table 2). High peel mass and peel thickness values are undesirable, as they do not contribute to percentage of pulp, especially in fruits for industrial processing (Medeiros et al. 2009).

Seed-free pulp mass and percentage of pulp are attractive traits to the juice industry, which establishes a minimum acceptable yield of 33% (Nascimento et al. 1999). For the PUM trait, the hybrids found to have the highest genetic gains were BRS *Gigante Amarelo*, HFOP-09, GP09-02, H09-09, and BRS *Sol do Cerrado*, with results ranging from 11.34 to 4.57 (Table 2). Those materials exceeded the general experimental average (68.24 g), ranging from 72.81 to 79.61 for the new predicted average. Regarding percentage of pulp, the best genotypes were FB200, BRS *Sol do Cerrado*, FB300, GP09-02, and H09-09, varying from 30.33 to 30.19 (Table 2). Genotypes H09-09, BRS *Sol do Cerrado*, and GP09-02 also showed large gains for pulp mass and percentage of pulp.

For the soluble solids, the genotypic average was similar to the new predicted average. This is possibly related to the lower gains, from 0.0 to 0.39 (Table 3). Still, genotypes H09-30, FB200, H09-07, BRS *Sol do Cerrado* and HFOP-09 had a certain level gain for this trait (Table 3). Fruit pulp acidity is desirable for the agroindustry since higher acidity avoids microbiological deterioration, allowing better product conservation as well as reducing the need for artificial acid addition (Freitas et al. 2011). For titratable acidity, the genotypes with best genetic gain were H09-30, BRS *Gigante Amarelo*, HFOP-09, HFOP-08, and H09-14 (Table 3). The values observed for TA were above the limit established by the Ministry of Agriculture, of 2.5% (Brasil 2003). From the SS/TA ratio, it is possible to evaluate fruit flavors, since it expresses the ratio between sugars and acids (Freitas et al. 2011). The five best hybrids were H09-07, BRS *Rubi do Cerrado*, FB200,

Table 3. Estimates of the genotypic effect, average predicted value in the environments, gain and new average in 14 yellow passion fruit hybrids evaluated in three locations in the State of Bahia for chemical traits of fruits: soluble solids (SS), titratable acidity (TA), SS/TA ratio

Genotypes	Soluble solids (SS)				Titratable acidity (TA)				SS/TA			
	g	μ+g	Gain	New Average	g	μ+g	Gain	New Average	g	μ+g	Gain	New Average
GP09-03	0.03	13.06	0.16	13.18	0.04	3.82	0.19	3.96	-0.05	3.57	0.12	3.74
HFOP-08	-0.03	13.00	0.14	13.16	0.12	3.90	0.26	4.03	-0.16	3.46	0.05	3.67
HFOP-09	0.08	13.10	0.23	13.26	0.18	3.96	0.30	4.08	-0.14	3.48	0.09	3.71
H09-30	0.39	13.41	0.39	13.41	0.43	4.20	0.43	4.20	-0.22	3.40	0.03	3.65
FB300	0.03	13.06	0.17	13.20	-0.09	3.68	0.10	3.88	0.05	3.67	0.16	3.79
H09-09	0.03	13.06	0.20	13.23	0.03	3.81	0.16	3.94	-0.04	3.58	0.14	3.76
BRS-SC	0.10	13.13	0.27	13.30	-0.02	3.75	0.13	3.90	0.06	3.68	0.21	3.83
H09-14	-0.16	12.87	0.11	13.13	0.05	3.83	0.21	3.99	-0.16	3.47	0.07	3.69
FB200	0.37	13.40	0.38	13.41	0.01	3.79	0.15	3.92	0.22	3.84	0.26	3.88
BRS-GA	-0.35	12.68	0.00	13.03	0.29	4.07	0.36	4.14	-0.37	3.25	0.00	3.62
BRS- Rubi	-0.18	12.85	0.08	13.11	-0.32	3.45	0.00	3.78	0.25	3.87	0.28	3.90
H09-07	0.21	13.24	0.32	13.35	-0.24	3.54	0.05	3.83	0.30	3.93	0.30	3.93
H09-02	-0.31	12.72	0.03	13.06	-0.19	3.59	0.08	3.85	0.05	3.67	0.18	3.81
GP09-02	-0.21	12.82	0.06	13.08	-0.29	3.49	0.02	3.80	0.22	3.84	0.25	3.87
Average (μ)				13.03				3.78				3.62

g, genotypic effect; μ+g, average genotypic effect predicted in the environments; μ, general experiment average. BRS Sol do Cerrado (BRS-SC); BRS Gigante Amarelo (BRS-GA). The five best hybrids for each evaluated trait are underlined.

GP09-02, BRS *Sol do Cerrado*, with a new predicted average ranging from 3.93 to 3.83 (Table 3).

Significant differences at the levels of 1 and 5% were observed in the deviance analysis among genotypes for traits fruit length and diameter, peel thickness, peel mass, titratable acidity, SS/TA ratio, total cumulative productivity and pulp mass. Regarding the GxE interaction, only variables fruit length, percentage of pulp, soluble solids, titratable acidity, and SS/TA ratio, were not significant, which indicates low interaction effects for those traits (Table 4). This can also be observed by low values of σ^2_{int} (Table 1). A study conducted with half-sibling passion fruit progenies evaluated in two environments in Rio de Janeiro also found no GxE interaction for the same traits (Oliveira et al. 2008).

The detailed study of the GxE interaction allows selecting the best genotypes for the various environmental conditions, and it results from the estimated phenotypic adaptability and stability (Silva et al. 2014), which enable identifying genotypes with behaviors that can be predicted according to environmental variations. For selection of the most stable and adaptable hybrid, it was the simultaneous selection method based on performance of genetic values (HMRPGV). Depending on the evaluated trait, the genotypes were classified differently through HMRPGV values (Table 4). However, in general considering all the studied traits, hybrids BRS *Gigante Amarelo*, HFOP-09, H09-09, GP09-02 and GP09-03, and BRS *Sol do Cerrado* stood out (Table 4). The most relevant traits for recommending the release of a hybrid are productivity, fruit mass, fruit length and diameter, and percentage of pulp. For the productivity trait, the most stable and adaptable in various environments were the hybrids BRS *Gigante Amarelo*, GP09-03, H09-09, BRS *Sol do Cerrado*, and H09-14. For fruit mass, the following genotypes were highlighted: BRS *Gigante Amarelo*, H09-30, HFOP-09, GP09-02, and H09-09. Fruits with larger lengths and/or diameters tend to have larger peel mass and thickness, which are not appreciated especially by the juice industry, because they negatively influence pulp mass and percentage of pulp (Negreiros et al. 2007, Freitas et al. 2011). However, hybrids BRS *Gigante Amarelo*, HFOP-09, H09-09, GP09-02, and BRS *Sol do Cerrado* were also found to have higher pulp mass, which indicates that this is not always a valid association (Table 4).

The chemical properties of fruits, such as soluble solids, titratable acidity and SS/TA ratio are important to the juice industry. Passion fruit pulp with higher sugar levels results in a smaller number of fruits required to obtain concentrated juice at 50° Brix (Oliveira et al. 2008, Freitas et al. 2011). High acidity in the passion fruit juice is an important characteristic

***Table 4**.* Deviance analysis and stability and adaptability (HMRPGV x μ) in 14 yellow passion fruit hybrids evaluated in three locations in the State of Bahia for traits of fruits: total cumulative productivity (TCP), fruit mass (FM), fruit length (FL), fruit diameter (FD), peel thickness (PT), peel mass (PM), pulp mass (PUM), percentage of pulp (PP), soluble solids (SS), titratable acidity (TA), SS/TA ratio

	MHPRVG x Avarage (μ)										
Genotypes	**TCP**	**FM** (g)	**FL** (cm)	**FD** (mm)	**PT** (mm)	**PM** (g)	**PUM** (g)	**PP** (%)	**SS** (ºbrix)	**TA** (%)	**SS/TA**
GP09-03	29.44	223.74	10.21	8.37	6.68	121.79	67.56	29.02	13.03	3.81	3.55
HFOP-08	25.46	221.48	9.44	8.37	7.30	123.00	62.78	27.85	13.03	3.93	3.44
HFOP-09	23.87	244.08	9.82	8.37	7.37	129.03	74.38	29.90	13.16	3.97	3.44
H09-30	20.43	246.34	9.15	9.12	8.06	138.67	68.93	28.15	13.55	4.27	3.37
FB300	23.34	216.96	9.63	8.12	7.02	110.94	70.29	30.49	13.03	3.66	3.69
H09-09	28.91	228.26	9.92	8.37	6.47	118.17	70.97	30.20	13.03	3.81	3.59
BRS-SC	28.38	223.74	9.24	8.54	6.54	115.76	70.29	30.49	13.16	3.74	3.69
H09-14	27.85	226.00	9.34	8.54	7.44	126.61	64.15	27.85	12.77	3.85	3.44
FB200	22.55	194.36	9.15	7.78	6.47	97.67	62.10	30.49	13.55	3.78	3.88
BRS-GA	30.24	275.72	10.30	9.04	7.16	148.32	83.26	29.61	12.51	4.12	3.19
BRS- Rubi	27.85	216.96	9.73	8.12	7.23	127.82	57.32	25.80	12.77	3.40	3.91
H09-07	27.32	205.66	9.24	8.03	6.27	104.91	64.83	29.90	13.29	3.51	3.98
H09-02	25.46	207.92	9.53	8.03	5.99	106.11	64.83	29.90	12.64	3.55	3.69
GP09-02	27.85	228.26	10.02	8.37	6.40	115.76	72.34	30.20	12.77	3.44	3.88
Average (μ)	26,53	226,00	9,63	8,37	6,89	120,58	68,24	29,32	13,03	3,78	3,62
Deviance analysis											
Genotypes	614,23*	939,79	31,24**	-14,98**	121,24**	826,93**	742,45*	454,27	113,02	-64,96**	-49,13**
GxE	618,00*	940,74**	25,54	-16,19**	117,00*	824,73**	743,44*	453,37	112,37	-72,08	-53,83

BRS *Sol do Cerrado* (BRS-SC); BRS *Gigante Amarelo* (BRS-GA). μ, general experiment average. The five best hybrids for each evaluated trait are underlined. ** and * significant at 1% and 5%, respectively.

for the processing, due to the possibility of reducing the addition of acidifiers (Nascimento et al. 1999). The SS/TA ratio is considered one of the most practical ways to evaluate the flavor of fruits. Acidity is decisive for the SS/TA ratio, because high levels of acidity decrease the value of this ratio (Freitas et al. 2011). Inexpressive gains were observed for soluble solids, titratable acidity and SS/TA ratio, which explains the similarity between the genotypic values free of interaction (μ+g) and the new average. Besides that, the chemical characteristics were not significant for the GxE interaction (Table 4), so the genotypes presented similar behavior in the evaluated environments.

According to the results, the average heritability of the genotypes for the six evaluated traits in the three environments was found to have medium to high values, which also indicates the existence of enough variability for the selection of superior genotypes, even if they are hybrids in the last validation stage. There were GxE interactions for all traits except for fruit length, percentage of pulp, soluble solids, titratable acidity and SS/TA ratio. The hybrids that were the most stable and adaptable to the evaluated environments, and hence can be recommended for commercial planting in Bahia, are BRS *Gigante Amarelo*, BRS *Sol do Cerrado*, HFOP-09, H09-09, GP09-02 and GP09-03.

ACKNOWLEDGEMENTS

The authors would like to thank the Conselho Nacional de Desenvolvimento Científico e Tecnológico (CNPq) e Coordenação de Aperfeiçoamento de Pessoal de Nível Superior (CAPES) for the scholarships and the Fundação de Amparo à Pesquisa do Estado da Bahia (FAPESB: RED0004/2012) for the financial support and Bioenergia Orgânicos (Agreement FUNARBE 7697) for making experimental fields available.

REFERENCES

Atroch AL, Nascimento FJ and Resende MDV (2013) Seleção genética simultânea de progênies de guaranazeiro para produção, adaptabilidade e estabilidade temporal. **Revista de Ciências Agrárias 56**: 347-352.

Borges V, Ferreira PV, Soares L, Santos GM and Santos AMM (2010) Seleção de clones de batata-doce pelo procedimento REML/BLUP. **Acta Scientiarum Agronomy 25**: 643-649.

Brasil - Ministério da Agricultura e do Abastecimento (2003) Instrução Normativa Nº 12 de 4 de Setembro de 2003. Diário Oficial. Brasília, Seção1: 72-76.

Cargnelutti Filho A and Storck L (2009) Medidas do grau de precisão experimental em ensaios de competição de cultivares de milho. **Pesquisa Agropecuária Brasileira 44**: 111-117.

Carvalho AD, Fritsche Neto R and Geraldi IO (2008) Estimation and prediction of parameters and breeding values in soybean using REML/BLUP and Least Squares. **Crop Breeding and Applied Biotechnology 8**: 230-235.

Carvalho LP, Farias FJC, Morello CL and Teodoro PE (2016) Uso da metodologia REML/BLUP para seleção de genótipos de algodoeiro com maior adaptabilidade e estabilidade produtiva. **Bragantia 75**: 314-321.

Costa RB, Resende MDV, Gonçalves PS and Silva MA (2002) Individual multivariate REML/BLUP in the presence of genotype x environment interaction in rubber tree (*Hevea*) breeding. **Crop Breeding and Applied Biotechnology 2**: 131-140.

Cruz CD, Regazzi AJ and Carneiro PCS (2004) **Modelos Biométricos Aplicados ao Melhoramento Genético**. 2nd edn, Imprensa Universitária, Viçosa, 480p.

Farias Neto JT and Resende MDV (2001) Aplicação da metodologia de modelos mistos (REML/BLUP) na estimação de componentes de variância e predição de valores genéticos em pupunheira (*Bactris gasipaes*). **Revista Brasileira Fruticultura 23**: 320-324.

Freitas JPX, Oliveira EJ, Cruz Neto AJ and Santos LR (2011) Avaliação de recursos genéticos do maracujazeiro-amarelo. **Pesquisa Agropecuária Brasileira 46**: 1013-1020.

Gonçalves GM, Pio Viana A, Bezerra Neto, FV, Pereira MG and Pereira TNS (2007) Seleção e herdabilidade na predição de ganhos genéticos em maracujá-amarelo. **Pesquisa Agropecuária Brasileira 42**: 193-198.

Hallauer AR and Miranda Filho JB (1988) **Quantitative genetics in maize breeding**. Iowa State University Press, Ames, 664p.

IBGE (2015) Banco de dados agregados: produção agrícola municipal. Sistema IBGE de Recuperação Automática - SIDRA. Available at <http://www.sidra.ibge.gov.br/>. Accessed on 06 Out, 2016.

Maia MCC, Resende MDV, Paiva JR, Cavalcanti JJV and Barros LMB (2009) Seleção simultânea para produção, adaptabilidade e estabilidade genotípicas em clones de cajueiro, via modelos mistos. **Pesquisa Agropecuária Tropical 39**: 43-50.

Medeiros SAF, Yamanishi OK, Peixoto JR, Pires MC, Junqueira NTV and Ribeiro JGBL (2009) Caracterização físico-química de progênies de maracujá-roxo e maracujá-azedo cultivados no Distrito Federal. **Revista Brasileira Fruticultura 31**: 492-499.

Moraes MC, Geraldi IO, Matta FP and Vieira MLC (2005) Genetic and phenotypic parameter estimates for yield and fruit quality traits from a single wide cross in yellow passion fruit. **Hort Science 40**: 1978-1981.

Nascimento TB, Ramos JD and Menezes JB (1999) Características físicas do maracujá-amarelo produzido em diferentes épocas. **Pesquisa**

Agropecuária Brasileira 34: 2353-2358.

Negreiros JRS, Araújo Neto SE, Álvares VS, Lima VA and Oliveira TK (2007) Caracterização de frutos de progênies de meios-irmãos de maracujazeiro-amarelo em Rio Branco - Acre. **Revista Brasileira Fruticultura 30**: 431-437.

Neves CG, Nunes OJ, Ledo CAS and Oliveira EJ (2013) Avaliação agronômica de parentais e híbridos de maracujazeiro- amarelo. **Revista Brasileira Fruticultura 35**: 191-198.

Oliveira EJ, Fraife Filho GA and Freitas JPX (2014) Desempenho produtivo e interação genótipo x ambiente em híbridos e linhagens de mamoeiro. **Bioscience Journal 30**: 402-410

Oliveira EJ, Santos VS, Lima, DS, Machado MD, Lucena RS, Motta TBN and Castellen MS (2008) Seleção em progênies de maracujazeiro-amarelo com base em índices multivariados. **Pesquisa Agropecuaria Brasileira 43**: 1543-1549.

Pio Viana A, Silva FHL, Gonçalves GM, Silva MGM, Ferreira RT, Pereira TNS, Pereira MG, Amaral Júnior AT and Carvalho GF (2016) UENF Rio Dourado: a new passion fruit cultivar with high yield potential. **Crop Breeding and Applied Biotechnology 16**: 250-253.

Resende MDV and Dias LAS (2000) Aplicação da metodologia de modelos mistos (REML/BLUP) na estimação de parâmetros genéticos e predição de valores genéticos aditivos e genotípicos em espécies frutíferas. **Revista Brasileira Fruticultura 22**: 44-52.

Resende MDV (2004) **Métodos estatísticos ótimos na análise de experimentos de campo**. Embrapa Florestas, Brasília, 25p.

Resende MDV (2007) **Software SELEGEM – REML/BLUP**: sistema estatístico e seleção genética computadorizada via modelos lineares mistos. Embrapa Florestas, Brasília, 359p.

Resende MDV (2002) **Genética biométrica e estatística no melhoramento de plantas perenes**. Embrapa Informação Tecnológica, Brasília, 975p.

Resende MDV and Duarte JB (2007) Precisão e controle de qualidade em experimentos de avaliação de cultivares. **Pesquisa Agropecuária Tropical 37**: 182-194.

Rosado AM, Rosado TB, Alves AA, Laviola BG and Bhering LL (2012) Seleção simultânea de clones de eucalipto de acordo com produtividade, estabilidade e adaptabilidade. **Pesquisa Agropecuária Brasileira 47**: 964-971.

Santos EA, Viana AP, Freitas JCO, Rodrigues DL, Tavares RF, Paiva CL and Souza MM (2015) Genotype selection by REML/BLUP methodology in a segregating population from na interspecific *Passiflora* spp. crossing. **Euphytica 208**: 493-507.

Silva PR, Bisognin DA, Locatelli AB and Storck L (2014) Adaptability and stability of corn hybrids grown for high grain yield. **Acta Scientiarum Agronomy 36**: 175-181.

Silva FL, Barbosa MHP, Resende MDV, Peternelli LA and Pedrozo CA (2015) Efficiency of selection within sugarcane families via simulated individual BLUP. **Crop Breeding and Applied Biotechnology 15**: 1-9.

Viana AP, Pereira TNS, Pereira MG, Amaral Júnior AT, Souza MM and Maldonado JFM (2003) Parâmetros genéticos em populações de maracujazeiro-amarelo. **Revista Ceres 51**: 541-551.

PERMISSIONS

All chapters in this book were first published in CBAB, by Brazilian Society of Plant Breeding; hereby published with permission under the Creative Commons Attribution License or equivalent. Every chapter published in this book has been scrutinized by our experts. Their significance has been extensively debated. The topics covered herein carry significant findings which will fuel the growth of the discipline. They may even be implemented as practical applications or may be referred to as a beginning point for another development.

The contributors of this book come from diverse backgrounds, making this book a truly international effort. This book will bring forth new frontiers with its revolutionizing research information and detailed analysis of the nascent developments around the world.

We would like to thank all the contributing authors for lending their expertise to make the book truly unique. They have played a crucial role in the development of this book. Without their invaluable contributions this book wouldn't have been possible. They have made vital efforts to compile up to date information on the varied aspects of this subject to make this book a valuable addition to the collection of many professionals and students.

This book was conceptualized with the vision of imparting up-to-date information and advanced data in this field. To ensure the same, a matchless editorial board was set up. Every individual on the board went through rigorous rounds of assessment to prove their worth. After which they invested a large part of their time researching and compiling the most relevant data for our readers.

The editorial board has been involved in producing this book since its inception. They have spent rigorous hours researching and exploring the diverse topics which have resulted in the successful publishing of this book. They have passed on their knowledge of decades through this book. To expedite this challenging task, the publisher supported the team at every step. A small team of assistant editors was also appointed to further simplify the editing procedure and attain best results for the readers.

Apart from the editorial board, the designing team has also invested a significant amount of their time in understanding the subject and creating the most relevant covers. They scrutinized every image to scout for the most suitable representation of the subject and create an appropriate cover for the book.

The publishing team has been an ardent support to the editorial, designing and production team. Their endless efforts to recruit the best for this project, has resulted in the accomplishment of this book. They are a veteran in the field of academics and their pool of knowledge is as vast as their experience in printing. Their expertise and guidance has proved useful at every step. Their uncompromising quality standards have made this book an exceptional effort. Their encouragement from time to time has been an inspiration for everyone.

The publisher and the editorial board hope that this book will prove to be a valuable piece of knowledge for researchers, students, practitioners and scholars across the globe.

LIST OF CONTRIBUTORS

Lo Thi Mai Thu and Vi Thi Xuan Thuy
Taybac University, Viet Nam

Le Hoang Duc, Le Van Son and Chu Hoang Ha
Institute of Biotechnology, VAST, Viet Nam

Chu Hoang Mau
Thainguyen University of Education, Viet Nam

Lia Maris Orth Ritter Antiqueira
Federal University of Technology - Paraná, Av Monteiro Lobato, km 4, 84.021-216, Ponta Grossa, PR, Brazil

Gabriel Dequigiovanni, Jucelene Fernandes Rodrigues and Elizabeth Ann Veasey
University of São Paulo, Luiz de Queiroz College of Agriculture, Av. Padua Dias, 11, PO Box 9, 13418-900, Piracicaba, SP, Brazil

Evandro Vagner Tambarussi
State University of Central West, PR 153, km 7, 84.500-000, Irati, PR, Brazil

Alessandro Nicoli
Universidade Federal Rural de Pernambuco (UFRPE), Departamento de Agronomia, 52.171-900, Recife, PE, Brazil

Renata Oliveira Batista
Universidade Federal Rural de Pernambuco (UFRPE), Departamento de Agronomia, 52.171-900, Recife, PE, Brazil
Universidade Federal de Viçosa (UFV), Departamento de Fitotecnia, 36.570-900, Viçosa, MG, Brazil

Ana Maria Cruz e Oliveira, Johnn Lennon Oliveira Silva and José Eustáquio de Sousa Carneiro
Universidade Federal de Viçosa (UFV), Departamento de Fitotecnia, 36.570-900, Viçosa, MG, Brazil

Pedro Crescêncio Sousa Carneiro
UFV, Departamento de Biologia Geral

Trazilbo José de Paula Júnior
Empresa de Pesquisa Agropecuária de Minas Gerais (EPAMIG), 36.570-900, Viçosa, MG, Brazil

Marisa Vieira de Queiroz
UFV, Departamento de Microbiologia

Samara Rayane Pereira de Morais, Laura Cristina da Silva Almeida and Guimarães Santos Melo
Universidade Federal de Goiás (UFG), 74.690-900, Goiânia, GO, Brazil

Ariadna Faria Vieira
Universidade Federal de Minas Gerais (UFMG), 39.404-547, Montes Claros, MG, Brazil

Luana Alves Rodrigues, Patrícia Luís Cláudio de Faria, Leonardo Cunha Melo, Helton Santos Pereira and Thiago Lívio Pessoa Oliveira de Souza
Embrapa Arroz e Feijão, 75.375-000, Santo Antônio de Goiás, GO, Brazil

Roberto Fritsche-Neto and Filipe Inácio Matias
USP/ESALQ - Genetics, Av. Pádua Dias, 11, CP 83, 13418-900, Piracicaba, SP, Brazil

Volmir Sergio Marchioro, Francisco de Assis Franco, Ivan Schuster, Tatiane Dalla Nora Montecelli, Mateus Polo, Fábio Junior Alcântara de Lima, Adriel Evangelista and Diego Augusto dos Santos
COODETEC - Desenvolvimento, Produção e Comercialização Agrícola Ltda, BR 467, km 98, 85.813-450, Cascavel, PR, Brazil

Alexandre Pio Viana, Fernando Higino de Lima e Silva, Rulfe Tavares Ferreira, Telma Nair Santana Pereira, Messias Gonzaga Pereira, Antonio Teixeira do Amaral Júnior and Ge¬raldo Franscisco de Carvalho
Universidade Estadual do Norte Fluminense Darcy Ribeiro (UENF), Laboratório de Melhoramento Genético Vegetal, Av. Alberto Lamego, 2000, Parque Califórnia, 28.013-602, Campos dos Goytacazes, RJ, Brazil

Gustavo Menezes Gonçalves
Petrobras-Biofuel, Av. República do Chile, 65, 20.031-912, Rio de Janeiro, RJ, Brazil

Marcelo Geraldo de Morais Silva
Instituto Federal Fluminense (IFF), Campus Avançado Cambuci, Estrada Cambuci, km 05, s\n, Três Irmãos, 28430-000, Cambuci, RJ, Brazil

Itamar C. Nava, Marcelo T. Pacheco and Luiz C. Federizzi
Universidade Federal do Rio Grande do Sul (UFRGS), Faculdade de Agronomia, Departamento de Plantas de Lavoura, Av. Bento Gonçalves, 7712, 91501-970, Porto Alegre, RS, Brazil

Mohammed A. A. Bari and Marcelo J. Carena
North Dakota State University, Department of Plant Sciences, Dep. #7670, Fargo, ND 58108-6050, United States of America

Messias G. Pereira
Universidade Estadual do Norte Fluminense, Departamento de Genética e Melhoramento de Plantas, Av. Alberto Lamego, 2000, 28.013- 602, Campos, RJ, Brazil

Marcos Deon Vilela de Resende
Embrapa Florestas, Universidade Federal de Viçosa, Departamento de Engenharia Florestal, 36.570-000, Viçosa, MG, Brazil

Rosana Gonçalves Pires Matias, Danielle Fabíola Pereira da Silva, Priscila Maria Dias Miranda, João Alison Alves Oliveira, Leonardo Duarte Pimentel and Cláudio Horst Bruckner
Universidade Federal de Viçosa (UFV), Departamento de Fitotecnia, Avenida P.H. Rolfs, 36.570-900, Viçosa, MG, Brazil

Olga Alikina
Russian Academy of Sciences, Branch of Shemyakin and Ovchinnikov Institute of Bioorganic Chemistry, Science Ave 6, Pushchino, Moscow Region, 142290, Russian Federation

Sergey Dolgov and Dmitry Miroshnichenko
Russian Academy of Sciences, Branch of Shemyakin and Ovchinnikov Institute of Bioorganic Chemistry, Science Ave 6, Pushchino, Moscow Region, 142290, Russian Federation
All Russian Research Institute of Agricultural Biotechnology, Timiryazevskaja 42, Moscow, 127550, Russian Federation

Mariya Chernobrovkina
All Russian Research Institute of Agricultural Biotechnology, Timiryazevskaja 42, Moscow, 127550, Russian Federation

Josiane Isabela da Silva Rodrigues and Newton Deniz Piovesan
Universidade Federal de Viçosa (UFV), Departamento de Bioquímica e Biologia Molecular, Instituto de Biotecnologia Aplicada à Agropecuária (BIOAGRO), 36.571-000, Viçosa, MG, Brazil.

Fábio Demolinari de Miranda and Marcia Flores da Silva Ferreira
Universidade Federal do Espírito Santo (UFES), Departamento de Biologia, 29.500-000, Alegre, ES, Brazil

Adésio Ferreira
UFES, Departamento de Produção Vegetal, 29.500-000, Alegre, ES, Brazil

Cosme Damião Cruz
UFV, Departamento de Biologia Geral, BIOAGRO

Everaldo Gonçalves de Barros
Universidade Católica de Brasília, Quadra SGAN 916, módulo B, bloco C, sala 213, 70.790-160, Brasília, DF, Brazil

Maurilio Alves Moreira
In memoriam

Karine Cristina Krycki, Carine Simioni and Miguel Dall'Agnol
Universidade Federal do Rio Grande do Sul (UFRGS), Departamento de Plantas Forrageiras e Agrometeorologia, Avenida Bento Gonçalves, 7712, 91.501-970, Porto Alegre, RS, Brazil

Pereira da Costa JH, Rodríguez GR and Pratta GR
IICAR-CONICET (Instituto de Investigaciones en Ciencias Agrarias de Rosario - Consejo
Nacional de Investigaciones Científicas y Técnicas),
Universidad Nacional de Rosario,
Facultad de Ciencias Agrarias, Campo Experimental
JF Villarino CC 14 (S2125ZAA), Zavalla,
Santa Fe, Argentina
Universidad Nacional de Rosario

Liberatti DR, Mahuad SL and Marchionni Basté E
Universidad Nacional de Rosario

Picardi LA and Zorzoli R
Universidad Nacional de Rosario
IICAR-CONICET, CIUNR (Consejo de Investigaciones de la Universidad Nacional de Rosario), Universidad Nacional de Rosario

Luiz Paulo de Carvalho, Francisco José Correia Farias and Camilo de Lellis Morello
Embrapa Algodão, Centro Nacional de Pesquisa de Algodão, 58.428-095, Campina Grande, PB, Brazil

Paulo Eduardo Teodoro
Universidade Federal de Viçosa, Departamento de Biologia Geral, 36.571-000, Viçosa, MG, Brazil

Magno Antonio Patto Ramalho
Universidade Federal de Lavras (UFLA), Departamento de Biologia, C.P. 3037, 37.200- 000, Lavras, MG, Brazil

Ângela de Fátima Barbosa Abreu, Leonardo Cunha Melo, Helton Santos Pereira and Maria José Del Peloso
Embrapa Arroz e Feijão, Rod. GO-462, km 12, Zona Rural, C.P. 179, 75.375-000, Santo Antônio de Goiás, GO, Brazil

José Eustáquio de Souza Carneiro and Marcos Paiva Del Giúdice
Universidade Federal de Viçosa (UFV), Departamento de Fitotecnia, 36.570-000, Viçosa, MG, Brazil

Trazilbo José de Paula Júnior and Rogério Faria Vieira
Empresa de Pesquisa Agropecuária de Minas Gerais (Epamig), Av. José Cândido da Silveira, 1647, União, 31.170-495, Belo Horizonte, MG, Brazil

Israel Alexandre Pereira Filho
Embrapa Milho e Sorgo, Rod. MG-424, km 45, Sete Lagoas, C.P. 285, 35.701-970, Sete Lagoas, MG, Brazil

Maurício Martins
Universidade Federal de Uberlândia (UFU), 38.400-902, Uberlândia, MG, Brazil

Luís Cláudio Inácio da Silveira and Márcio Henrique Pereira Barbosa
Universidade Federal de Viçosa (UFV), Departamento de Fitotecnia, 36.570-900, Viçosa, MG, Brazil

Bruno Portela Brasileiro, Edelclaiton Daros and Heroldo Weber
Universidade Federal do Paraná (UFPR), Departamento de Fitotecnia e Fitossanitarismo, 80.035-050, Curitiba, PR, Brazil

Volmir Kist and Luiz Alexandre Peternelli
Instituto Federal Catarinense, 89.703-720, Concórdia, SC, Brazil

Francisco Tiago Cunha Dias and Cândida Hermínia Campos de Magalhães Bertini
Universidade Federal do Ceará, Av. Mister Hull, 2977, Antônio Bezerra, 60.021-970, Fortaleza, CE, Brazil

Francisco Rodrigues Freire Filho
Embrapa Amazônia Oriental, Trav. Dr. Enéas Pinheiro, s/n, Bairro Marco, CP 48, 66.095-903, Belém, PA, Brazil

Rosana Mendes de Moura Oliveira, Ângela Celis de Almeida Lopes, Karla Annielle da Silva Bernardo and Akemi Suzuki Cruzio
Universidade Federal do Piauí, Campus da Socopo, 64.049-550, Teresina, PI, Brazil

Francisco Rodrigues Freire Filho
Embrapa Amazônia Oriental, Travessa Doutor Enéas Pinheiro, s/n, Bairro Marco, 66.095-100, Belém, PA, Brazil

Valdenir Queiroz Ribeiro
Embrapa Meio-Norte, Av. Duque de Caxias, n° 5.650, Bairro Buenos Aires, 64.006-220, Teresina, PI, Brazil

Gustavo H.F. Klabunde, Sarah Z.A. Tenfen and Rubens O. Nodari
UFSC, Rodovia Admar Gonzaga, 1346, 88.034-000, Florianópolis, SC, Brazil

Camila F.O. Junkes
UFPel, Rua Gomes Carneiro, 1, 96.010-610, Pelotas, RS, Brazil

Adriana C.M. Dantas
UERGS, Rua Júlio de Castilhos, 3947, 95.010- 005, Caxias do Sul, RS, Brazil

Carla R.C. Furlan
Faculdades Integradas FACVEST, Avenida Marechal Floriano, 947, 88.501-103, Lages, SC, Brazil

Adelar Mantovani
UDESC, Avenida Luis de Camões, 2090, 88.520-000, Lages, SC, Brazil

Frederico Denardi
EPAGRI, Rua Abílio Franco, 1500, 89.500-000, Caçador, SC, Brazil

José I. Boneti
EPAGRI, Rua Araújo Lima, 102, 88.600-000,São Joaquim SC, Brazil

Amadeu Regitano Neto
Embrapa Semiárido, Rodovia BR-428, km 152, CP 23, 56.302-970, Petrolina, PE, Brazil

Ana Maria Rauen de Oliveira Miguel, Anna Lúcia Mourad, Ercília Aparecida Henriques and Rosa Maria Vercelino Alves
Instituto de Tecnologia de Alimentos, ITAL, Av. Brasil, 2880, CP 139, 13.070-178, Campinas, SP, Brazil

Paulo Vitor Dutra de Souza, Sérgio Francisco Schwarz, Claudio André Werlang and Pedro Augusto Veit
Universidade Federal do Rio Grande do Sul (UFRGS), Faculdade de Agronomia, Departamento de Horticultura e Silvicultura, Avenida Bento Gonçalves, 7712, 91.501-970, Porto Alegre, RS, Brazil

Maria Teresa Schifino-Wittmann
UFRGS, Faculdade de Agronomia, Departamento de Plantas Forrageiras e Agrometeorologia

Divanilde Guerra
Universidade Federal do Rio Grande do Sul (UFRGS), Faculdade de Agronomia, Departamento de Horticultura e Silvicultura, Avenida Bento Gonçalves, 7712, 91.501-970, Porto Alegre, RS, Brazil
Universidade Estadual do Rio Grande do Sul (UERGS), Curso de Agronomia, Rua Cipriano Barata, 47, 98.600-000, Três Passos, RS, Brazil

María Fernanda Guindon, Eugenia Martin, Aldana Zayas, Enrique Cointry and Vanina Cravero
IICAR-CONICET, Instituto de Investigaciones en Ciencias Agrarias de Rosario, Zavalla, Argentina

Karunamurthy Dhivya, Sundararajan Sathish, Natarajan Balakrishnan, Varatharajalu Udayasuriyan and Duraialagaraja Sudhakar
Tamil Nadu Agricultural University, Department of Plant Biotechnology, Centre for Plant Molecular Biology and Biotechnology, Coimbatore, India 641 003

Lenise Castilho Monteiro, Gleiciane de Lima Benteo and Cláudia Barrios de Libório
Instituto Federal Goiano-Campus Rio Verde, Av. Sul Goiânia, km 1, Zona Rural, 75.901-970, Rio Verde, GO, Brazil

Jaqueline Rosemeire Verzignassi, Sanzio Carvalho Lima Barrios, Cacilda Borges do Valle and Celso Dornelas Fernandes
Embrapa Gado de Corte, Av. Rádio Maia, 830, Zona Rural, 79.106-550, Campo Grande, MS, Brazil

Carlos Roberto Riede, Luiz Alberto Cogrossi Campos, Klever Márcio Antunes Arruda and Deoclécio Domingos Garbuglio
Instituto Agronômico do Paraná (IAPAR), Rod. Celso Garcia Cid, km 375, CP 10.030, 86.047-902, Londrina, PR, Brazil

Avahy Carlos da Silva
IAPAR, Av. Euzébio de Queirós, s/no, Uvaranas, CP 129, 84.001-970, Ponta Grossa, PR, Brazil

Monalisa Sampaio Carneiro and Fernanda Zatti Barreto
Universidade Federal de São Carlos (UFSCar), 13.600-970, Araras, SP, Brazil
UFSCar, Departamento de Biotecnologia e Produção Vegetal e Animal

Roberto Giacomini Chapola, Danilo Eduardo Cursi, Thiago Willian Almeida Balsalobre and Hermann Paulo Hoffmann
UFSCar, Departamento de Biotecnologia e Produção Vegetal e Animal

Antonio Ribeiro Fernandes Junior
UFSCar, Estação Experimental de Valparaíso, 16.880-000, Valparaíso, SP, Brazil

Alírio José da Cruz Neto and Adriana Rodrigues Passos
Universidade Estadual de Feira de Santana, 44.036-900, Feira de Santana, BA, Brazil

Raul Castro Carriello Rosa, Eder Jorge de Oliveira and Onildo Nunes de Jesus
Embrapa Mandioca e Fruticultura, 44.380- 000, Cruz das Almas, BA, Brazil

Sidnara Ribeiro Sampaio, Idália Souza dos Santos and Plácido Ulisses Souza
Universidade Federal do Recôncavo da Bahia, Centro de Ciências Agrárias, Ambientais e Biológicas, 44.380-000, Cruz das Almas, BA, Brazil

Index

CPSIA information can be obtained
at www.ICGtesting.com
Printed in the USA
BVHW011619180619
551322BV00002B/28/P